WE'LL TRACE THE RAINBOW

We'll trace the Rainbow

O joy that seekest me through pain,
I cannot close my heart to thee;
I trace the rainbow through the rain
And feel the promise is not vain,
That morn shall tearless be.

George Matheson (1842–1906)

JEAN BROWN

Illustrated by Pauline and Joe Walters

We'll trace the Rainbow
© Jean Brown 1998, 2008

First Published in 1998 by
H H Brown Currer Laithe, Moss Carr, Keighley BD21 4SL

Second edition published 2008 by Palatine Books
an imprint of Carnegie Publishing Ltd
Carnegie House, Chatsworth Road, Lancaster, LA1 4SL

ISBN 13: 978-1-874181-49-1

All rights reserved
No part of this publication may be reproduced, stored in a retrieval system, or transmitted in any form or by any means mechanical, electronic, photocopying, recording or otherwise, without the prior written permission of the author and publishers.

Cataloguing-in-publication data
A catalogue record for this book is available from the British Library

Typeset by Carnegie Book Production, Lancaster
Printed and bound by Biddles Ltd, King's Lynn

Reading from Jean's book is like sitting by a cool stream of water in the heat of the day. It's refreshing, because she shares some of the things in life that we all long for.

Ernestine Aberle – New York USA

Acknowledgements

'Peace of the River'
from The Ditty Bag © 1956 *by Janet E Tobitt*
Used by permission of Girl Scouts of the United States of America

'Going Home'
from New World Symphony by Anton Dvorak
words by William Arms Fisher
Reprinted by kind permission of Alfred Lengnick & Co
(A division of Complete Music Ltd)

Introduction

This is an autobiography 1987–1996, but it is not about me, it is about us and the sixteenth-century farm Grandfather bought in 1929. It is not the beginning of the story nor the end. To find the beginning you must put this book back on the shelf and lift down *We'll see the Cuckoo*, and to read the end you must wait a while, I hope. You may quite happily begin in the middle or even open the book at random, for every dawn is a new beginning and though storm may rage each day, every sundown has a peaceful ending. There are few things you need to know about us for atmosphere and drama can be enjoyed without sound or sight of the stage-managers.

The history of the place in which we live is rooted in the Middle Ages when stone walls were built and fields tilled by the monks of Rievaulx Abbey. In 1571 we know that the beamed and mullioned farmhouse, with inglenook fireplace and beehive oven, was bought from the Paslewes of East Riddlesden Hall for £30. Almost 400 years later it was without windows, water or electricity, without sanitation and road, and we came to live in it. Every moment since has been a challenge.

There is an atmosphere here, hard to define. From whence it came we will never know. What generation created it, we cannot say. Was it monk or yeoman, tenant or farmer, weaver or bard? Whether it was ploughman or shepherd or milkmaid will remain a secret but whatever the presence it is benign and warm and radiates happiness. It has been our roof now for longer than it has sheltered any other family and we propose to live here until the day we 'give back the life we owe'. Had we not come here this would not have been our story for you to read. Make of it what you will.

We are farmers and will always be. We have a love affair with the land, the great outdoors, the animals we rear and the wild things which are not too shy to be our neighbours. When Father died I chose to retire from a happy, twenty-one-year-long headship of a village school up the dale, so that my sister Margaret, the competent farmer, could keep our promise to him that the farm would continue in perpetuity. Through The National Trust we have done so. It is now our good fortune, and I believe theirs, to share the life we lead with hundreds of people, from all corners of the world, who elect to spend their holidays at The Currer.

We are ordinary people to whom extraordinary things happen. Perhaps it is normal and other ordinary people have similar encounters and it is arrogance on my behalf to write ours down and to think other people will be interested, but it is basic to all conversation that listening to the experience of another excites the memory. If I can do no more than make the reader remember, it is enough.

Jean Brown, April 1998

Chapter One

I went to the window where the morning was.

Gerald Bullett from 'White Frost'

It is not a new experience to wake early, to get up immediately and start a new day. I have been doing it all my life, the biggest half of which has been spent at The Currer. Exposed, as we are, near the summit of this Pennine ridge, our closest neighbour, the weather, communicates its mood and season daily in the few seconds before I rise.

If there is a wuthering in the chimney of the fireplace in my bedroom, I know there is a wind. If it is rattling the wooden screen on the hearth, I know a gale is blowing with unstoppable force and skimming the tops of the hilly range we proudly call the backbone of England. I know that, if dawn is breaking when I open my eyes, clouds will be scurrying across our uninterrupted horizon. Inevitably the day will be fraught with the danger of loosened corrugated tins hurling themselves into space as do the hang gliders on Baildon Moor. I know that carrying bales of hay will pose a problem, that straw will drift on the shed floor like flotsam on a windy beach and that the cattle will prefer to stay indoors. I prophesy that it will be a struggle to get the human 'oldies' and the dogs into the Range Rover for, unlike the cattle, they will choose to go and it will take superhuman strength to shut the car door.

In a few seconds before I rise I can calculate the degree of rain, heavy clouds may be throwing on the window-pane and I know that the yard will be cleansed and that the dyke following the road down Currer Lane will be full of dancing, leaping water. I know that the gateway mud and the manure the cattle have walked out of the sheds will have held the water and will be tenacious of wellingtons.

If, when I awake, daylight is already several hours old, I know open windows will be letting in the early sounds of summer and that the air will be scented with a myriad of wild flowers and new mown hay; that the chattering I hear will be gossiping tits and finches in the silver birches, and blackbirds in the sycamore. If the sun is already sending its image through the mullioned windows onto the white bedroom wall, I fear the cockerel may be already hoarse, that cattle will have been up and grazing long ago and if I fail to get up quickly and let the dogs out it will be impossible to get them back indoors before the arrival of the post van. How they love to give chase! For seconds only I absorb the pastoral peace before I remember our converted barn is full of holiday-makers who require breakfast before their daily exploration of the broad acres of the Yorkshire Dales.

If, when I awake, the open window is letting in a misty dampness and the fungoid smell is of earth and wet bracken, I know the blackberries will be ripe in Jimmy's Wood and down the almost vertical Marley Banks, that sycamore leaves will soon be littering the yard and that the cattle will ripple with autumn fatness.

But when I awake to complete and utter silence I know that snow has fallen. I do not need to open my eyes. When there is stillness and a tangible silence I intuitively know that drifted snow curtains the windows, lies clean and unbroken across the farmyard and is sculptured against doors and walls. I cannot explain the waking awareness that snow has fallen. It muffles the wind and quietens the birds and animals wintering in straw. It distances the town and the A650 in the valley. It isolates us in a way which can only be understood by those whose neighbours are several fields away. No matter how heavily it is falling it does so in complete silence and it takes away the urgency to get up. It slows the clock for nothing needs to be on time. The pattern of the day must be changed and a different routine followed.

I can rise confident in the knowledge that logs piled on the fire last night will still be aglow and that the kettle I place on the Aga, before stepping out to check the cattle, will be boiling merrily on my return. I know there will be only semi-darkness for the valley lights will still be burning and the white carpet will itself illuminate the landscape long before the late dawn comes. The cattle will still be sleeping and one at least, will be stretched out to frighten me into poking it and saying, 'Hey, are you alright?' If it responds favourably to my fearful whisper, the dogs and I will return to the kitchen and snow will fall dry onto the porch floor.

I have so little time to be in retrospective mood it is a pleasure to tarry over a cup of coffee. When the golden opportunity arrives I sit with pen in hand so that, in remembering, I can trap it on paper and preserve it in perpetuity. To do so seems as important as our Covenancy with the National Trust which preserves The Currer for posterity.

Some things we have changed since we first came, when Margaret was only nineteen and I had not yet discovered the village school up the dale where I was to be headmistress for so many years. But when snow falls all our creativity disappears. Things look as they used to do. No cars litter the yard. There are no foreign footprints, no strangers' dogs or children. Drifts pile up in the yard and conceal the mullioned windows so that the ancient laithe looks again what it was for four hundred years, a mistal and a loft within a hay barn. What we have done, the road we sledge-hammered for ourselves, physically and metaphorically, is temporarily hidden. All hard surfaces disappear, clear outlines are softened and isolation is comfortable and agreeably acceptable. No one has access except on foot. There will be no postal delivery, no dustbin collection; what we need from the village we must carry as we, and our predecessors used to do, and should our water-pipe freeze we will be as dependent on the spring as all who have lived here before.

I know too, as I descend the stairs, that the kitchen will echo a past we occasionally grieve for. It was infinitely precious for a host of reasons. Our guests, I know, associate pastoral peace and quiet with The Currer. With reason, for here they find old-fashioned countryside, hospitality, ways and fare. They believe serenity reigns and a quietness the outside world has lost. They do not know that, not so very long ago, there was a greater one when our dogs did not bark at the frequent arrival of a car, that twentieth-century menace man can no longer do without; when jogging was not fashionable and heavily breathing man did not rouse our canine friends and when our over-worked washing machine did not spin an almost perpetual background noise.

Snow brings a welcome return to that quieter life we used to live and as I come down the stairs I know that our wellingtons will stand in pools on the porch floor, our coats will smell of cow hair and dried wetness and the dogs will be unusually clean. There will be the comforting, selfish belief that no one will come, no neighbour, no guest, no holiday-maker, no traveller. If the phone rings and I am urged to put in pay phones, buy a particular brand of coffee or advertise somewhere, I can say honestly 'No thank you. We are only farmers.'

I can take longer over my coffee, turning over my still wet gloves on the Aga. With the minimum of imagination I can believe I must cross the fields, not easily for drifted snow will have buried the walls to the top stones and field gates will no longer open, and trudge with other fellow workers into town for the bus which used to take me up the valley to my school. I was Head and

Joan was my colleague and, now long retired, memories are still evoked. Whilst I struggle to remember who our guests were last week and what I served for dinner yesterday, I have no difficulty in remembering the location and atmosphere of my village school, especially on a day when the playground blanket of snow would be completely disturbed by sliding, snowballing children. Their happy faces, red noses and cold fingers are not forgotten though they are grown with children of their own.

If I plumb the depths of my memory still further back into my own childhood I am again the little milk-girl, teetering on snow-padded clogs, carrying the heavy can of milk. I can remember Dick the black horse, and Father, my best friend and workmate. I can hear the ice on the creamy milk in the churns and Grandmother reminding me to keep changing hands, or I'd be lopsided, like Arthur!

It is snow, not wind, or rain, or sunshine which takes me so forcefully down the channels of my memory. It gives me time to wallow in the joy of living here. The peopled outside world has no say in what we do. Blessed snow! It isolates us from the present and, whichever way we look, it shows us the beautiful place in which we live.

* * * * *

Let me be honest. We think there is no better place to live than the Yorkshire Dales. To make such an assumption we should be widely travelled and we are not. Admittedly we have spent our holidays in very beautiful places for such abound in Scotland, The Lake District, Northumbria, Wales and Ireland. My sister Margaret and I have taken hundreds of children, annually and joyfully, to camp on Hebridean Islands for nearly three decades. When early sun has brought a shimmering God-given day and finally painted red the western horizon before dipping into a glittering sea, we have thought, with reason, we had found paradise prematurely. Of the wider world, however, we must confess to having seen little, having only once, briefly, left the British Isles. So I have no right to make such a definite statement but I have a gut feeling that, for me, I am right and that there is no better place to live than the Yorkshire Dales. To live at The Currer, for all its back-breaking demand on our lives, is a bonus indeed.

If poverty is shortage of money we have been poor, but it is no such thing. It is the absence of opportunity, or the inability to take advantage of it, or the lack of survival genes to compensate for the lack of finance. The Currer was our opportunity, we snatched at it and we are survivors. Whatever the 'winter' we will live to 'see the cuckoo'.

The Currer was bought by my grandfather when my disabled brother,

Harry, was just two years old and I had not been born. Its now fertile acres were then neglected, its farmhouse and barn were rapidly becoming a ruin, exposed, roadless, without water or sanitation or electric. There are other such farmhouses and steadings scattered alone and abandoned, but not many for farmers are tenacious of property and of a way of life and refuse to give up easily. They are battered by storm and misfortune but are still on the hillside when disaster and winter have passed.

To wake and be aware of snow is intuitive but no more so than to wake and know that it is Christmas. There is something uncannily holy about cattle sheds on the Eve which celebrates Nativity, and magic in the air when dawn comes. There was no snow when I awakened on Christmas morning 1987 but the silence was equally profound. I lay quietly trying to anticipate a new experience. We were about to be totally selfish. Lying there I could not remember not having to leap out of bed, drag out my sister and begin those chores necessary to every day and add to them those peculiar to Christmas.

From early childhood the festive season meant more work. Almost everyone will claim that it does. When I was a child milk was still being delivered every day of the year. Cattle still have to be milked and fed but nowadays people can buy in advance and store in the fridge. Always in my childhood, after the milk round, we feasted at Grandmother's house with aunts and uncles and cousins galore. By 1980 the extended family had shrunk but when we converted the barn a score of guests booked in every year. Two large turkeys were needed and several of Mother's clouted puddings had to bounce in the pot. Feeding so many, there had been neither the time nor the place to eat ourselves.

It was therefore quite unusual for me to be able to lie, even just a moment, in bed on Christmas morning, but in 1987 I was able to do just that. There was no urgency to enjoy this relatively selfish day. It was still dark and quiet outside. There was no post van to send the dogs into a frenzy. There had been a Christmas Day delivery in my student days when I'd been seasonally employed by the Post Office. Not any more, and the dogs were asleep and the geese docile and the donkeys awed into silence. Perhaps something hereditary in their genes fills them with the wonder of The Nativity. The hundred Hereford heifers were still curled up in last night's straw and they never begin to bawl until daylight is hours old and hope of food begins to recede. The cockerel was probably already hoarse but he never wakens me. Just the dogs, and there was evidently no wandering fox to fill the air with alien scent and disturb Lusky, and Jess, the pup, was too little, anyway.

The thought of the pup made me rise. Jess was only three months old and, although house-trained toilet-wise she was still not socially acceptable. She was naughty when awake, eating chair legs, gnawing hair-brush handles negligently left on the window seat and causing havoc with Auntie Mary's knitting. I will never forget the horror of finding a newly sewn-up jumper almost in shreds and will forgive Auntie Mary, for always, for the admirable way she took the news. So Jess and I went into the seven o'clock darkness to check the herd. I flicked the torch to spotlight each recumbent animal. All being well we returned to the kitchen where the kettle was boiling. We huddled against the Aga. I drank my coffee, tickling Jess's warm belly with my stockinged foot, anticipating this quite rare, selfish day ahead.

Oh boy, had we had our fill of feeding people these past seven years? Last Christmas we had said to Mother, 'Someday we will close for Christmas Day and free ourselves to enjoy it.'

She had said, 'Someday is no good to me!' Indeed at ninety-one she could not afford to procrastinate. 'OK,' we had answered. 'Next year we will close.' We had kept our promise and refused all bookings for Christmas Day. It was quite a serious loss of income, but we were beginning to feel less insecure and we were fed up of feeding people. Fed up of food, for that matter. The animals had to be fed, of course, but not one of us could bear the thought of turkey and all its trimmings. Neither could we think of sitting watching inappropriate programmes on television. In fact we had had a complete overdose of The Currer. It had been an exhaustive run-up to our Covenancy with The National Trust. It had been a strain and now all we wanted to do was to get into the Range Rover and go. All we needed was a flask and a few ham sandwiches. We'd planned this heavenly day for a full year.

It wasn't quite unselfish for, a week ago, William had phoned. He had been coming, on and off, for six years paying for his accommodation with his allowance from Social Security. Neither we, nor our other guests, minded William. Shabby, down at heel, he needed a haircut as much as he needed a good meal and a roof over his head. He was Irish, nervous, intelligent and excitable. His eyesight was very poor which may have accounted for his out of-work status. If we had been in the business of providing long-term accommodation William would have stayed, but we are pre-booked sometimes twelve months ahead. I had answered the phone and my heart had disappeared into my boots. 'No, William,' I had said, 'we are closed over Christmas. Definitely closed. We have said so to scores already. We really mean it. We are closed!'

Well, could he come for a week? Just until Christmas and then he would go? He promised. I was gently inhospitable suggesting he would be much better finding accommodation in town. To be at The Currer in winter was foolish. We could be snow-bound any day. He had no car, we had no other guests, it was a non-starter. Full stop! But the Irish are persuasive. He'd come for a week, he insisted, then he really would go. He didn't, of course. Who'd have William who didn't know how nice he was, how kind and unassuming and grateful.

Mother wouldn't hear of us turning him out, but I wouldn't relent on the Christmas dinner business. We were postponing it until Boxing Day when Auntie Dorothy was coming, as she had for sixty years or so. Since Uncle Willie's death twenty years ago she'd come and stayed the week. William would be provided with ham sandwiches and left to get a headache in front of the television and we'd have Christmas dinner on Boxing Day. 'Lovely,' he said. Mother was not quite sure about the rightness of our decision, but she was addicted to the back middle seat of the Range Rover. It was her special brand of treacle and she'd lick.

Margaret went out to feed the cattle. Auntie Mary put on her Christmas apron and slipped a cassette of carols into her recorder and the family had breakfast. Then Mother and Harry washed the dishes. Auntie Mary made the sandwiches, and I fed William and lit his log fire. Then we climbed into the Range Rover and left.

It was a strangely beautiful day of empty roads and winter sunshine. Traffic on Christmas Day keeps to the fast roads and those winding country lanes we love so well, are deserted. Upper Wharfedale had recently been given to The National Trust. We drove along the valley floor in the comfortable knowledge that, come May, we would be putting our signatures to the Covenancy.

Gliding round the Dales, at no more than 20 mph, we met hardly anyone. There were no farmers in the frosty fields, feeding mangolds to sheep or hay to out-wintering cattle. There were no bread vans or corn trucks and the milk tanker must have already been. In all the world there seemed to be only us, singing untunefully as we meandered along. 'O come all ye faithful,' we sang to the blackbirds in the hedgerows. 'While shepherds watched their flocks,' we carolled to the grazing sheep. 'Away in the manger,' we croaked and tears blurred my sight. The long, lovely years with children were no more.

There was no-one by the river, no-one scaling the crag at Kilnsey, no-one in the car park at Kettlewell. No-one tried to pass us or came towards us on the narrow road through Starbotton to deserted Buckden.

We alighted to visit the toilet and peer through the closed shop windows and we saw wild flower seeds for sale. We do not need them. If Margaret closes field gates and keeps the cattle on the high ground and allows the low pastures to grow unnibbled, hundreds of wild flowers colour the hillside. If spring is late and we have to let cattle graze the whole 170 acres, the meadow flowers are eaten before they flower, but we know they are still there. Nor have we any doubt about the multitude of different grasses for a professor from Alabama came to study them at the Turf Research Institute nearby and the many varieties adorned his bedroom for several weeks. He was studying fungi on them and oblivious of time. He would be sitting on the seat at the top of Currer Lane, watching the sunset and quite forgetful of his evening meal, and one day he phoned his wife at breakfast-time unaware that it would be 2 a.m. at home!

It is difficult to do anything these days which does not stir up a memory. As we pushed our ageing family back into the car I could not be unaware that, not many yards away, was Heber Cottage where Julia lives. She was a pupil, a Guide and a Sea Ranger who was so wonderfully competent with the disabled at White Windows Cheshire Home. She won a Juliette Low award and represented England for six weeks in America, yet was satisfied to marry a farmer and rear two little boys in remote Buckden. Not far away, too, were her parents who had found a fourteenth century chest in an old barn and given it to us to polish and treasure.

But we were not visiting. Not on Christmas Day, so we went further up the dale and ate our sandwiches in the shelter of Hubberholme church thinking about the last time we had driven up our neighbouring vale. Had it really been nine years since we had followed the Wharfe from Grassington to Buckden, just before Father died? Nine short, interesting years, so full of activity we had had no time to do so again. It had been autumn then, and Margaret had stayed at home to do the work even a Sunday demands. Harry says it was November 5th and he is always right. We never argue with him about dates. That year autumn was aglow and the leaves were still on the trees, richly colourful. We had had tea with Julia and then she had taken Father round the farm and he had had a wonderful time. It proved to be his last outing before that appalling winter which followed and the deep snow which covered the hillside during the February that he died.

Sitting there, the five remaining members of our family, eating a meagre Christmas lunch on a calm, crisp day, we felt some of the weight of the last nine years drift away. How busy we had been, farming our inheritance and providing hospitality for the thousands of guests the tide of tourism left briefly on our doorstep! We felt exhausted. The winter sun was as weak as we were but managing, nevertheless, to light up the lovely landscape, the bare trees, the old stone field walls and church, but casting only delicate shadows. In spite of the crispness of the December air no ice fringed the river flowing sedately down the dale.

It wasn't a carol which repeated itself again and again in my head as we sat close together waiting for Harry, our handicapped brother, to finish his sandwiches and the cream cake I'd brought to follow. It wasn't The Nativity, it was the river which was my hypnotist. I wasn't with my school-children sitting round the Crib, I was back in the camp-fire circle with my Guides. There I had found real joy and peace, learned what to do with solitude, how to admire beauty, to recognise timelessness and acknowledge an extraordinary Creator somewhere out in the space we all need.

I am a shocking singer. I do it least well of the many things to which I admit incompetence. Even so, I can't help singing. It comes to me as naturally as breathing. I've just got to sing and if it amuses people so be it. Life has been one song of an experience. Sometimes sad, sometimes hysterically happy, often grateful, many times reverent but never coarse or vulgar. I have no time for today's noise. It is strange, perverted, repetitive and meaningless to me. More than that it is offensive and I turn it off. I know enough songs for every day of the year, every season, every part of each day, every mood and every activity.

The river excited one on that lovely first New Christmas Day of the many that have since followed. I began and Margaret joined in for she was reared in the camp-fire circle too.

Chapter Two

Peace I ask of thee o' River,
Peace, peace, peace
When I learn to live serenely,
Cares will cease.
From the hills I gather courage,
Visions of the day to be,
Strength to lead and faith to follow
All are given unto me.
Peace I ask of thee o' River,
Peace, peace, peace.

Glendora Gosling

I find it necessary to define for myself the word 'serenity'. Surely it only comes with increasing age and is spiritual rather than physical. If it is a gentler way of life, if it is indeed freedom from care, I haven't found it yet, but I do not interpret it so. I think serenity comes simply from just being able to cope with all the twists and turns of even the busiest sort of life. It is the ability to remain calm in extraordinary situations. Joe Walters, wheelchair bound, says it is being able to cope when you can't cope. Another of our guests gave me this definition. He was a nurse in charge of a group of mentally handicapped people brought here on holiday. For me it had been a day of domestic mishaps. Little, annoying, frustrating incidents had crowded my day. I said to this understanding man, 'I seem to have been saying, "Oh dear," all day!' Life with the disabled is full of spills, grazed knees, broken crockery, sickness, incontinence, epilepsy, infuriating wheelchairs, medications, lost articles of clothing and continual delays. Each minute of the young man's day was filled with exasperating problems but he remained composed. 'I say that all the time,' he told me, 'but I follow it with "Never mind."'

I think that is serenity. The ability to say, 'Never mind', or as Father used to say, 'Something will turn up,' or Mike's, 'It'll pass.' I find myself saying, 'I'll get used to it.' With this outlook on life serenity is within reach. The social disease of the closing millennium is human inability to cope. We are failing if we do not teach our children to do so. To be able to handle birth and death, family and marriage, children and elderly, prosperity and adversity, health and illness, morals and beliefs is not just important, it is essential and when individuals or families or governments lose control, then we need to worry!

We were very relaxed during the spring of 1988. We had worked so hard on our run-up to our Covenancy with the National Trust that we could put away the paint brush, ease up on the wall building and pause for the first time since coming to The Currer thirty years before. We heard each other mutter an incredulous, 'Phew!' and saw each other smiling and shaking a head in a daze as life, for a brief period, slowed and we recharged our batteries.

Jess was our spring entertainment. We had brought her from a Lakeland farm hidden away at Hubbersty, in the Lyth Valley. She had been born in a litter of seven and had lived in a clean cell in the dark recesses of the barn. Her mother had a wriggly backend and was all over us when we first saw her pups during our Grange October holiday. We returned a month later to collect the only bitch pup left of the litter. Our friend, who keeps beautiful collies and shows them at Crufts, will take a long time choosing and pay a sizeable sum. We had no choice, paid £15 and took what we were given, the only dog we have ever had which was car sick.

Margaret mopped up for her all the way home. I drove and Mother, bit by bit, used up a whole toilet roll for Jess drooled continually. Poor, miserable baby, saliva poured out of her mouth in pints. When we reached home she was trembling and cold. She had never been in a farmhouse and we had reared all our pups in a barn. Only old age had brought our dogs into the house, but it was a bitterly cold November day and we hadn't the heart to turn her out.

Harry was more interested in Jess than he had been in any other. We turned off the television just to watch her. It was too dangerous to have her running riot in the kitchen for fear of the infirm falling over her. In the sitting-room we tied her to the furniture and she performed saucily for us. Unsupervised she would steal anything. Returning from town, one day, we bought fish and chips from a relative's shop, a thing we rarely do. We were late, hungry and cold. We lifted hot plates from the Aga and emptied the fish from the paper, preparing to eat inelegantly huddled to the blazing fire. My plate was on the coffee table when Jess appeared from nowhere and the whole battered fish went down in one. Harry laughed, and has been laughing at Jess ever since.

Other Collie pups had been born with a herding instinct. They practised on ducks, geese, donkeys, goats, cars, people and cattle. Jess was not interested. 'She's going to be no good at all,' we predicted. That she was sick every time we went in the car was a disaster. Dorothy brings all her dogs with her when she comes as caretaker to The Currer. When we went on holiday Jess would have to travel all the way to the Hebrides. Friends showered advice. Hazel said, 'Put music on. That works!' We don't doubt her wisdom but we have never had a car with a radio. We tried singing when we set off. I'm confident this taught Jess to compete. When the car started she persecuted us with a piercing, ear-splitting bark. Lusky, our Hebridean Collie, joined in and Mother, with hearing problems, protested loudly.

One day the exodus from The Currer was fraught with more stress than usual. I was staying at home. It was possible for the non-driver to hold Jess's mouth firmly closed but Mother couldn't do that. Harry only laughed and there was such noisy havoc in the car that Margaret stopped at the Moor Gate and said she was returning to put out the dogs and leave them with me. She couldn't cope, she said, with all that hassle and pick up Auntie Mary. Mother screamed that they were too late already. Margaret shouted a reply and suddenly the ninety-two- year-old lady lifted her soft, small palm and left a weal-mark across Margaret's face. Now several years later, Margaret remembers with affection the dear old lady who thought she was still the boss. The sickness persisted. We tried dog pills, human pills, front seat, back seat and still Jess was sick. In desperation Margaret bought a chain to conduct static electricity and that was almost the end of our problem.

Jess was not interested in cattle at all, or ducks, or anything which moved. We despaired. Lusky wasn't always a good example but she did not even mimic him. There was no question of Lusky's ability to move cattle. He herded everything from cattle to car wheels and empty plastic containers but he chose to do it his own way and was deaf to instructions and praise alike. He would manage to get every calf or heavy bullock into the shed, but one. On purpose he would take the last one away, just for fun, and no matter how we yelled at him, it made no difference. Jess didn't accompany or copy him. If she saw an ice-cream carton being herded she merely retrieved it and would not let it go. But we put up with all the barking and sickness in the car and the uselessness outside because she loved Harry so. A muddle over bookings taught us she would fly to his defence. Guests leaving the cottages often ask to re-book for next year. I like to have a chart ready for it would be disastrous to double book a cottage. The phone rang, one August and Margaret was nearest and picked it up. I heard her say, 'The Loft. A week on Saturday Mrs ... ' (I can't even remember the lady's name.) Margaret was running her finger on the chart but the lady's name was not there. 'Don't worry. Please don't worry,' she comforted the woman. 'If a job comes these days you have to take it. No we won't charge. We'll fill it, I'm sure. You can't help it.'

'I feel so awful,' the caller was apologising, 'I booked the cottage two years ago!' Margaret put the phone down trembling. It was the nearest we had ever been to double booking.

The Pratts, when they were leaving, asked me to hold the same cottage for them the following year, before my new chart was made. I know I told everyone. I definitely told two other families who, on leaving, tried to bid for that earlier week. 'Sorry,' I said, 'It's already been spoken for.'

The phone rang. Harry and I were in the front room. Margaret answered and came to us to announce she had just booked the week in question to someone who hadn't been before, who would send a deposit and an address. There followed a heated quarrel. I insisted she had been told. She said she hadn't, and Harry, who sides with whoever is in trouble, said accusingly, 'You didn't tell her!'

We both turned on him and chorused, 'You keep out of this. It's nothing to do with you. You don't have to sort out this mess!'

From somewhere Jess came flying to his rescue, leaping on his knee and licking his face. That was the day we learned Harry had an ally, a protector. Always if we shout at him, which we do for he is our brother, she comes flying to his defence. Her peace-keeping ability is profound. But she was no good with cattle until our neighbour let out his dairy herd at the beginning of May. Gaining freedom after winter, even old bovine ladies get giddy. Margaret and Jess were walking across the moor when Jess spotted the frolicking cuddies and suddenly knew what to do. She leapt over the wall into George's field. She set the newly milked herd galloping and their empty udders were swinging alarmingly. Margaret was totally shocked. It is possible, if you are tolerant, to have a dog which barks when the ignition is pressed, is sick in the car, steals your dinner and intervenes in a quarrel, even one that jumps on the window seat and frightens the life out of the postman, but there is no way you can have a dog that chases a neighbour's cattle.

It wasn't easy to catch the errant pup, but when she did Margaret held her collar and beat her soundly. I don't think Margaret had ever done that before, but Jess received a spanking several times. She was put on a long rope and her training began. Having once chased cattle she was all set to chase ours as well. It was difficult for Margaret to teach and for Jess to learn, but the alternative was unthinkable. Jess must behave or go. It was as simple as that. Jess learned and became the best cow dog we have had since Jed in the 1960s.

* * * * *

We signed the Covenancy with the National Trust on 13th May 1988. I remember it in detail. The Range Rover was full as usual, Harry, Margaret, Mother, Auntie Mary, both dogs and me. The solicitor's office was ten miles away. The day was gloriously sunny and we were in holiday mood, driving with wheels on the ground and heads in the clouds. The unbelievable was about to happen. The once-ruined farmhouse, which was our home, was to be preserved in perpetuity. How we loved the word! Wouldn't those who had built it 400 years ago have been proud? Wouldn't Father have smiled? We were sure our friends would be pleased and neighbours would be glad we'd held back the concrete jungle of the twentieth century!

We parked the Range Rover and, each taking an arm of Harry, we marched confidently into the solicitor's office, our smiles stretching from ear to ear. We must wait, the girl said. There was no-one else, but we sat in trance-like silence, still wearing the beam of pleasure, the grin which had been our hallmark since November. The moment had arrived, but the solicitor was otherwise occupied. I saw the minute hand of the clock move slowly towards his dinner hour. Surely, I thought, the good man will be delighted for us, and happy to forgo a little of his lunch break for such a momentous occasion. He had not been introduced to Harry. That in itself was worth missing lunch altogether!

There was a pile of literature for those wishing to buy a new house, but it didn't interest us. We were going nowhere! Half-an-hour passed. Fearing we had been forgotten, Margaret went to the girl in the office who picked up the intercom. 'He won't be long,' she said, so Margaret came back and walked all round the room looking at the pictures on the wall and wearing a smile generally associated with the cat which has been at the cream.

It was almost midday when we were summoned. The steps to the first floor were steep. Margaret pulled Harry from the front and I pushed him from the rear. It is impossible to enter a room with dignity. I remember everything so well. The solicitor's expressionless face, his formal handshake, his disinterested, 'Oh.' His complete lack of enthusiasm dampened our excitement. Men! Had he no idea whatsoever what the signing of this agreement meant to us or how we had had to work to arrive at this moment? Apparently not! Thank goodness his approval had not been necessary for his interest was purely academic and legal.

How often we have watched a cloud temporarily dull the hillside across the valley. The shadow spreads gradually from the west, creeping quietly over the medieval pattern of walled fields climbing the hill to the wastes of Ilkley Moor. So did the shadow of this solicitor dull the atmosphere

and suppress the laughter we wanted to release. Silently we signed the most precious thing we had into safe keeping, shaking our heads at the remoteness of the man behind the desk and shrugging our shoulders at each other. I risked just one little wink in Margaret's direction as we picked up our copy of the agreement and stepped out into the sunshine of the street. We had half a century of dedicated slog behind us, we'd been too busy with animals and other people's children to produce our own heirs, and yet we had secured future custodians for our earthly kingdom. Oh boy, had we something to celebrate! What ailed the solicitor?

We had expected something more, but his detachment was no doubt good for our ego. We had responded in kind, echoing his silence with our own. Even Harry was quiet as we hurried him to the car park and bundled him into the Range Rover. Only then did we laugh. Joyous, not derogatory, laughter for the fact that the man could not relate to our enthusiasm was funny, very funny, of cartoon quality in fact. We did not bear him any grudge for he had never been to The Currer and didn't know anything about us at all.

We were to discover that no-one knew anything about us either. There was an urgency to go back home and get busy again. We'd laid back long enough and the pre-holiday week of madness and catastrophe was about to begin. We had a burning desire to get home, but on an impulse, I turned off the main road and drove through the village of my school. Twenty-one years I had been Head and I thought it might be nice if I saw someone to tell for we had kept our secret for six months. Sitting outside her creeper-covered house was Pam whose mother had been our school dinner lady. Her father had been a school manager and the competent butcher we needed when pork had to go into the freezer and sides of bacon had to be salted. She, too, took bed and breakfast guests, but was sitting sunbathing and was as brown as if she had been abroad. I drew up before her door admiring the colourful hanging baskets, the polished entrance and the windows sparkling with cleanliness.

'Hello,' she said. 'Where have you been?'

'You'll never guess! We've just signed a Covenant with The National Trust!'

'Oh,' she said, 'Is that good or bad?'

Our news was received with the same lack of understanding almost everywhere. We thought our local MP might be interested, but received no acknowledgement of our letter. I thought the Metropolitan Council should be told but only when I remarked to a local councillor, several months later, that I'd had no letter from them either, did I receive a brief note of approval from the Leader of the Council. A relative said in disgust, 'I don't believe in that sort of thing!' We didn't care. After a week of pre-holiday fiasco, which no-one understands either, we were off to The Hebrides. 'Tiree is the place for me!' one of my camping children had written in the sand two decades ago. Well, Tiree was the place for us just then.

That small western island, so frequently in the sun, was Mother's favourite. Its roads have no hair-raising bends, no precipitous plunging towards the sea, no blind corners and, should you meet an oncoming car, you can pull off on to the grass. You can drive onto almost every beach, right to the water's edge, to watch seals play a few yards away. There are no heights to climb, no rough machair or dunes to cross. It is like walking on a bowling green and it is very, very beautiful. It was definitely Mother's favourite island for all those reasons. She was ninety-two years old and game for anything. The engines on the boat kept Auntie Mary awake, but Mother curled up on the lower bunk and slept like one of the Guides.

We were approaching Tobermory before the family finally entered the spacious lounge. Mother liked the open deck better and we had plenty of rugs for later on, but 9 a.m. is too early. In the corner of the lounge of the *RMS St Columba* there was a writing desk with a strip of fluorescent lighting on the wall above. I have seldom seen it being used. Most of the armchairs were occupied for the early morning was still chill. Mother, whose eyesight was failing, wanted to know, 'What's that white thing in the corner?' She was also a little deaf, so Margaret answered quite loudly, 'It's only a light, Mum. Over the desk. Look, I'll show you.' She heaved herself out of the comfortable chair rather stiffly, for hard work shows most when we rise and take the first few steps. The longer we sit the more we hobble until the oil runs freely in our joints. Margaret worried about it until she heard a ballet dancer, Wayne Sleep, confess to hobbling to the bathroom in the middle of the night.

Everyone in the boat lounge looked at Margaret, grasping any opportunity for momentary entertainment. 'Watch, Mother,' she said reaching for the string pull. Immediately she pulled a voice called, 'Attention! Attention! We are approaching Tobermory. Will passengers wishing to leave the boat please assemble ...' Thinking she had activated the loudspeaker announcement Margaret emitted a strangled cry and re-pulled the light string. The lounge echoed with spontaneous laughter. 'Do it again' someone called and there was an instant affinity amongst the passengers.

Mother's eyesight had been troubling her for some time. Our faces blurred, the television was unclear, distances a problem and reading impossible. We tried everything. The optician, a brighter light, an adjustable table lamp, a large magnifying glass. Nothing helped. We moved the television this way and that and exchanged it for a bigger screen to no avail. She saw curtains and crowds of people and entertained us continually with descriptions of the distortions. The optician said he'd never heard of anyone with Mother's complaint which was not surprising for there never was anyone

quite like Mother. She had worn glasses for seventy years. Now she took them off, seeing no better with them on. We had a great fear that she would go blind, but we need not have worried for she never did.

Tiree posed her a serious problem. There, where flat grassland becomes shining shore and sparkling sea becomes infinite sky, how can ageing eyes tell one from the other. On Tiree there are no trees or mountains to give height to the flatness and mark the horizon. Mother had been going to Tiree for over twenty years and she knew the houses stood apart, distanced by well eaten pasture. When her eyes multiplied a house into a cluster, it troubled her and after a day in the sun, black blobs danced on the white walls of Charlie's cottage. We'd known Charlie since 1951 and had been using his mother's Salum cottage after she had died.

It was a pity Mother's vision upset her for, otherwise, she was 'in her usual' as the islanders say. We had a lovely holiday and The Currer seemed to be on another planet. We have this wonderful ability to distance ourselves from our almost eighteen hour a day routine and it is a blessing. Whenever we all get into the Range Rover we leave work behind. We do not miss making an evening meal if the opportunity to do so presents itself. We eat a simple Yorkshire tea of bread and jam by the fire and it is as normal as if we never do evening meals at all. I find this phenomenon difficult to understand and therefore impossible to explain, but it alone, keeps us sane. The yard without cars does not strike us as empty. Life without people is not lonely as most of our guests believe it must be. Without the chore of making meals for many we do not feel a tangible freedom. The pre-1980 life is still the dominant one. If we could take a blackboard rubber and erase the bed and breakfast I don't think we would miss it. Stop farming and we'd soon be dead.

The ability to distance ourselves from the continual river of guests which flows through our door is a good thing. Because we cannot

remember who came last week, even who left yesterday, caring for them is never a burden. It is a pleasure for the thousand is never more than the few. We can go to Tiree and only the island exists. We can unload a wagon of hay or sit in the Auction Ring and believe we never diversified. We can follow an empty road and we are in limbo. We can entertain our own friends in the sitting-room and forget we have meat in the oven for twenty. People think we have already gone round the bend, but I do not think we ever will.

There is, however, a certain shock returning to a task having completely forgotten it, but I prefer the trauma of returning home to a converted barn full of holiday-makers, whose needs and appetites must be catered for, than the alternative which would be to remember it and anticipate it all holiday.

Mother used to say, 'I couldn't get to sleep last night thinking about what I was going to bake.' So she baked twice when once was enough. I can't do that. Asked what we are having for the evening meal I do not know until the moment I begin to prepare. It means I can tackle a big job without stress which is a modern enemy. We do not live entirely without it, but we can turn it off. When panic is past we can breathe out long and gratefully and say, 'Phew!'

Of course, if we are honest with ourselves, we thoroughly enjoy our guests. 'You must love meeting so many people,' everyone says and it is true, we do. A lady chatting to us one day many miles from home, opened her bag and gave us a religious tract. It told us of the many depressing ways individuals described life. It's a pain, it's a groan, it's a race, it's a fight, it's a toil, it's a giggle. We were on holiday and life for us was a joy. The tract set me thinking what I would have said if I had been asked to identify life. Surely life is an encounter, an experience and long may it be so. A return from spring holiday means an encounter with Isebrook School for children with special needs. We try to plan it this way for it guarantees that the first week home will be one of the happiest of summer. They love their annual holiday at The Currer and we like having them. When the Sunshine bus arrives in the yard there are whoops of joy, leaps of delight and hugs all round.

I cannot honestly say I enjoy shopping or that I delight in cooking, but I love to walk into the dining-room with laden, steaming plates when those around the table gasp with pleasure, lick their lips and cannot wait for their plates to be filled. I am sure Isebrook children are no more satisfied than their peers or our older appreciative guests but they are far more vocal and say such nice things. 'You are a good cook.' 'It's my favourite.' 'I like it here.' Well, so do I. I glow with pleasure at their praise. I love to have them under my feet wanting to help set or clear the table. I respond to their description of the exciting day they have just had. I answer their, 'Goodnight', with, 'Sleep well,' and if someone dares to offer a kiss, I present my cheek. I like it best when ordinary holiday-makers are here at the same time. Modern opinion of children is low. When guests come here with Isebrook, or Kingsley, or Forest Gate, or Fairlawn, or North Gate they have their faith renewed, and certainly their admiration of the teaching profession receives a boost. Those who remain after the children have gone feel sorry that they have left and await letters of thanks as eagerly as we do for they are carefully written and wildly enthusiastic.

May is a pleasantly green month and I'm happy to mow the lawn for it means that grass is growing and cattle will be content. The daffodils are over but it is dandelion time and every roadside verge is a sea of yellow. If we do not let cattle into our approach road, that too will become golden, their many-petalled heads reflecting the sun above. We approve of dandelions! I find myself pushing the lawnmower round them on the lawn rather than beheading them just as one year we vacuumed round a colony of butterflies choosing to hibernate on the family bedroom passage carpet. The dust was not taken from that patch until, on a warm day, they awakened and flew away.

Unexpectedly we received an invitation to go to the Dedication of Fountains Abbey as a World Heritage Site. Be it wedding or funeral, Open Day or Reunion we have a big dilemma. What have we to wear? Unlike other ladies whose dilemma is to choose something suitable, ours is a different problem and easier to solve. If our one outfit is not suitable we can't go. It's pathetic, but true. We habitually wear trousers for the nature of our work demands freedom to bend and climb unimpeded and discreetly. There are no hours left in the day, now, to bring out the sewing machine which was my very good friend for so many years. Consequently I never learned to 'try on' and to buy.

In any case, we said, we hadn't time to go to any Dedication however important. How on earth could we get through our chores to get to Fountains Abbey even if, when we hauled our one dress out of the wardrobe, we deemed it suitable?

We looked at our colourful chart on the wall which shows who the day's new arrivals will be. This visual picture of the year's bookings invariably shows every bed space full. I stick it behind the phone, to lose it would be catastrophic. One year, in Arnside, Margaret bought a card bearing the reminder, 'The impossible we can do. Miracles take a little longer.' We decided to attempt the impossible.

My association with Studley Royal and Fountains Abbey spans nearly half a century. As a student, whose friends went home for the weekend, I spent long, blissfully solitary hours walking in the park and Abbey grounds. There was a time when every gnarled tree bordering the path to the Studley Royal church was familiar, when I knew each of the ornamental lakes and

the view of the Abbey from every angle. In the years that followed I took school children there, and Guides when we were camping in the Dales. In the early days of owning a Land Rover station wagon I used to take the whole family on College Reunion days, leave them and the vehicle in the park and walk briskly back to Ripon. In those days I had time to go to the Cathedral service with the other five hundred past students. Now the Range Rover drops me tardily at the College gates and takes the family to Fountains without me.

The day of the World Heritage Site Dedication was utterly beautiful, confirming my long experience of glorious weather if I return to Ripon. We drove into the crowded car park and left the family to picnic whilst Margaret and I ran to the entrance with no time to spare. Incredibly everyone seemed as casually dressed as we were. It was too hot to be dressed otherwise. There were chairs but many guests were just sitting on the grass. We joined them. It was lovely. I think, in retrospect, it was very important we went to that very special ceremony. Everything about it was so right and so easy. It was such a perfect day and I am sensitive to atmosphere and find historic places peaceful. They teach us whatever toil and stress, danger or tragedy, pain or pestilence, it passes and there is peace. There is a tranquillity about National Trust places which is soul restoring.

We had a comfortable feeling of being known, of having done the right thing whatever friends and relatives thought. We know too, for our guests have told us, that a feeling of monastic peace, of silence, of distance from the rest of the world pervades at The Currer. Of course we feel it too, but we are living in the present, a very busy and demanding present, and it is only early in the morning, or when we pause for coffee in the sunshine, or watch the red splendour of sunset, or look up to a canopy of stars that we are aware. The solicitor, our local MP, the Metropolitan Council, friends and relatives became a family joke after our Fountains Abbey experience, and have remained so ever since.

One of our long-stay guests, a credit manager at a factory in town, was one of the few who applauded our decision. He bought a house on the lakeside at Foulridge and his wife wanted us to drive over one weekend to view it and have tea. We said we couldn't go on a Saturday but we would call on Sunday bringing Angie and Ian, two frequent guests who live so admirably from wheel-chairs.

Saturday is a changeover day for cottages and some who come for bed and breakfast, and there is the daily evening meal for so many it is a no-go day for any other activity. I'd bought a new pack of fish and had pulled off the continuous strands of flat plastic the night before, to serve plaice for the Friday meal. I should have cut the strands. More than that I should have put them on the fire instead of leaving them to fall onto the Freezer 'Ole floor. In those days the Freezer 'Ole was an absolute mess. Dating back to the sixteenth century it had once been a curing kitchen where hams had been smoked. When we came in 1958 it was a ruin we wrote off and walled up, but later we re-roofed it to store camp equipment. When we built another tent store we used it for coke and then, when we diversified, we opened the walled up entry into our living kitchen and used it to house three freezers. The walls were crumbling, the beams were rough and we never let anyone in. Inquisitive guests who tried to follow us had the door firmly shut in their faces. We had hidden the three steps to this glory-hole by enclosing them and the storage space under the bedroom staircase so that two doors kept back prying eyes. It was pitch dark unless the light was switched on, but my arms were too full to do so on that Saturday before we went to Foulridge. I was about to descend the three steps when my foot stepped on a flat plastic band from the fish box. It stood erect and my other foot tripped in it. Incapable of saving myself I fell heavily, hitting the jamb of the second door six inches from the floor. I was convinced I'd done irreparable damage. I called for help thinking I would black out any minute and felt my forehead. Without exaggeration there was a lump as big as a lemon. It didn't hurt so I got up and when Margaret saw it she ran for a bag of frozen peas. Guests surrounded me, insisting I must go to the hospital, but it didn't hurt, I didn't feel faint, my sight was normal and, what's more, I had an evening meal to make. The frozen pack brought down the lump and by the time I was serving all the evidence of a fall had disappeared. If that had been the end of the story it would not have been worth the telling.

Early on Sunday afternoon, twenty-four hours later, we were preparing to go to Foulridge and I glanced in the mirror, a thing I seldom do. I saw two blood spots one in the corner of each eye. They could have been put there with a felt tipped pen and by the time we reached the Cobbys' I had two very black eyes. The purple spectacles got bigger and bigger and blood began to collect in bags on the summit of each cheek. They were heavy and when I moved they swished about like cleansings from a newly calved cow. When I put on my glasses they perched on the moving bags of liquid and my sight was distorted. Without glasses I didn't even have to look down to see them and the weight of them bent my head.

Amused smiles surrounded me everywhere. When I served Sunday joint everyone was laughing. Next day I went to the doctor and asked him to drain away the liquid. He took one look at me and said, 'Good God!'

'Just prick them and get rid of the blood', I begged.

'I can't do that,' he was laughing too. 'I can give you some pills.'

That's all they ever do these days. It's a pill or a jab which is expected to cure everything. The pill emptied the sacks by distributing blood to the rest of my face. I looked like some alien monster for a month and it was impossible to evade comments. I tried to choose the same checkout lady at the supermarket to avoid having to tell the same story over again and I sent Margaret to greet new guests for fear I would frighten them away.

* * * * *

We thought we might just be being heard by the Council, re Altar Lane, that two mile stretch of the green lane half of which runs along our southern boundary. We had been shouting for attention since the 1960s when the GEB had put up new pylons and the YEB had taken a row of poles just inside our perfectly secure, dry stone wall. Between them they had made a dreadful mess of the lane. It had become deeply rutted and very nearly impassable. Each had blamed the other and The Council had failed to insist both must contribute to its repair and we were the losers.

Stranded car drivers habitually took stones from our wall to help themselves out of difficulty. We had complained annually and bitterly for The Council, who owned the road, had an obligation to maintain it in kind. They never did. The eroded wall became insecure and there was constant danger of cattle finding a sudden gap and straying. Our neighbour's cattle had wandered onto the St Ives estate and left footprints on the golf course. They had been asked to pay a sum of money we could not have afforded. We secured an insurance to cover us against this happening and endeavoured to maintain our wall constantly. As more and more stones found their way into the ruts, to repair the wall became impossible and Margaret and I fenced half-a-mile way back in the 1970s.

The lane had deteriorated until it was only passable by off-road, four-wheel-drive vehicles. Anything else had to be hauled out by tractor. Occasionally the fence had been cut to allow drivers to use our dry land to by-pass the quagmire and, in the late 1980s, the fence needed renewing.

Fencing is expensive even if you do it yourselves and, unexpectedly, the government was giving grants for the building and repairing of dry-stone walls. We realised that, grant aided, such would cost little more than half-a-mile of fence. We made enquiries from The Ministry and found our share of the £12,000 would be £5,000. We could not spend that amount of money on a wall bordering a deeply rutted and waterlogged road, so we sent for officials from the Council who came and agreed we could not and should not. They admitted funds could not be found to repair the road and suggested it should be closed to all traffic. Dirt track enthusiasts would have to go elsewhere. It would take a month or two for legal closure, the two men said. They asked us if we would pay half the legal costs, and we said, 'Yes'.

It seemed as easy as that! After twenty years trying to attract attention we had, at last, found two men who would listen. So we went ahead with the wall building. We hired a professional and it took him three years, cost the Government £7,000 and us £5,000 and nothing whatsoever was done about the green lane. Men! If we'd had less to do we would have protested more frequently. Even so we phoned occasionally and wrote many letters and got nowhere. We had great joy seeing that beautiful wall grow daily in a perfectly straight line but knew that, if the old road was not closed soon, the new wall would disappear into the ruts as the old one had.

We find men, the gender we mainly have to deal with in our male dominated profession, quite exasperating, which makes us appreciative of a story told to us by a Hebridean friend whose husband had been ill and it had been her lot to attend to the outside lambing and calving and feed the bull housed for the winter.

'It'll have to be you, dear,' her husband had said, one very wet night. Having great sympathy for the bovine lady whose lot was not enviable whatever the weather, she had got up, dressed and crossed the field to attend to a long and arduous calving. Soaking to the skin our friend had, at last, been able to leave a baby suckling its mother and, as day was breaking, she returned to the house. The huge bull poked his head over the half door as she passed. She told us she had stood in front of him and shouted, 'MEN!' Good for her.

* * * * *

We went to Grange-over-Sands as had been our custom for several years, and we left The Currer in the good hands of Dorothy and George Winup. All day, on the Thursday before our holiday, Auntie Mary, in her usual, baked biscuits to eat at the close of each lovely autumn day in Lakeland. She also made chocolate coconut slices and Weetabix loaf to butter. Her perfections filled several tins. We collected these when we brought her and her luggage to spend Friday night at The Currer, ready to go to Grange next morning. We put the pile of tins on the table in the entrance to the porch, ready to pack into the Range Rover.

Next day we drove the short sixty miles to the nicest side of Morecambe Bay, in the most appalling weather. We arrived in a cloudburst and everything we had was under cover on the roof rack. If we'd have been able to make even a cuppa we'd have waited, but it would have made no difference for the deluge continued all night. So we pushed Harry and Mother nearer to the fire and stationed Auntie Mary in the kitchen for it was her bi-annual job to stow everything away. Margaret's job on the roof rack was the least enviable. We were soon both soaking. Fortunately we were able, always, to drive very close to 'The Studio', our autumn retreat. We are used to getting wet and to working fast and were happy to be on holiday. Everyone was amenable and relaxed. We had a whole fortnight to enjoy just being together. What more could we want? The rain would pass. We always had sunshine on holidays! By morning everything would be fine. We slung the last articles into the cottage and lifted down the dripping wheel-chairs. Our vehicle was too high to sleep in the garage but we always put the wheel-chairs there. We shut the door firmly behind us and said, 'That's everything!'

'Where are my tins of baking?' Auntie Mary asked. Where indeed!

'They must have come in,' we said with conviction, but we knew darn well they had not. We braved the storm outside and went to the empty Range Rover. Margaret went back on top though we could see there was nothing. We looked inside and under the seats even though it is impossible to push tins there. Eventually, empty-handed we returned to the tense atmosphere of the sitting-room. If my baking, or Mother's, had been left we would have laughed, phoned home and told Dorothy and George to have a feast. We couldn't possibly do that to an elderly aunt who had baked all day.

Harry was as if he had been electrified, every nerve trembling. We couldn't begin a holiday like this. The rain cloud would pass and so would the metaphorical one, for Auntie Mary is no ogre, but Harry would get into a state and Mother would not sleep for worrying. The wonderful affinity we were used to on holiday would be shattered. Margaret and I are prepared to grovel for family peace. We will do anything to preserve it.

'We'll slip home for the biscuits tomorrow.' I promised. 'We are only an hour and a half away. We'll have our first day of the holidays in the Dales and pick up the biscuits.' Harry relaxed.

'It doesn't matter, really,' Auntie Mary said. 'It's not necessary'. But Mother was emphatic. 'That's what we'll do,' she declared, 'but we're not driving on the Top Road, letting everyone see us come home. We'll go down to Marley, have our lunch there and you can walk up the hillside and get them!' No mean feat, but Mother didn't bring up her children soft.

Beauty so often follows rain. The newly washed morning was incredibly lovely. Margaret walked the dogs on an empty promenade whilst we packed sandwiches to eat on our neighbour's land which climbs from the valley floor, steeply, until our hill-top perch is reached. Long ago Marley, too, was owned by the Currer family. William and Isobel were living there in 1571 when the Paslewes of East Riddlesden Hall sold The Currer to their son, Arthur, for £30. At the first telephone box we alerted Dorothy to the fact that we were coming home and would be scrambling up the hillside because Mother didn't want all the village tongues wagging. Somewhere on the A65 we were flashed and hooted down by George and his grandchildren who'd come out to meet us and the biscuit tins were handed over. There was a great deal of laughter all round. We happily turned off the main road and went to Slaidburn. The sun was so hot we all had ice-cream and then we followed the road through the Trough of Bowland and had our picnic there.

Mother's big toe was beginning to cause trouble during the night. The Studio was an ideal place for sleeplessness. The bedroom I had begun to share with her led straight into the sitting-room. A few strides from the bed and I could switch on the gas fire. It was a luxury. Only when she lay back did it trouble her. One or other of us would sit in comfort beside her until the pain left and she could take the few strides back to bed. During the day everything was fine and we had a lovely holiday. On our return we called the old doctor and he put Mother on the first pill, other than the occasional antibiotic, she had had in over ninety years, but it did little to help. We put a bed in the sitting-room and Margaret and I took turns sleeping on the floor. The doctor prescribed a cage over her foot, a sheepskin rug under her feet; we bought an electric overblanket, but nothing helped and we were up and down all night. During the day everything was normal. Mother was baking as much as usual, was as eager as ever to accompany us every time we went out in the Range Rover and was still the perfect hostess to all our guests. We cut a hole in her shoe and she never complained, but when she went to bed the pain returned.

* * * * *

November was wet. The earth accepted what it could and then standing water appeared in the fields and the valley floor was flooded. We had recently completed a drainage scheme in the sloping fields rising to the moor. It was disappointing to see water gushing down from the undrained pastures which had once been moor and had remained so named. The rough track following the wall of the intakes became a river weeping tributaries down the supposedly drained eight-acre. Margaret sent for John, who does all our contract tractor work, and together they walked the quarter of a mile of waterlogged track and he said the only solution was to dig a dyke on the top side and channel the water to the moor gate and into the one running down Currer Lane. With a few loads of quarry bottoms the track could be made passable again. It seemed a good solution to our problem.

'How much would it cost?' she queried.

'Well,' said John, 'We'd be thinking on £2,000.' Twice as much as Grandfather had paid for the house and barn and 170 acres just sixty years ago!

The two came into the kitchen and told me. Our chart for December customers was empty. We had forfeited any hope of a seasonal bonanza by closing. Our income on the run-up to Christmas would be Mother's pension. Our winter expenses of fodder and fuel and insurances would be exorbitant. Our overdraft would soar again. However we decided we had no alternative and John was told to start work at once. The phone rang and Margaret took a one night booking for five for Sunday night. The group of young people wanted an early breakfast at 6.15 a.m. She took the booking because we were desperate for income. 'One day won't kill us,' she said. Grandmother Smith had frequently said, 'You can stand on your head for a fortnight if you have to!' We've proved you could keep it up much longer than that. We were tired with sleepless nights and the winter feeding of cattle, but we are survivors.

The young guests said they would be in town until Christmas, staying at a guest house in the centre for they were to train employees at the newly built McDonald's burger restaurant. On Monday they left for permanent, more expensive accommodation, but the manager of the new venture phoned to book in two other trainers for a fortnight. We smiled with pleasure. We were back in business, albeit in a very small way. At the town guest house the five young people met opposition to their request for a 6.15 a.m. breakfast for the first shift and to lie in until dinner time for the second. 'No,' the proprietress said with good reason. 'You will not get that deal anywhere!'

'We did this morning,' they said, but when asked to say where, were unprepared to do so.

The good lady, however, guessed correctly and said, 'I've heard of them. They'll do owt for money!'

'Too true!' we said when the five young ones came back and told us.

Our cheaper rates left money for taxis so we got all the McDonald staff, fourteen of them, right up to and past Christmas. We could also provide them with an evening meal which suited them fine and drove us nearly round the bend. I never knew how many there would be, for their different shifts had all of us confused. Some would want early breakfast and some would want it at 10.30 a.m. and some would sleep until noon. Those on the night shift would come home in the early hours. Taxis sallied forth every few hours and we earned every last penny of the money we made to pay for what has since been known as the McDonald Ditch. The water, which frequently leaps noisily down beside Currer Lane every winter, proves the immense value of that December project.

Happily the new young doctor came to see Mother's foot and told her to take two of the tablets and the pain went away and never troubled her again. We all went back upstairs to bed and we began to think positively about Christmas. Life for everyone is some good, some bad, some up, some down. I'm just not sure that it changes so frequently for other folks as it does for us. We never have to wait long for excitement of the positive or negative variety. One thing follows another and we scarcely have time to take breath.

This opportunity to take a long day out at Christmas lasted just three years. On this second of our Yuletide picnics we fed 100 Hereford heifers and, on a gentle, pleasant day we went to Byland Abbey. It was a long way to go but the empty roads could be meandered along slowly and with the dignity demanded by a day so holy. I remember we were almost happy. Mother's pain had gone. We were all well. We'd made a packet of money to nearly pay John. If we'd flogged our aching, ageing bones to do so, it didn't matter a scrap!

Byland Abbey was deserted. We parked in view of it. If we are not careful, these days, we are in danger of taking Divinity for granted. We have no Sunday rest in which to worship and when the opportunity arises to do so I tend to be disappointed. I appreciate it is necessary for mankind to be reverent. I have not forgotten, though I am no longer singing, 'Heavenly Father may Thy blessing,' with my school children in Assembly or, 'Just as each new day is dawning,' with Guides camping on island shores but I know I have lost something of infinite value.

I am deprived of inactivity. That's what's wrong and I don't know what on earth to do about it – yet. On Christmas Day 1988 all five of us ate our sandwiches in the bliss of inactivity, in view of the Abbey. Monks had tilled The Currer before us. People had not lived as long in those days but they had had more space, more time to stand and stare. Mother, sitting eating ham sandwiches, was still active every waking hour of every day of her ninety-third year.

Margaret and I left them to finish drinking tea and walked, with the dogs, quietly among the Abbey ruins looking for an inglenook fireplace similar to our own. The only one we have found remotely like it, is a 13th century one in Skipton Castle. We returned to the Range Rover via the front of the pub and two recumbent stone dogs guarded the door. Jess, who doesn't like her own species, nearly went berserk. On subsequent years duty has called us home earlier, before darkness lends magic to lighted Christmas trees and the windows of shops and inns in village streets. Our journey home from Byland must have been a record one for decorations. We drove entranced all the way home.

When you are only almost happy there must be some dark cloud somewhere. Lurking in the minds of all of us was unease about Thomas. His family had looked after The Currer for several years releasing us to go on holiday. Dot, his mother, had been one of my Guides and her family is very dear to us though they live far away in Suffolk.

One year, when Thomas was only eleven, we'd returned from holiday to learn he was going home to a doctor's appointment. They were worried because his balance seemed poor. Immediately he was rushed into Addenbrooks Hospital for a brain tumour to be removed. There had followed the agony of post operation, then recovery and heaps of high-apple-pie-in the-sky hope. For sixteen-year-old Thomas the cancer was now terminal and his parents promised he should spend the New Year at The Currer and they would celebrate with a huge family party. Dot had booked the whole house for this great occasion which everyone knew was a farewell party. There was no mistaking the fact that though deeply drugged, the boy was still in great pain, but there was a special something about him peculiar to so many cancer sufferers.

What a party that was! Dot was one of six children and they all came with their wives and husbands and children. There was magic in the air for they were musicians and singers and instruments and voices filled the The Currer as never before. Mother stayed as long as she could and, having taken her to bed, I should have returned to the party, but I was so tired my head cried for the pillow. Harry and Margaret stayed and Harry hasn't stopped talking about it since. The old house resounded to the music of fiddle and flute, ballad and song and when the mid-night hour was reached the boy, in whose honour it was, picked up a brass pan from the hearth and collected a substantial amount of money for Addenbrooks. The Currer has provided the rendezvous for so many social occasions but none more extraordinary than this, none more beautiful, none more poignant than this farewell to a teenage boy going home too soon.

We had a letter from him on his return saying he planned to make a parachute jump in aid of charity and that he was as happy as 'the King of England'. Towards the end of January he became very ill and died at home amongst his family.

We tried, in the quiet weeks that followed, to make as few evening meals as possible. The pre-Christmas period had been lucrative but had worn us out. Our chart was filling for the summer and the days were beginning to lengthen. If workmen rang for accommodation we said bed and breakfast was £8 and omitted to add we did an evening meal unless asked. Naughty, but no-one grumbled. Then snow came. Really heavy, good old fashioned snow. The man who had been with us several weeks decided to stay in town rather than walk up and down a drifted Currer Lane. 'I'll be back when it's gone,' he said.

I cannot over emphasise the wonderful feeling of freedom isolation brings. It only comes with the roughest of weather, but we love it. We made no attempt to dig out the road. Auntie Mary stayed, we stoked up the fire, shut the dividing door between us and the converted barn and prepared to thoroughly enjoy being cut off from the rest of the world.

It was already dark on Monday evening when the phone rang. "Ave yer room fer four fer tonight, Missus?' I was asked.

'Well, yes,' I replied, confident our accommodation would not be acceptable. 'We're empty, as a matter of fact. Our road is completely blocked. No one can get here.'

"Ow far is it?' the man asked.

'No-one can get to within a quarter-of-a-mile of us,' I announced gratefully. 'The one man staying with us has decided to stay in town.'

'We can walk a quarter-of-a-mile,' said the persistent man. 'We are working on the reservoir and are absolutely frozen. We don't mind walking the last 500 yards.'

'It's deep,' I said, doing my utmost to put him off. 'And we're only doing bed and breakfast at the moment because we have a lot of cattle to feed and umpteen problems.'

But the man would not take no for an answer and said they'd be here at 8 p.m. We were feeding the cattle and I had come to answer the phone bringing well-manured snow into the front porch. I lit a huge log fire in the dining-room and returned to help Margaret in the cattle sheds loosening the bales she had carried into the mangers. Eventually we saw car headlights at the top of the road and anticipated the shock of the men when they stepped out onto our deeply buried road. We could see their unsteady torch and knew they were floundering. I decided to climb the hill myself to lead them to safety. In places the drifts were three feet deep.

'Bloody hell!' said one man as I reached them.

'I did say,' I chided. They were dressed for bad weather, booted and waterproofed, but they were already bedraggled. They followed me silently clutching battered holdalls. It's not fair, I thought. Why can't folk just leave us alone?

The fire was blazing cheerfully when the four men shook off the snow and left their boots on the mat. They clustered round the burning logs completely disinterested in my efforts to show them to their rooms. All they wanted was the fire. I think I knew before I asked them. 'You haven't eaten. Have you?'

They shook their heads.

'I did say!' I sighed.

'We were too cold and wet,' they said. Indeed their sodden kagoules were festooned on the chair backs dripping pools onto the parquet floor. So I opened some baked beans and made some chips and the dining-room windows steamed with drying clothes and the stockinged feet of the workmen cluttered the hearth rug.

I firmly believed they'd only stay one day, but the warmth of our fire induced them to walk up and down that buried road for the rest of the week, to eat nothing more exciting than egg and chips, sausage and chips, fish and chips until their job at the reservoir was finished. Peace was not to be!

* * * * *

Margaret had a bee in her bonnet. She would argue that it had a legitimate right to be there and I wholeheartedly agree. It's a good job that I really do believe in the same things as Margaret for I fear she would take me by the scruff of the neck and make me if I did not. She was alerted by the local newspaper to the proposition of the Metropolitan Council to allow the building of four hundred houses on the hillside across the valley. Such development would destroy, for ever, a sizeable woodland and many acres of traditional walled pasture alongside the Leeds/Liverpool canal. She motivated me into protesting by letter and attending the March Enquiry. I became heartily sick of that development project in the spring of 1989. If someone had told me that it would be years before the decision of the Enquiry could be implemented, and that, even as I write this, ten years later, little action has been taken, or any guarantee that it ever will be, I might have enjoyed the fight. No-one did and I hated the whole thing. When the scheme was agreed it was shelved and I, for one, did not want to think of it again.

Dutch elm disease had killed eleven of our trees. We were aggressive as well as sad. Several of them had helped the line of beech trees at the foot of the Five Acre to obscure the ribbon development along the valley road. Besides being old and beautiful they did a worthwhile job and would be even more necessary should there be further development on the far hillside. To have them felled troubled us. We got a local firm to cut them down and saw them into discs. John, our contract tractor man, collected them and a mountain filled the old bullock shed, partially disused since we had built new. Thereafter, for two years, Margaret and I split them with a woodman's axe and we did not have to buy which was a great saving on our ever empty purse.

The Council offered to replace the trees and planted a screen of fifteen saplings along the west wall of the Five Acre. Knowing that cattle and goats eat young trees we accepted the offer on condition the small plantation was fenced. The men instructed to do so came via Altar Lane. The green, walled lane looks quite passable on the Ordnance Survey map and appears to offer a usable short-cut. The men were swearing mad when they eventually arrived but I'm sure they didn't vent their anger on us by erecting the weakest fence I've ever seen, because they were nice men and promised us a lovely silver birch in memory of Thomas. The weak fence lasted no time at all. Cattle broke in before breakfast and they'd all the trees eaten before we knew.

For years we had gone, every second year, to Harris, in the Outer Hebrides. We had spent a fortnight on the islands every spring for nigh on forty years. We wondered whether it really was too far for Mother and agreed we ought to go somewhere only one day's journey away. Dot's parents came to see us and suggested we went to Southwold, in Suffolk. We had been hesitating because, with the elderly and infirm we believe it is necessary for us to be among friends who will come to our assistance if we have an emergency. We think this is far more important than anything else and the wealth of that care, demonstrated by people on the islands, has compensated for the tremendous distance and the long sea crossing. Southwold seemed to offer us that necessary reassurance for Dot would be only a short distance away and her husband, Peter, was headmaster of Southwold Primary School. We asked him to find us a self-catering cottage and he found no difficulty in doing so.

We had had a hard winter. Snow, frozen pipes, gales and heavy rain. Since converting the barn we had had to build a stack outside. It was always unsatisfactory. Waterproof until disturbed, it initially made us feel secure against shortage in winter. When we began to use it, it became a liability vulnerable to wind and rain. After a storm bales would be wet and heavy. After a gale the plastic sheet would be torn. We'd carted a hundred and more old tyres from a garage in town in an effort to weight down the cover but hollows and creases collected water and became miniature lakes which were apt to cascade over us when we moved bales. We had a love/hate relationship with that stack. On the one hand it was essential fodder and on the other it was our enemy. As we will not, and cannot afford to waste, Margaret dragged the heavy bales free, weighted down the cover and she and I dodged the waterfalls believing anything was preferable to being short of food.

As March gave way to April, with no let up in the weather, we began to tire of winter and long for spring. Auntie Mary had a small cottage on the other side of town. She had spent at least three days a week at The Currer since Auntie Janie died in 1975. One day, just before spring, I said to her, 'Don't spend another winter alone. Come and live with us before the next one comes.'

Unexpectedly she said, 'Do I have to wait so long, until next backend?'

'No!' I said. 'You can come tomorrow,' and there followed a month of extraordinary activity. We carried Margaret's bed to join ours in Mother's room and turned our largest bedroom into what we thereafter called Auntie Mary's boudoir. Our builder put in a south wall mullion to lighten it and we pulled out old cupboards and put in new. We decorated and carpeted it until it became a luxury apartment.

Auntie Mary's neighbour said, 'It's foolish to give up your home before you really need to,' but Auntie Mary was not deterred.

'I think I should go before, whilst I am still able to help them. As I feel now, I think I can do that.' It is a pity that such a philosophy is not shared by millions of old folks, living alone, with nothing to do, who would be quite capable, as was eighty-two-year-old Auntie Mary, of purposeful activity. As I have said before, nothing is more important in life than to be needed. She brought everything from her cottage and we distributed it around The Currer and her house was put up for sale.

Easter came before we were ready for it and winter was disinclined to go.

One of our annual Easter guests did not come for holiday but as a volunteer at the Worth Valley Steam Railway. Our guests all called him John. Almost everyone is so called and he was special so we never called him anything but Mr Skeggs. He came laden with jigsaws for Harry, who is a winter addict and expert, and with photographs methodically catalogued in albums which was his addiction and at which he was an expert. Mr Skeggs, the Railway Man, became well-known and loved by all our regular guests. Because he worked all Easter he brought his daughter and her family at Spring Bank and they all had a good holiday. He always arrived on Maundy Thursday and those arriving on Good Friday and Easter Saturday would greet us and then say, 'Is John here?' and we would say, 'Of course.' Sadly he died early in 1996 shortly after making his booking.

The Barbault family came for Easter, Bernard, Yveline, Denis and Emilie who was only four. Normally they had been coming for three weeks in the summer. Coming in April for two weeks made them known to nearly all our regular guests. They were becoming quite a part of the family even though only Bernard could converse. The two children were delightful but we were always too busy to encourage their English. They used to come from the dining-room with empty glass jugs and say, 'More water, please.'

Richard, our Down's syndrome friend and regular Easter visitor, brought his dungarees and, whilst his parents toured The Dales he painted our field gates in the yard, and our seat and the garage doors. He would have painted us if we had stood still. Sadly there is no opportunity to do that. Richard is very proud of his work and shows off his handiwork to all returning for evening meal. He is a more employable person than most and we pay him for his services because he does a good job. In the evenings he would sit with Mother in our front room, asking many questions and showing his medal for his achievements in the Special Olympics. He loved to say, 'Bonjour,' to Yveline and 'gute Nacht,' to Marlene, an annual Easter guest from Germany.

Richard's bedroom is always tidy, far more so than other young people's these days. He can stack dirty plates and assemble cups sensibly on trays far more responsibly than most. He clears everything from the table and resets it almost correctly and having done a job once he can repeat it exactly right. He, like the Isebrook and Fairlawn children (he was educated at Fairlawn), not only enjoys his food but says so. He is a comedian who will sometimes don an apron and help me serve. He wears a beaming smile and has everyone in hysterics. In 1988, he helped competently on the farm and promised, at the end of his stay, to come and visit us in Southwold. His father, Terry, put together some of Auntie Mary's furniture and Bernard, who stayed a fortnight after his wife and children left, put up a strong fence on the bottom of the First Intake.

Before he left Bernard said he thought Denis would soon be old enough to come alone and learn English properly. The boy was only twelve. We chose not to hear. How could we possibly deal with a lone teenager, a boy who couldn't speak English, whose home was in the city of Lyon? What on earth would he find in common with four old ladies and a sixty-two-year old disabled brother? If, or when, Bernard asked us outright we would have to say, 'No!' We did not relish the task for the French are totally charming and Bernard could be very persuasive.

We had previously taken our spring holiday at the end of May but the calf-rearer who supplied our calves, a hundred three-month-old Hereford heifers, anticipated a need to buy earlier than usual. It meant we had to go to Southwold at the end of April and we thought it might be quite cold. We were used to hot sunshine in the Hebrides.

A few days before we left I received a worrying phone call. For the twenty-one years that I had been head of the happiest village school in the dale, Joan had been infant teacher extraodinaire.

Her contribution had been enormous. There had been a working relationship between us of tremendous value and we had remained friends, having both retired on the same day. Sadly the nature of our work makes a social life impossible, but inasmuch as it was possible, Joan and I had kept contact. The phone call was from her son alerting me to the worrying news that his mother was in hospital undergoing an emergency operation. I went at once to visit her. 'It was malignant,' she announced. 'They are confident they have got it, so not to worry.' Not to worry indeed! I was trembling as I left the hospital and anxious all holiday.

* * * * *

On a cold morning at the end of April we clambered into the Range Rover in poor fettle. The wind was bitter. There was an inadequate growth of grass and last year's heifers must have an evening feed. Dorothy and George, who would have to administer this, arrived early, before we had packed the roof rack. In our haste one of us, (I was blamed), knocked off the inside windscreen mirror and we had insufficient means of seeing what was behind us. As we pulled out of the yard, somewhat apprehensively, I was still wearing my overall and my shoes were untied. Projecting from the roof rack was the silver birch to plant in Suffolk. I always drive first because we know that leg of the journey. When we have to follow a map in detail, Margaret drives. I drove slowly to the Leeds Ring Road and pulled off in a lay-by to eat a late breakfast.

'It will only take you five hours to get to Southwold,' we had been told. At 10 a.m., two hours after leaving home we were still on the Ring Road, having been delayed by breakfast, toilets and Charlie Brown's for a new mirror for the Range Rover! We didn't reach Suffolk until seven. We were shamefaced. No, we hadn't taken a wrong turning, no we hadn't had an accident or a hold up. Why did it take so long? Only those with elderly, handicapped and dogs really understand.

Dot was waiting for us at the self-catering house. We are used to cottages and this proved to be a mansion, in a beautiful garden, near the sea. My, were we impressed! It had five bedrooms and the central heating was extremely efficient. This was a blessing for the wind howled and rain slanted down. We took off sweaters and wallowed in luxury. What's more, we really believed we deserved it! No one year is really harder than the other, but the winter just past always seems to have been the worst. Mother's toe, the McDonalds, snow, workmen, Thomas's death, Joan's operation, Auntie Mary's removal! We flopped into armchairs and let Dot serve us tea. It was wonderful! All Sunday it rained and the wind blew and we, who work in winter conditions annually, without feeling cold, shivered and froze and felt we couldn't stand it. We took Harry out in the wheel-chair festooned in waterproofs and were wet and perished no matter how much we exerted ourselves.

'Is it weakness, or is this a colder place than the Hebrides?' Margaret wondered. Nostalgia for the islands could not be tolerated, we decided, as we pushed Harry back to the comfort of the warm house. Mother was cosy but she would venture out any day rather than stay in doing nothing. She could neither read nor watch television and if we'd said, 'Come for a drive,' she'd have hopped in gratefully. Dot and four-year-old Hannah were coming to dinner and the rest of the day we closeted ourselves indoors and talked about Thomas to ease the pain of mourning.

Most families, where the members were all over fifty, would have no problem doing little for days on end; especially farmers who sit too infrequently and have too little sleep and too few luxuries. It is on holiday that the job we always call our third one proves to be our principal one. At home we might say farming is our major role. Asked for our profession, we do not hesitate to say so. Animals come before people, we tell our bed and breakfast guests who have to wait five minutes for their evening meal. Our role as carers seems relatively insignificant which is a totally unfair judgement on all the millions of carers all over the world. It is only on holiday that we actually notice how time consuming that job really is. For every ignored disabled person there is an ignored carer which should give some people food for thought.

If the wild wet weather had persisted in Southwold we would have found being prisoners in the house quite difficult to cope with, but the sun, our loyal companion, took pity on Mother, Harry and Auntie Mary who did not want to spend their holiday indoors (Margaret and I would willingly have hibernated), and the rest of the holiday we sun-bathed. How wonderfully warm it became.

We liked Southwold. It was old-fashioned in the most pleasing way, quietly sleepy and definitely out of season. The unique, continuous line of beach huts were being painted for the summer. Few were being used but we decided to hire one for the fortnight. Should we need it, it would give us a windbreak and we could cook lunch there and eat in comfort. It was a different holiday from our usual island ones but there was something very nice about it. Mother accepted the wheel-chair and the long hut-lined promenade was an easy push. Harry would go from Land's End to John O'Groats in his. The daily sun was hot but low and we lost it soon after lunch so we would then cruise around the unspoiled countryside. We found it pleasantly monotonous and we never quite knew where we were or if we had been that way before so we spent less time than usual in the

Range Rover and sat on the high, cliff-top promenade which was still in sunshine. We spent a lot of time gazing out to sea. The beach was almost empty and the sea always near and we relaxed.

Mother said the sea was 'uneventful' which was true but we liked its gentleness and were happy with Dot and Hannah, getting full value for money sitting outside our beach hut. It was, after all, only like the store tent of our left-behind camping days.

We had a memorable day out, visiting Joe and Pauline Walters in Fourncett St Mary. Both are in wheel-chairs but their ordered home is run more competently than many. It was a brilliantly sunny day. Pauline had made a hot lunch but we ate it in their cared-for garden and then went indoors to admire their pictures for both are skilled artists. We were exceedingly impressed by their lovely bungalow in its delightful setting. We so seldom have a chance to visit our friends.

Cynthia, Terry and Richard visited us in Southwold as they had promised and, because we had a spare room, they stayed overnight. It was ridiculous to be doing bed and breakfast on holiday. What ails us? When we are on islands we are farming! On the Sunday night Cynthia, Auntie Mary, Richard and I went to evensong at the very lovely Southwold Church. It was an old fashioned service and memories of my Ripon College days came flooding back. I felt sudden nostalgia for The Church of England for which I had had no recent time. My cathedral had a ceiling of blue and my God was the God of the Open Air. He accompanied us as we tended cattle and people and welcomed the many disabled who take holidays at The Currer but for years I had not had the opportunity to go looking for Him in Church. This is quite, quite lovely, I thought. The evening sun was streaming through the clear, leaded windows and a profound peace crept in and wanted desperately to stay. I was hypnotised into remorse that I could not do this more often. The prayers and responses came easily from my lips for I had learned them well as a child. I allowed the memories of College to possess me. There were two services every day. The morning one was obligatory and the evening compline voluntary. The traditional chapel of St Margaret at Ripon has become a library and a new one has been built in the grounds to accommodate the mushrooming of the student population.

The Southwold Church was both big and beautiful and the visiting clergyman led an old-fashioned service. I think we should have tiptoed out before the sermon. It was a shame to spoil what had been so pleasurable. The message of the visiting preacher was commendable. I approved his theory that, in some cases, it was better to do the job yourself rather than employ someone else. He was getting round to a complaint that congregations employed him to sing daily Matins and seldom came along to sing for themselves. His first parable was that an overweight person might employ a dietician to help him but it was far better to lock the pantry door. There was no bigger nuisance than the mole, he said. You could employ a professional catcher if you wished but far more effective was to put your motor bike exhaust down the hole yourself! Had I been alone I would have walked out. I partially appeased my anger by ignoring the preacher on my way out through the vestry and I spent a restless night mentally telling him just what I thought of the inhumane solution he had recommended in the Church of the Creator. I rose early and wrote him a letter and said I was a teacher who had been trained at a C of E College and that I had been head of a C of E School which qualified me to criticise. A few days after our return I received a reply addressed to Mr Brown. He'd been a teacher, too, and was sorry, old boy, if he'd offended but his congregation were country folk and he was sure they would be fed up with moles. Some you win and some you lose!

* * * * *

Some you rear and some you don't, especially if you are a duck. One was 'sitting' on our return. Usually a broody duck goes off into the nettles and is in danger of becoming dinner for spring fox cubs. This one had the sense to stay in the Duck 'Ole. To do so was to her credit but to sit on fourteen eggs was surely insanity. She had been there far too long for Margaret to remove some so she was left to her fate. To our astonishment all fourteen eggs hatched. Margaret hurriedly made a vermin proof pen and every duckling survived. They surely cannot all do so in the wild. They are so small. I prefer not to picnic at Bolton Abbey in early spring for minute babies are squeaking mid-stream all the time. Within a month of our duck population doubling their pond dried up for the summer and I had to dig a catchment area below the spring or the flotilla would have had nowhere to sail.

Joan recovered well from her operation. Sometimes, on her visits to The Currer, we would go bilberrying. Once we all went in the Range Rover and had a picnic at Newton in the Forest of Bowland where she and Jimmy had had a weekend retreat and where she had often taken school-children to play by the river. There was nothing I loved more than to spend a while with Joan remembering our twenty-one years together at the village school and telling each other which grown child we had individually, recently seen. Joan was necessary to me because only to her could I talk of my teaching days. Good days should never be totally left behind. Life is a jigsaw and one missing piece spoils the whole. Joan was my connection with a past neither of us had really wanted to leave behind and just a word, a smell, an atmosphere, or an encounter triggered off a shared memory which needed no explanation. Ours had been an experience of children and parents. There were memories only the two of us shared, situations over which only the two of us had laughed, things only the two of us had overheard and perhaps I, more than she, needed to share a look-back because my life was so overflowing with novelty.

Equally important was it for me to watch the film taken in camp and to return annually to The Islands. Nothing about my Guiding must be lost. This was infinitely easier to cope with, for Margaret shared all those memories. Together we could go back to Harris and Tiree, to Barra and Skye and past and present would be united.

With school it was different. We did not revisit. Not because we did not wish to but because the link between the old and the new was not there. We were strangers to the new staff. My only real links with the school, in which I taught for so many years, were the children and Joan and all had left. It was terribly important to me that Joan made a good recovery. It was no surprise to me that she coped with the shattering blow as well as she did. I had long experience of her courage. Her sense of humour was the biggest asset she had. She thought one of the most amusing ways to spend an hour was to sit in our kitchen and have hysterics at our efforts to produce and serve an evening meal in a limited time. We were always late starting on the days she visited us and so we had to work at greet speed. It is disconcerting how many guests like to sit in our kitchen and watch us cook. We never minded Joan because she never sat in silence. She dared to laugh at us, which few do, when we are cooking. It is silent watching which unnerves. One frequent visitor sat at my table watching me prepare breakfast. It was so early Margaret was still in bed. Unaware of where she was and what she was doing, this lady said, 'I don't like people in my kitchen when I'm cooking.'

'Don't you?' I asked.

'No,' she said. 'I always turn my husband out,' and she watched me put out the bacon and fry the eggs for those who'd ordered early!

Joan and I not only shared the same attitude towards children and learning and discipline. The same feather tickled us. She might have laughed at my 7 a.m. audience, but I dare not.

Chapter Three

'Make new friends and keep the old,
One is silver and the other gold.'

Camp-fire Round

Present society poses many problems not least the new generation. What to do with children not disciplined at home, how to cope with them in school, how to teach them to be socially acceptable? Margaret says that until adults clean up their act, until visual media stops polluting and money stops corrupting there is no hope. I see the faintest glimmer if the Government would do a U-turn on small schools. We had no discipline problems. Only in small numbers could the direction of society be changed and disaster diverted.

Equally I like to rub shoulders again, now and then, with the Guide Movement. Mary Johnson in Denver, Colorado, USA wrote to say she had an eight-strong troop of senior Girl Scouts planning to come to England in July. Their visit would be part of a challenge. In return for hospitality they would provide a weekend of adventure for forty Guides. They would like the venue to be The Currer. One girl had visited us with her parents and knew our Guiding history. I was alight with enthusiasm!

We arranged for the local Guides to camp in the Five Acre. The cattle had been kept on the hill for there was plenty of lush grass that spring and the field was a mass of flowers. I counted a hundred different species on the hillside. I still had the company tents in the camp store. I was in my element. If Rat liked messing about with boats, so help me, I like messing about with tents. The smell of canvas turns me on. Silver, Hebridean sand fell from them as we emptied them from their bags. Oh, boy, do I love camping! Hazel, who replaced Margaret as my able assistant when Father grew too old to be left with farm chores, brought me an oven glove on which is written, 'I like camp food!' I have never used it. I see it with tremendous pleasure every time I open the front room sideboard drawer. With a few helpers we pitched our tents in a perfect horseshoe, on the Thursday afternoon, and with poles and string we made gadgets to hold kit. It should have been the responsibility of the Guides to do this but they were at school until late afternoon and a weekend was too short to do everything those enthusiastic Girl Scouts planned to do.

It was a hot, wonderful weekend. The Guides enjoyed it and our holiday-makers had such a field-day they didn't disappear into The Dales but stayed around and joined in. The Guides cooked all the meals on an open fire, outside, except the evening one which I cooked. As we couldn't seat everyone in the dining-rooms we put a table outside the big doors. We had beef and Yorkshire puddings for we like being traditional. Afterwards we took chairs into the yard and became an audience for the Girl Scout Road Show. The theme continued the day's Early Settlers' activities. Pioneering skills had been taught and now old songs were sung and traditional entertainment enjoyed.

I love balmy summer evenings when guests sit talking outside until well after dark, when the owl screeches from the lane and the swirl of snipe and bat lend magic to the night. The cattle repose on the hill and moon and stars assure us that there are unsolved mysteries in Heaven and on Earth.

In retrospect I am sure we were taking everything possible from 1989. Every opportunity we had, we jumped into the Range Rover to explore God's Acres. Even a visit to the household tip became an outing. Having dumped our rubbish we found a viewpoint to drink from our flask of coffee. Mother seemed incredibly well. Cosseted and cared for, adored and busy, she was the most precious thing we had. She was far more active than frail Aunt Dorothy, her junior, who lived alone. We never wanted to leave Mother. We took her everywhere. We loved the way she, so small, stood so straight and said such funny things. 'We like your food, Mrs Brown,' an appreciative guest said.

'We don't do anything fancy,' she replied, 'We just serve farmyard food!'

I know with absolute certainty that if Mother had been Queen, today's royal family would not have behaved so badly. Margaret declared Mother's look should be bottled and sold in Mothercare shops.

On principal and for choice we serve only traditional Yorkshire fare. We had a French boy, Henri, staying for several weeks as a management trainee at Peter Black's factory in town.

He complained, 'Six weeks and no spaghetti!' I am sorry but no-one gets pasta or curry here. The Mistal cottage was empty so I advised him to buy some and cook for himself. Having never cooked spaghetti he did not bargain for it swelling. When it began to overflow the pan he removed some of it, then some more and some more until the kitchen was full of it. Then it burnt on the bottom of the pan which, when he emptied it and ate, was left in cold water in the sink. Like most men he left me to clean up and neither he nor I suggested he repeat the experiment.

* * * * *

Having sold her house Auntie Mary gave a small lump sum to all her nephews and nieces. We pooled ours and bought an ATV Yamaha four-wheeled bike which is an occasional god-send. The growth of our boundary wall pleased us but the Council did absolutely nothing about the road. They had given us a spoonful of sugar, keeping us quiet with promises but neglecting to do anything at all. For three years there was always a vulnerable spot where the man was building. Between the old and new wall, we leaned an iron gate which fell if insecurely tied or was removed by vandals. Then the cattle would get out, a hundred of them charging about the estate, endangering walkers or dashing downwards towards the trunk road two miles away. We complained to the Council and to the Police.

'There's going to be a dreadful accident. The road is impassable,' we yelled down the phone. 'Do something! Quick! You promised closure! Temporarily close it now, before all our cattle gallop down the A650.'

'We can't do that,' both replied. 'You can't just close a road. It's illegal!'

'You can close a dangerous road temporarily,' I insisted. 'Anyway, it isn't a road! It's a river! It's a lake! It's a gully, a maze of two-foot-deep clay ruts full of water! It's a death trap for unsuspecting motorists following a local map! We'll buy gates and pay for them to be installed if you will let us!'

'If you gate it,' we were warned, 'you'll be prosecuted.'

We were afraid of someone being killed and put it in writing to use as evidence should necessity arise and we became untiringly vigilant. We had a problem. When the cattle got out, which they frequently did, the dangerous task of bringing them home could not be done without the responsibility of taking Harry with us, across the moor to the boundary, in the Range Rover. We could not leave him at home with the two old ladies. If he fell they could not get him up. So many stampeding cattle are frightening even if you've been handling such since childhood. They behave differently out of bounds. The friendly, placid animal becomes a bucking bronco, a wild thing blindly and unstoppably following the leader. It isn't fun!

We continued to shout our predicament over the phone to the Highways Department, the Road Traffic Control, the Council Legal Services, the Police, all to no avail and then, as had happened for thirty years, we tired of banging our heads against a brick wall and gave up until the next time. It sadly looked as if we were going to spend £12,000 on a wall which would be thrown into the ruts by stranded drivers. We had a busy summer. I think it was a hot one. I cannot remember further than the heat wave of the Denver Girl Scouts and the fact that the duck pond was dry all summer.

There was an alarming occasion when I had planned turkey for the evening meal. The Aga cooks slowly and very competently. When guests are appreciative of food we say, 'It's the Aga!' I put an 18 lb turkey into the oven before going to the Cash and Carry which is on this side of the city. I could not possibly use it if I had to drive through the centre. Margaret was going to check the cattle and the wall, a daily must. We all returned for a late afternoon snack, putting the kettle on the Aga to boil whilst we stowed away £400 worth of food. I am not a Cash and Carry fan like Auntie Mary is. She loves it and will not yet be left behind. Mother and Harry only went for the ride and to enjoy the activity outside the super-store whilst calmly drinking coffee.

When we first converted the barn and began taking holiday visitors our eighty-four-year-old Mother had thought nothing of coming inside. Mother had always been the shopper and had always held the family purse. Before we had our own transport she had walked the half mile to the bus twice weekly. When we bought an £80 Austin Gypsy she was seventy, and Margaret took her into town but it was Mother who did the buying. She would never have trusted us with that job any more than she would hand over the baking. She was eighty-four when we began the evening meal lark which might be the death of us, who knows, but she was still cake and pastry cook until at nearing ninety-four, she bowed out honourably. She frequently said, 'One day I'll teach you. When I have time to stand over you.' She thought nothing of baking eight sponge cakes before elevenses, six of which she would put in the freezer. There was a memorable occasion when I was still teaching. It was a Christmas Fayre day and I took eight cakes out of the freezer to cream and sell. Mother came down early and saw the array of cakes and presumed I had been baking. She was put out and snapped, 'Yer won't need me in a bit!'

Having stowed away our purchases, I opened the Aga door expecting to hear an encouraging sizzle coming from the turkey I had put in hours ago. There was silence. I took off the foil and exposed a very white, uncooked fowl. The Aga was out! There followed the usual outburst of disbelief. Margaret said it was my fault for not noticing the oven was cool when I put the turkey in and I blamed her, for it was her job to refuel the Aga and she had forgotten. We blamed an unknown person who had put a plastic bag on the front room fire, carelessly so that the draught had blown it upwards and blocked the flue. It was easy to lay blame but we'd eighteen to feed for the evening meal and an enormous bird to cook in less than an hour and a half. We were in a mess! We have no knowledge of vegetarian cooking but there was no way we could raise the oven temperature enough to properly cook.

We had a Baby Belling electric cooker but the turkey was too big so we brought a camp billy from the store in the Five Acre and a pillowcase from the linen cupboard and we put the turkey in it and dropped it in the billy and boiled it furiously. I remember popping in an onion, an apple, some chicken oxo cubes and plenty of seasoning to help make the meat a fraction tasty. The very well cooked fowl was easily lifted from the pot in the pillowcase. We hung it to drain before carving the tender breasts and arranging the meat decoratively for serving. There was no way I could have taken it out whole and white and sagging. It was beautifully tender and as tasty as turkey ever is but apparently the guests had not been able to christen it.

'That lovely meat we had last night, Jean,' I was asked, 'What was it?'

For a short while I kept a large tin of corned beef in the larder and called it my panic tin. It was never used for that purpose. 'We couldn't have fashioned!' Mother declared. I used it for making sandwiches for a school party and now we have no panic tin and the Aga must not go out. Visiting schools give us great pleasure. They are invariably Special, catering for the physically handicapped or educationally sub-normal. Government policy has leaned to the closure of these excellent schools so now we have fewer of these wonderfully pleasant weeks when we join the singsong round the piano before starting the dish washing. One group of children added a verse to 'He's got the whole world in His hands,' singing lustily, 'He's got Currer Laithe in His hands.' Indeed He has.

* * * * *

The month of the cattle sales approached. Just when Hereford heifers looked so beautiful they had to go to market for their buyers needed to run them with the bull in order that calves should be born in the spring. Selling a hundred was a gigantic task and we spread it over the month selling at two markets. Every year the margin of profit narrowed. Inflation increased the purchase price of our calves, their feed and veterinary bills, but the selling price remained constant. We could see that the margin of profit might soon become the margin of loss. We knew the reason, of course. It was because the diversification meant we couldn't buy calves at a week old and rear ourselves. We were happy to let the bed and breakfast be the bread-winner subsidising the exorbitant cost of buying calves at three months old providing they made some profit, albeit small. We were not happy to do all that work and make a loss.

It was such a wonderfully workable programme. We realised the sense in buying all the spring calves at once. We could worm them in one batch, inject for copper deficiency, and against blackleg, all at the same time. We could graze them and winter-house them together and, in August sell them. Nothing could be simpler but the overheads were too great. However there seemed nothing much we could do about it. bed and breakfast was flourishing. That we preferred solvency to be achieved by farming was, for the moment a pipe dream. We were committed to ten years' bed and breakfast and evening meals simply because we had accepted the small ETB grant in 1980 and we couldn't feed people and rear calves. No way!

We went for our autumn holiday to Grange, confident that all was well. Having sold all hundred eighteen-month-old heifers we left Dorothy and George with only those six months old. We had a lovely, sunny holiday. We pushed the two wheel-chairs on Grange promenade. Jess saw another wheel-chair approaching, thought it was Harry and jumped on the strange man's knee. Lusky quarrelled with a native collie, bit and was bitten much to our consternation. We spent one enjoyable day with Margaret Watson at Wall End in Deepdale where The National Trust idea had been born. We were more worried about her, alone, almost blind and a long way from her nearest neighbour, than we were about Mother, though we remember now, that just before bed she would say, 'I've got stomach ache.'

'Go to sleep and it will go,' we promised her. That she was restless we blamed on the bed. She had no problem in her more comfortable one at home. We never suspected anything. Mother was going to live to a hundred. Easy!

The one to die was the billy goat, similarly named. He had been born in 1980, in a November snowstorm. The nanny kid had died and the billy kid was so beautiful we had neither the heart or the guts to destroy it. An old Irish dealer said good luck prevailed where there was a billy goat and had castrated the little fellow for us. We have found that the visiting public knows nothing at all about farming, little about the countryside and hardly anything about the birds and the bees. It is small wonder the country is in such a bonny pickle. Taught sex education at school Joe Bloggs believes all black cows are bulls, has no idea what a bullock is, firmly believes a large bellied billy goat is in kid, thinks all eggs will hatch and calls a fully grown hen a chicken. Joe Blogg's wife thinks milk cows just give milk without having a calf every year and her vegetarian children think they can live on yogurt, cheese and milk and have no idea that they'd have to kill the calf at birth if unprepared to rear it for beef. Ramblers think animals can be taught not to walk on footpaths. Joggers think they can run through a farmyard, panting and sweating and that collies can be advised not to give chase. Almost everyone thinks that the countryside would look just the same if not grazed and animal lovers who fight for freedom believe that cattle are better left out in winter rather than housed. Tell that to the cattle who will not go out even if the shed door is opened. And, most pitiful of all, people believe that fellow human beings who are evil are 'behaving like animals'.

We have plenty of proof that all these suppositions are wrong but we have no proof that having a billy goat brings good luck! We believe we have had, perhaps, more than our share. We cannot believe in a God who singles out some for luck and some for misfortune, so we are left with four possible solutions to our ability to survive and 'see the cuckoo'. The first is that we work very hard indeed. The second is that we can trace the rainbow through the rain. The third is that we have a guardian angel in heaven in the shape of Father and the fourth is we have a billy goat. As we cannot afford to ignore any one of the four it was imperative that we found a new billy quickly. It was yet another excuse to put our oldies into the Range Rover and roam the countryside looking for a field with a herd of goats in it.

It was actually from our vet's assistant that we got Charlie. She had wanted to keep him as a pet but kids and lambs, piglets and calves remain so for only a brief season and then become too big. Charlie had had to be boarded out at a smallholding, up above Oxenhope. The lady owner kept goats but didn't want a useless billy who was a pest anyway. His reputation came with him. 'You'll never keep him at home. He can climb and jump anything,' we were warned. An OAP came with him, too. Isobel had been famous having starred on the television soap which portrays local lifestyle so appallingly that Harry will not watch. Nor will we, for that matter. No wonder Joe Bloggs gets it all wrong. Her owner did not really want to part with Isobel for she genuinely loved goats and preferred to pension them out rather than have them put down. Isobel was still giving some milk and, being goat milk fans, Mother and Harry wore delighted smiles.

So Isobel and Charlie joined the other two old ladies of the species. Charlie grew and grew until he was enormous and added to the myth that billy goats have kids. Isobel gave us milk for about a year before becoming dry. There are plenty of non-profitable animals on every farm. With our good luck assured we went confidently into winter. There were fewer workmen needing accommodation. The recession was beginning. It made little difference to the tourist trade as far as we were concerned but workmen stopped coming in any great numbers. We had previously been as full in winter as we had been in summer. Building and shop fitting halted and local people secured the few jobs available. There was no movement of the work-force. Of the few outsiders who succeeded in getting a job in town, several had to forfeit it and return home finding it impossible to sell their houses.

* * * * *

Mother made all the Christmas puddings just as she had always done for nigh on ninety years. Eldest in her family I am sure she was helping to stir the mixture long before she was ten. She would not hand over the ritual job of baking to anyone. She needed us to lift down the biggest boiling pan, to carry the scales onto the table in front of the Aga and bring out the yellow glazed earthenware bowl, only used at Christmas since her bread making days were over. It was she who grated the suet, fresh from the butcher's, she who cleaned the fruit and whisked the eggs, she who weighed all the ingredients and who called us all individually to the bowl for the need for each to stir was traditional, a ritual necessary to well being in the year ahead.

She tore up the linen for the cloths and wetted each square thoroughly and put them, one at a time, into a bowl and dropped the right amount of mixture into it with a wooden spoon. Then she drew the corners together and I was needed to tie the string tightly. There had always to be a broken plate at the bottom of the pan and the water must be boiling merrily before the puddings were introduced and it must continue to cover them completely. She would make five puddings, three in the big pot and two in a smaller one and eventually the boiling pans would be put in the top oven of the Aga and later transferred to the simmering oven to be left there all night. The unmistakable smell of cloutie dumpling would fill the kitchen. Next morning I caught the strings and threaded them onto a walking stick to dangle over the big kitchen sink and when cool I hung them from the beams to dry.

Mother would never buy mincemeat for the Christmas pies and tartlets. Once again she grated the suet and cut the peel in the days when it was bought in one big sugary piece. It was she who squeezed and grated the lemon. She was very proud of the row of 1 lb jars which adorned the Welsh dresser.

We decided to go to Downham to eat our Christmas picnic. Auntie Mary made the sandwiches and Margaret put in a stale loaf to feed the mallard in the clean stream near which we would park for lunch. However, travelling west, we saw rain clouds approaching and changed our course and went north-east to Studley Royal Park to see the deer and feed the ducks there. We had sunshine all day and Mother enjoyed herself enormously. Feeding the nine-month-old heifers, before we left, had not been time consuming and to get home, to give them their evening feed and bed them with straw, was not urgent. We could relax and be unconscious of time all day. It was lovely! It was also the last time we could go out on Christmas Day relatively free of the chains of The Currer.

On Boxing Day Aunt Dorothy came to spend the best part of a week with us, and Craig and his mother dropped in. I had taught him a dozen years ago, at least. Craig was physically handicapped and mentally more than average alert. After graduation he was currently a personnel officer at St George's Hospital in London. He shook hands with me but put his arms round Mother and kissed her. She was so pleased. I think most people loved Mother just for herself. They did not necessarily think of her in the role of Harry's mother, perhaps because society chooses not to see handicap or accept any responsibility. To say you do not notice Craig's disability is an excuse and an insult to his extraordinary ability to achieve. It is also to deprive his mother of the accolade she deserves for limitless care and support. When relatives and friends have professed not to notice Harry's cerebral palsy I could only pronounce them blind and feel sorry for Mother whose dedication must have gone unseen. But Craig, my pupil, knew and Mother knew he did. We had a lovely afternoon!

Cousin Freda had bought Mother a dressing gown, but it was too big, it enveloped her. Auntie Mary bought it from her and gave Mother the money to spend. 'I need some new nightdresses,' she said, so we went to town for this specific purpose. Mother had not recently been shopping, for the new car-free precincts may be less hazardous traffic-wise, but they are paved and cobbled so unevenly as to be a danger to the elderly. We are strangers in the new shopping centre and the covered market, but Mother was out to enjoy herself. We suggested she used her wheel-chair which she quite liked on holiday, but would have none of it in her home town, and she seemed so well and agile she didn't need it. She was on her feet much of the day at The Currer. What was the difference?

That shopping expedition boosted her morale. She loved markets, always had. When we were in Grange she wanted a day at the open air one in Morecambe. She would never go to Skipton without a trip down the High Street poking round the stalls. I'm not like her in that respect, at all. She seldom bought anything except food, never having money for anything else. She liked buying a few kitchen plates or utensils. She loved to buy a vest or some underpants for Harry and she really enjoyed buying the nightdresses. In our family such articles will last twenty years and more. People buy too many clothes! A farmer friend's very old mother died, and Harry, who is capable of the most extraordinary calendar calculations, said to him, 'It is thirty-three years since your father died.'

'Aye,' he replied, 'An' Ah went t't'funeral in t'same suit.'

The clothes we wear on the farm are invariably hand-downs from affluent friends. Our best clothes last for ever. I am worried in case my college contemporaries notice I come to The Reunion in the same dress every time. Mother couldn't remember the last time she had bought nightdresses.

'Wear them,' I said. 'Don't just keep them for holidays!'

We had booked a cottage on Arran for May and were confident Mother would get there. Harris may be too far but Arran was only one day's journey away, and we had re-booked The Studio for October. Mother was stepping into the new decade as positively as she had stepped into the new century ninety years ago. She had survived ten years of bed and breakfast and the killer evening meal. She had become a legendary figure to thousands of holiday makers. 'We'll go t'T'Currer' she had said thirty-three years ago when it was still a ruin without windows, road and sanitation. She had braved all the privations, hardship, work, worry and financial struggle saying, 'Ah'd worry if Ah'd nowt t'worry about!' Mother was The Currer.

As we said good-bye to the old year and welcomed the new we did so with confidence. A lot of struggle was behind us. We had greatly reduced the bank overdraft, we had sorted out the work problem and eliminated a lot of the danger. We were organised, happy, healthy and we'd adjusted ourselves to having Auntie Mary with us permanently. She was a wonderful help in the house. Things were bound to be easier, we thought. 1990 here we come!

Chapter Four

Mother's there 'spectin' me,
Father's waitin' too;
Lots o' folk gather'd there,
All the friends I knew.
Home, home, I'm goin' home!

William Arms Fisher

We had barely entered January when Mother was sick. Bugs are unwelcome in a catering establishment. However, our New Year guests had gone and we only had an odd workman. Janet and Michael were coming next day but none of the rest of us seemed to have the bug and Mother was over it, whatever it was, so we told no-one. It must have been the richer food we had eaten over Christmas or that delicious Irish stew Mother had so enjoyed. Perhaps I had left too much fat on the mutton and Mother, who liked it so, had taken more of the tasty bits than was her share. Failing that, it could have been a chill she had caught in the market. Nothing to worry about!

Two days later she was sick again and then came the unmistakable signs of jaundice. We didn't immediately recognise them. A chill on her kidneys, we diagnosed and sent for the doctor. 'Bring her down to the surgery for a blood test,' he said.

We began to notice the frequent, 'Ah've got stomach ache,' we'd previously ignored and suddenly she was very yellow. The doctor was coming with the results of the blood test and Mother decided to stay in bed, just until he'd been. She was very cheerful, laughing at the funny colour everything was. We came downstairs with the doctor after his thorough examination of the lively lady and joined Harry in the sitting-room. The doctor gently told us the blood test and examination had revealed cancer of the pancreas and that he could take Mother to hospital for a scan to prove it but, in a ninety-three-year-old lady, treatment would not be advisable and an unnecessary trial for an old lady whose life expectancy could be no more than a month.

Shock is the first, spine chilling reaction. Having gone through the same experience when the older doctor had told us Father's prostate trouble was malignant, we knew that almost immediately would come anger. Anger that this should happen to our adorable mother when she appeared so well, had all her faculties and everything to live for right up to and beyond her century. Anger soon gives way to fear. We were not afraid for Mother, that she would die, for our faith is, as my school children used to sing in Assembly, 'an oaken staff' and Father was already there, 'expecting her'. Mother's soul was in order alright but her body wasn't and we were petrified, anticipating the course of events and the suffering which might precede her dying.

Fortunately we need not tell her. There was no question of her going into hospital, an experience she had never had in nearly ninety-four years and we need not lie to her. She only had to look in the mirror to see she had jaundice to satisfy her with an explanation. To tell her the cause of the jaundice would have been impossible.

The most important thing, when faced with such a diagnosis in a loved one, is to learn how to treat the symptoms. There is advanced knowledge of how to control pain, sickness, irritation, thirst and whatever. The doctor told Mother she must avoid fats. 'No butter,' he said. No matter how poor we had been, Mother had always insisted on butter. She could taste margarine by the smell of it before it reached her mouth. She judged people to their disadvantage if the bread they served was spread with it and, should it be served in an hotel, which once happened, she was very derogatory. She ridiculed the television advertisement which declared you could not tell the difference. Of course she could! She used butter liberally. 'Let Sarah butter,' Grandfather used to say when crumpets were toasted.

For much of our pre-bed and breakfast life at The Currer, our parents had churned our own, painstakingly skimming the thick cream from the rich Jersey milk and turning the churn by hand. Father's prostate operation in 1975 had encouraged us to buy an electric churn for their Golden Wedding anniversary. An odd present, no doubt, but it helped them to continue the chore for almost four years before Father died, in 1979, when the snow was so deep at The Currer we were completely cut off. The doctor, walking our treacherous road to visit him one day, slipped and broke the fibula in his leg. In 1990, with January and February still ahead of us, we wondered if history would repeat itself, but it seldom does at The Currer. We are plagued with variety. Mother's fear of being ill was all anticipation. When she was ill she was a perfect patient. She accepted that she must not eat butter with no argument at all.

To treat the symptoms of cancer you need to consult a professional angel. Ours was a former pupil, Guide and Sea Ranger, another Joan, who had trained first as a nurse and was currently working in a hospice for terminally ill cancer patients. She knew enough remedies for distressing symptoms to make Mother's life bearable. She was the most extraordinary source of practical help. She knew just what to do about Mother's sickness, telling our doctor what to prescribe. She knew a relief for Mother's dryness in the form of what we jokingly called 'tin spit' and she knew the right ointment to ease the terrible itchiness with which Mother began to writhe. Once these distressing things were treated life began to be not so bad at all.

It was early in her illness that Mother had her vision. It was a Tuesday. Since her stroke a few years earlier, cousin Freda had been brought to lunch once a week. I had some shopping to do and would collect her so Mother said she would stay in bed until Margaret came in for coffee. Whilst they drank it, Margaret read some of my autobiographical manuscript aloud for Mother to enjoy. Suddenly the old lady became confused, saying it was wrong, that her mother wasn't dead and that she, herself, was young. Margaret stopped reading at once and put the offending script away. Mother didn't get up. She turned over and fell into such a deep sleep Margaret believed her to be in coma and when all of us came home she reported Mother strangely ill, asleep, maybe never to wake again. We tiptoed round the room not knowing quite what to do.

I kept going downstairs to prepare and serve the dinner and Margaret and I ate ours in the bedroom where Mother lay in this death-like sleep from which we did not expect her to wake. But she did, all agog with a new and wonderful experience. She hadn't been dreaming. Oh no, it was far more dramatic than that. She had had a vision, a lovely one and she wasn't one bit afraid of dying. For the next four months she continued to live on cloud nine telling everyone, old and young, great and small of the wonderful vision she had had. She had been, she said, at the foot of a staircase, amid beautiful flowers. At the top of the stairs had been Father, her mother and father and Jesus and she repeated again and again how beautiful it had been. She told Joan and the doctor and the district nurses. She told the Guiders and the Guides and the Brownies coming for weekend trainings and holding 'camp-fire', for her, in our sitting-room. She told paying guests, relatives and friends and she began a new lease of life.

Thanks to Joan she was not sick anymore. When she was dry we squirted 'tin spit' and when the itching was driving her crazy we had this remarkable remedy. She lived on Rice Krispies and a special sandwich she found so delicious she wanted all our visitors to try it. I used to pulp tomato, shred boiled ham and crush pineapple. I cut off crusts and this tasty sandwich kept Mother alive for four months. During this time she rose each morning for coffee time. I helped her dress. She could not fasten hooks but she could bend down to put on her tights and having got them to her thighs she could do a little bounce on the bed which got them on completely. I've tried but I cannot do so. We used to laugh and joke all the way to the top of and down the stairs to her waiting wheel-chair. In the sitting-room she sat erect, like a queen. She was very beautiful. The yellowing looked like sunburn when flushed by the fire. Margaret made an exceptionally good job of cutting her hair which was a miracle for Margaret is no hairdresser and has farmer's hands. A friend, Maureen, gave Mother a blue, crocheted shawl which was very attractive. She looked so fine we loved her almost to distraction.

I told her so, once, for we are not demonstrative as a family. We were walking to the foot of the staircase. Her head was on my shoulder and both arms linked in mine.

'I do love you, Mother,' I said, quite loudly for she had removed her hearing aid. She did not answer.

'Hey! Don't you love me?' I prompted.

'Huh!' she grunted. 'What d'yer think?'

* * * * *

I don't think we have ever been out so much in the Range Rover. It was comparable to being on holiday. Everyday we went somewhere. Mostly it was on business, to the Cash and Carry, the supermarket, household tip, vet or the agricultural merchant. Any excuse found us filling a flask and going places. We took Mother's medication with us and neglected the spring cleaning. If we had no legitimate reason for going we said it was a treat and went on Baildon Moor or over the tops at Kildwick or Lothersdale or simply up to Keighley Gate where the good road ends and the rest of the way down into Roman Ilkley is but a track. If we had little time we would go no further than St Ives to park where we could watch the squirrels and be grateful that a thousand-strong protest, which we had all supported, had prevented the Council from selling.

But we cancelled the holiday. We were convinced that the doctor had been wrong when he gave Mother one month to live. She wasn't going to die yet. She had the last furlong to go and was ready to run superbly to the winning post. We told her we'd written to Arran and she was relieved. She no longer had to worry about whether or not we would be able to go. Without a deadline on the calendar we were able to relax and enjoy the present. Everyone came to see her. She was never without flowers and she was always smiling and composed.

She was more composed than any of us one day in February when the fire alarm went. The only people we had in the house, at that time, were half-term holiday-makers in the Loft Cottage. They had gone out for the day. Having cleaned out the 'cold' embers from yesterday's fire, they had emptied the ash tray into a cardboard box and left them in the bathroom. They thought the day too windy to carry them outside and risk being covered in dust. They did not know that yesterday's ashes are never too cold to light cardboard.

At three o'clock, four hours after they had left on some sight-seeing expedition, the box had ignited, smouldered through the lino and the floorboards which, fortunately, covered a mass of electric wiring. It was an electric fault which had activated the fire alarm. All we knew was that smoke was pouring out of the overflow pipe from the bathroom and we dialled 999 at once.

Mother said to her sister, 'Come on, Mary, we'll both go to the toilet in case we have to evacuate!'

No-one knew just how capable Mother was. All she needed was an emergency and she coped. Margaret knew better than I for, working at home, she had been with her more. She wasn't all that surprised how well Mother had tackled the adventurous move to The Currer in 1958, the hardship, the lack of amenities, the financial struggle and the cruelty of the weather. Margaret was less amazed than I how well Mother adapted to the upheaval of the barn conversion and the thousands of guests which followed. I had gone out to teach each day for thirty-one years. I was far more surprised than Margaret how Mother remained composed through terminal cancer. Her performance was amazing. I use the word purposefully. Mother was a born hostess and knew, more than most, how to entertain, to welcome graciously and to say goodbye with dignity. She was an actress who warmed to applause and modestly took her bow.

The fire brigade of three engines came more quickly than it took me to digress. Fortunately the fire doors were shut and without oxygen the fire had been restricted. Even so, the smouldering floorboards had belched out enough smoke to blacken the bathroom walls and fill the whole cottage with an almost irremovable smell. Apart from that, damage was minimal. We had to have a new shower and our electrician had his work cut out to repair the wiring and Dorothy came and spent hours scrubbing the tiles. We accepted her help for February is a busy month feeding our quite big and hungry herd and looking after Mother was the only thing we wanted to do.

The fire alerted us to a serious fault in our fire certificate regulations. The alarm was activated in the cottages only by heat detectors in kitchen and lounge. There wasn't a single smoke detector anywhere. By a miracle the holiday-makers had put the cardboard box over the wiring. By another miracle the fire had been generated whilst we were at home. Had we gone to 'buy another' on that February Monday, with that wind, our home would have gone up in flames. The expensive, compulsory alarm system demanded and inspected by professional authority, was useless if we were not there to ring 999. We are too far from any other house for a neighbour to hear. Had the day not been so windy that we were afraid to leave the rattling tins on the dutch barn, all we have ever worked for would have gone. We sent immediately for Keybury Alarms and had our system connected to a central control which would automatically alert the Fire Station and they put smoke detectors in the cottages, outside each kitchen door. The solution was expensive but reliable. We soon found it to be a major liability.

Throughout March and April we continued the practice of taking Mother out whenever possible. She found the gentle motion of the Range Rover very soothing. She couldn't watch television or read and she couldn't bake any more. She had stopped drying the hundreds of pieces of cutlery, a job she had assumed when we started catering. She continued to fold serviettes, far more than we needed daily, so that eventually we had a cupboard full and I barely folded one all summer. When we returned after a jaunt Mother would jokingly say, 'Bring me my dinner to the Range Rover.' It was almost as good an anodyne as the strong pain killer she took.

We loved her so. So did the district nurses and all the many people who came to see her and brought the flowers she loved and the chocolates she could not eat. She worried a great deal about news of child abuse, neglect and divorce. We loved the way she felt she could have an opinion and the right to voice it. She would say things like, 'What would yer mother've thought about yer carry on?' She kept Harry in constant laughter at the quite daring things she said. She wept for people when floods and earthquakes brought trouble and was quite out-spoken against politicians. Every minute of Mother's last three months was precious to us.

We protected her so diligently we thought our vigilance was foolproof. We slept in her bedroom thinking that between us we would hear every movement and were horrified when she fell out of bed under our very noses, as it were. We both leapt out in horror. 'Don't be alarmed,' said the old lady, 'I'm alright.' We all laughed, Mother in amusement and Margaret and I with nervousness. We really were tired. Mother knew this and didn't like to wake us. If she had to, she preferred to wake only one. The weight of the farm work was falling almost totally on Margaret. It was she who did most of the foddering of cattle. It is a heavy winter job entailing the carrying of thousands of bales and loosening them in mangers. Mother tried only to waken me but almost always Margaret stirred as well.

One night, in April, Mother's pain was bad and she got out of bed to sit on the commode chair. Margaret hurried downstairs for the pain-killing pills and I knelt on the floor in front of Mother whilst she gripped my hands. I was unaware that the light was subdued or that Margaret and I looked decidedly old-fashioned in our nightgowns and our plain hairstyles, Margaret's cropped and mine drawn back in a spinsterly knot. I'm sure we looked worried. Mother looked up and saw Margaret standing behind me and said, 'What's she up for?'

'I've brought your pills,' Margaret answered.

'You do look silly,' Mother said. She paused and then a smile illuminated her face. 'You look like them lasses from Haworth!' she said. Never have the Brontë sisters been so maligned. Like Maria, Mother could always make us laugh.

She began to remember songs from way back in her youth. 'I can't stop singing this one,' she said one day.

> Don't you wait up for me dear, I won't be back 'til three,
> I've got a poor sick friend, dear, who needs my help you see.
> So leave a little candle in the hall tonight
> And hang out the front door key.

'I can just see my Dad and your Dad laughing at that on Morecambe Pier.'

The flow of guests continued. The annual spring decorating was abandoned. The weather was good and the daily outings seemed endless. 'Tin spit', ointment, pain-killers, a mixture of soda water, tonic water and glucose, Rice Krispies and pineapple-ham sandwich and life was not bad at all, but any missing ingredient in Mother's recipe for survival and she had problems. We had no difficulty getting the 'tin spit' at our own chemist but we found it was not regularly on pharmacy shelves. Our doctor had never heard of it and we were greeted with a blank stare when we tried to get some, on prescription, in Skipton. We had to go there on some errand or other. We had no reason to go to the cattle market in April. Perhaps we needed a caterers' bag of medium oatmeal for the Easter guests who always opt for porridge. It is immaterial. All I remember is we did not take the second wheel-chair because we did not expect Mother to want to get out. Harry's was permanently in the vehicle, in those days, and when we stopped in the car park Mother busied herself with a head square and insisted on going round the market. Her mouth was very dry. We were going to the chemist and she was determined to come with us.

I think, until then, we had not accepted how ill Mother really was. On that cold April day she looked ancient and very yellow. She was so desperate for the 'tin spit', I left Auntie Mary, whose pace was slow, and hurried with Mother in the wheel-chair to the chemist. Further behind still, Margaret was coping with a walking Harry. We had not noticed how increasingly disabled our sixty-three year old brother had become. He could not get further than the seats in front of the Town Hall. Margaret left him there to lend an arm to Auntie Mary. What a decrepit family we were!

The chemist had no 'tin spit'. We bought glucose sweets and lucozade but neither helped. We were not used to being miserable on an outing. I took Mother back to the car and retraced my steps back to the Town Hall, with the wheel-chair, to pick up Harry.

Margaret said, 'He could hardly walk at all. I'm sure he needs some physiotherapy.' He was becoming so bent it could not possibly help his hiatus hernia. The 'do not stare' theory has been interpreted as 'do not look'. It is society's way of avoiding responsibility. We found, when we applied for a disabled sticker for the car, that our brother had never been registered as such. It was not Mother's fault. She had tried valiantly to get him into school without success. When examined for National Service he should have been registered. Having asked for a sticker we were told he should have been having a mobility allowance. For sixty years he had required twenty-four hour care, but he was refused an attendance allowance.

We were struggling and nobody noticed that either. I remember feeling quite bolshie. We drove to our own chemist, a sad and cold family. The next day our kind district nurse told us we should get attendance allowance for Mother. We burst out laughing. 'You must be joking,' we said, 'we don't even get one for Harry!'

Suddenly someone had looked! She didn't believe us. She reported to the doctor who, as is the custom, sent a medical inspector to assess both Mother and Harry. He was shocked he had been called so late in the day. Mother died before the slow wheels of the National Health turned, but eventually, after she who had cared for him since birth without help, had died, Harry was finally recognised. There had been no family allowance for Mother, no education for Harry, no day care, no physiotherapy, no speech therapy, no hospital or medical care which could have improved his physical ability. Mentally Harry is sound, but the medical officer failed to notice that. He said Harry would be exempt from paying the current Poll Tax. I filled in the form he sent me without reading the small print. It came back to me exempting Harry from paying on the grounds of 'Severe Mental Impairment'. We were so angry we sent the form back to where it had come from and told that particular government department it could 'get stuffed!' For as long as the Poll Tax existed, Harry paid.

Had he been given a normal education, Harry might have been a mathematician. He handles the calendar better than the rest of the community. We have to ask him the date because we do not know and he always does. He could tell the time before he could walk. He knows the day we went on holiday, came home, bought calves, sold cattle, this year, last year and the year before that. He knows everyone's birthday, wedding anniversary, funeral, even the day of the month annual visitors came on holiday. He knows the dates of next year's Saturdays, what day of the week our birthdays, Christmas and Easter will be. He is equally as reliable as the calendar we hang on the wall. If he knows the date of a wedding he will tell the day of the week the couple will celebrate silver and golden weddings and if he knows your date of birth he will accurately give the day of the week. Guests who know this try to catch him out and we have been known to attempt to pose him a problem beyond even his ability.

Years ago Auntie Mary found the printed catalogue of the sale of Great-Grandfather's livery stables, coaches, horses, harness and tackle. There were 271 lots to go under the hammer. Fifteen valuable horses were advertised and named Albert, David, Harry, Billy, Jet, Maggie, Johnny, Mick, Nelly, Peter, Kitty, three called Dick and one un-named, other than the Brown Mare. There were twenty carriages, landaus, broughams, Clarence carriages, wagonettes of all sizes, a char-a-banc to carry twenty-three people, two hearses and a funeral coach.

Seeing the auction was billed for Monday 19th May 1919, Margaret, knowing Harry had not even seen the catalogue, asked what day of the week 19th May 1919 would have been. 'I can't do that,' he initially said. How could he? It was before he was born.

'Just a minute,' he said suddenly, realising there was a date reasonably close from which to start. We waited less than a minute before he laughingly but confidently said 'Monday!' We were unbelieving.

'Aw,' we all said, 'That was a good guess.'

'No, I did it,' he replied and went on to explain. He was laughing so much he could barely be understood. Between November and May, Harry said, there were six months and thirteen days which would place May 12th on the same day of the week as 11th November, Leap Years excepted. We did not know, but he did, that Armistice Day, on the 11th November 1918, had been on a Monday therefore May 12th and 19th must have been on Mondays in 1919. There was no way we were prepared to label Harry as mentally impaired.

* * * * *

The Barbault family arrived from France two weeks before Easter and stayed for a fortnight. Denis was thirteen and Emilie five. Each evening they left the table to come to the kitchen with empty water jugs, 'More water, please,' seemed to be the only English words they had. Margaret heard Donie try a few hesitant sentences to other children in the farmyard but we had time for no-one. The thing we found most difficult was playing hostess to a multitude of holiday-makers. Mother was doing so well we could not be anything but happy, but it was a tense happiness. It was the happiness of seeing a proud ship sail unerringly, flags flying, into safe harbour for the last time. It is impossible to be heart-broken watching someone keep so triumphantly on course, towards final anchorage. Mother was going willingly, happily and confidently. She knew where she was going and who would be there and that we would be alright and would care for Harry and Auntie Mary and in no way was she going to let the side down. Not she!

I remember one evening when she was in the Golden Wedding chair, so named because we had bought it fourteen years earlier for that occasion. Auntie Mary had gone to bed and Mother did not want any supper. Margaret and I were on the rug. I didn't think we were going to able to get Mother upstairs she lacked so much energy. She said, 'I think I'm dying. Don't try to keep me.' But the funny feeling passed and she ate some Rice Krispies and we wheeled her to the foot of the stairs and took her to bed.

On Maundy Thursday we went to Bolton Abbey for a picnic. The Yorkshire Dales were empty for Easter had not yet begun and in any case we never go by the main road. We climb the Pennine ridge through Eldwick and turn right at Dick Hudson's approaching Ilkley via the road which passes the Cow and Calf rocks. Then we take the single track road through Langbar, pause for coffee at the viewpoint below Beamsley Beacon and eventually enter Bolton Abbey, by the back door as it were.

It was a beautiful spring day. Sunshine bathed the dale and there was a holy quietness preparing mankind for Good Friday. How many would listen to the silence, I wondered? What ails everybody that we are all too often too busy for things which really matter. There were daffodils everywhere for Easter, that year, was quite late.

Mother was on top form. Never a singer she suddenly had voice. She amazed us by singing Father's song. We only ever heard the first two lines from him but we had heard it so often we had called it his song.

> I'm so happy, oh so happy,
> Don't you envy me?

Mother was sitting watching the ducks on the Wharfe, having eaten her sandwich and she suddenly gave us more.

> Try to picture me, upon my mother's knee,
> back home in Tennessee ...
> Father, mother, sister, brother,
> All are waiting there,
> And there's an empty chair ...

We applauded and she gave us an encore. Father had been dead eleven years. We do not pretend to know the mysteries of life but Margaret said she felt his presence many times during the next ten days. We did not want the precious day to end. Mother was encased in rugs and hot water bottles and there were few cars on the riverside. We bought expensive ice-creams and Mother licked so feebly hers wouldn't go and Margaret had to finish it for her. We had taken Father to Buckden for his last tour of his countryside. That day, too, will not be forgotten. We belong to the Dales and I hope someone will take me a drive, on a sunny day, when my life on earth is nearly done.

Next day the Barbaults were returning to Lyon. I told Bernard Mother had cancer and had already lived three months longer than the doctor had expected. Sadly she would not be here should they come next year. She had not risen when they came to see her, before leaving, so they went upstairs to her bedroom. She was sitting waiting with sweets under her pillow for the children. The French are so much better at this sort of thing than the English. Bernard did not say, 'Au revoir'. There was no point in concealing the fact that he would not see her again. 'Good-bye, Marm,' he said kissing her on both cheeks. She was all smiles. 'Good-bye, Marm,' Yveline said, doing likewise. They did not prompt the children but the thirteen-year-old boy did not hesitate. Denis leaned across the bed, kissed Mother twice and said, quite clearly, 'Good-bye, Marm' and the baby, only five years old, climbed, without being told, onto the bed and kissed and whispered, 'Good-bye, Marm.'

Easter guests all came to see Mother sitting by the fire in the front room. Sally in her wheel-chair, soon to be awarded the MBE, the Smiths with Richard who had brought his overalls as usual, all the usual crowd. On Wednesday Mother got up in the bedroom only and on Thursday she did not get up at all. She was sleeping in the afternoon and I remember dissecting boiled chickens, for the evening casserole, at the foot of her bed, angry that life and show must go on.

Guests were coming who did not have a car and I had promised to meet the bus in town. Richard came with me. Although it was rush hour I managed to park on the one way street opposite the bus station. The coach from London was an hour late. I phoned home to tell of the delay.

I was frantic. My Mother could die whilst I was sitting here waiting for people I had never seen before! Eventually the long awaited bus arrived and I greeted the couple, happily coming on holiday to one of the most hospitable farmhouses in the country. 'Hello! Lovely to see you. Your long journey is over. Come on and have a lovely holiday.'

I could not say, 'I do not want you or anyone else for that matter. I don't want to cook for you or laugh with you. I just want to shut the door and cope with the next few days.' How could I? Having transferred Richard to the back seat and put all the cases in the back of the Range Rover with Lusky, I climbed into the driving seat. Ours is a Range Rover van and I saw the road was clear through my rear windows but failed to see the small car crawling alongside. I edged slowly out from the curb, a foot, no more. The impact was so gentle for the old man in the small car was driving very slowly but very near. We both stopped. I was trembling. The two vehicles were touching. Either of us could have reversed a fraction but the man pulled away forward and a scratch appeared the length of his car.

I got out. There wasn't a mark on the Range Rover. 'I'm sorry, I'm sorry!'

I was near to tears. I wasn't fit to drive, I'd waited an hour, my Mother was dying at nearly ninety-four. I was a wreck! I drove the newcomers home so slowly I could have been arrested for loitering. However had we got ourselves into this predicament? We could not send people home. Many of them had re-booked last Easter. We could not criticise them for laughing in the dining-room and singing in the snug. There are things you cannot change and one was that it was Easter and ten years ago we had converted the barn and all the world and his neighbour liked to come.

On Friday Mother was really angry with me saying I wouldn't let her get up. She wanted her pinny and she wanted to bake. Joan said Mother's sudden temper was pain related. It passed and she was calm in the afternoon and welcomed Freda who came and Richard who wanted to see Mrs Brown.

Like the French, the Down's syndrome young man knew better than anyone what to do. He gave her a very gentle kiss and sat on the bedside and held her hand.

'How are you, Mrs Brown?' he said. 'You do look lovely. I think it's your hair. We're having fish and chips for dinner. You have some, Mrs Brown. It'll do you good.' He continued to chatter and to hold her hand. He was making a far better job of his visit than Freda who stood beside Auntie Mary, tears streaming down their faces and Freda's husband who came to collect her and couldn't find anything to say at all. I was preparing the evening meal and missed it but Margaret said Richard's performance should have been on tape and used to teach the public how to cope in these circumstances.

I made Mother her usual sandwich and at 7 p.m., when we served the evening meal, she remembered Richard's advice and wanted some fish and chips. I recall having to go and steal some from the serving tureens in the dining-room.

'Richard told Mother to have fish and chips,' I excused myself. 'He said it would do her good!' There was laughter and a feeling of goodwill around the table.

And then the pain came back and Mother wanted Harry, said he was missing and we could not convince her that everything was alright. We told the nurse what Joan had said about pain related stress. She told the doctor and he came and gave Mother an injection and the morphine took away the fears so she could sleep peacefully. Joan's long experience of working in the hospice was so comforting.

Saturday was the day we should have all gone to Arran. The sixteen Easter guests were leaving. Come to think of it, we'd never have managed to get to Ardrossan, even for the last ferry on Saturday night, not at the end of Easter week, even if Mother had been well and we'd had no reason to cancel. As it was Mother did not wake on the Saturday. We dared not leave her lest she leave us without someone being there. We had never left her. She had no experience of being alone. For more than sixty years where she had been there had always been Harry. I remember thinking, once when she had been quite ill, long before Father had died, that a lady who had never been on any journey alone couldn't possibly go to heaven all by herself. One has some very funny thoughts when someone very dear is ill.

There was a distraction in the yard. Someone was repeatedly pressing an unresponsive ignition. Some leaving guests could not do so because their car would not start. The suspense of Mother's coma left us intolerant of departing guests. I heard myself praying aloud for the engine to leap into activity but the repetitive grating noise went on and on. 'He'll run down the battery. The stupid man,' I groaned in accusation. It was increasingly obvious we were going to have to get out the Range Rover and the tow rope and drag him up the hill to the top road where, perhaps the down slope would help him to bump it off. We were so envious of those who don't have bed and breakfast guests and can get on with living and dying quietly, without intrusion.

Margaret was already untangling the tow rope when a Land Rover pulled into the yard. Way up on the moor, live the Ainleys. They can relate to us in more ways than one because they have converted an old farmhouse into comfortable accommodation and steadings into a workshop-studio. Tony is a commercial artist and Pauline his wife and work-mate. I have memories of the heavy snowfall of 1947 when Father, Jack Clay and I sledged hay from there just a few months before the Ainleys bought it. Tony was only seven. Nearly fifty years later he is still there so they know all about being isolated from the town but imprisoned by business. We see them infrequently for we are all busy people, but on that Saturday, at the end of Easter, as Margaret was pulling out the tow rope, wondering whether our vehicle would start, Pauline drove into the yard. She did not know Mother was ill. She only wanted a brochure for a friend and seeing our predicament she desperately wanted to do something useful for us.

'Give me the tow rope,' said that indomitable woman. 'The least I can do is tow out this car.'

Did anyone ever appear at a more welcome moment? Did anyone ever offer more acceptable help? Our gratitude was sincere. Our guests departed and new ones came, but Mother knew nothing of either. She lay peacefully unconscious for over three days. Someone was always with her and we sat up in bed all night afraid to sleep and critical of ourselves if we did, for we were very tired. Our dear lady never moved. We did not know whether she could hear when we talked to her. The district nurses came night and morning and once Mother gave the most radiant of smiles in our direction.

'Father, mother, sister, brother,' were all waiting where she was going. She had said once, early in her illness, 'I've always been loved. My mother loved me, then yer Dad and then you children.' I remember Auntie Mary visiting her cousin in hospital. She asked the elderly, lonely spinster, if there was anything she needed. Cousin Edith had unexpectedly said, 'I need a little love.' Our Mother had never been deprived of that. We tiptoed round the room afraid to waken her and for her to find she was still here and not with Father where she wanted to be.

New guests were in the house. I remember there was a man with a Chinese wife who was an opera singer. Had they entertained the guests with pop music, I could not have tolerated it, but her singing was quite beautiful. We had one lady who was in a wheel-chair. She came, alone, by train and we collected her from the station. We had warned her, when she booked, that it was impossible to propel herself up the steep Currer Lane to the village and that she would be farmyard bound. Although it was a beautiful sunny day on Tuesday morning and all other guests had gone sight-seeing, she was still at home.

The district nurse came to turn Mother but decided not to do so she was so still and, even while we whispered at the bedside, Margaret, the nurse and I, Mother's breathing changed and when it stopped altogether we kissed her tenderly and said goodbye. I remember the nurse pulled the sheet over Mother's face, as is the custom, and when she had gone we pulled it back. Our mother was so lovely. It was April 24th 1990 and Mother well into her ninety-fourth year.

We had done everything possible. Now the only thing we could do was send for the undertaker and wash the accumulation of breakfast dishes and re-set the dining-room table. We rang our friend in the village who had given Mother the lovely shawl and asked if she could take the lady in the wheel-chair out for the day. We wanted to take off the persistent phone for we were disinclined to talk to would-be guests, we did not even know, who wanted to book accommodation, but life has to go on even in bereavement.

We cannot experience even this without drama. Margaret always says that to film us straight would be a Soap with more variation than any fictional one. The undertaker had just left, perhaps it would have been two o'clock. Life wearies us with excitement and will not leave us alone for more than a few minutes at a time. The car which pulled into the yard surely must have passed the hearse at the top of the road. The unexpected comes round the corner to meet us. We have no 'hours of in-between'. We have to rapidly readjust, to double-de-clutch, change gear and direction without having even time to look at the map.

The strange car came into the yard on that beautiful morning. The day Father had died in 1979 there was so much snow we had to take the coffin in the Land Rover. Deep snow had fallen on April 24th 1981 causing sudden death to thousands of animals in the Dales. But our 'script writer' finds new drama for us. Never does the same thing happen twice.

The phone was ringing as the strange car pulled into the yard. I lifted it. My insurance man wanted me to invest. I told him, as politely as I could, that I had no funds and that we had just had a bereavement and that a stranger's car had just pulled into the yard. Out of it stepped a blonde nordic couple, vaguely familiar. They stepped out excitedly as only returning guests do.

'Hello,' they cried, beaming their pleasure at having found the place again, us at home, the sun in the sky and the cuckoo calling.

'Hello, you remember us? We stayed with you. We are from Finland. One night we stayed with you.' All guests expect never to be forgotten. I opened the dining-room door and welcomed them in, a fixed hurting smile on my face. I did not need to say anything they were so vocal.

'We came last year. We sent you photographs of the goats. Remember?' They stood turning in the dining-room. 'See, it's just as we remember!'

They ran up the slope into the sitting-room. They stood with their backs to the mullions and smiled with pleasure at the inglenook fireplace. 'It's just the same,' they said. They were so excited I did not know what to say so I excused myself and went for Margaret and Harry.

'We've brought you bon-bons because, this time we cannot stay,' they said displaying some attractively boxed Finnish confectionery. 'We are going south but we wanted to see your farm again.'

'That's lovely,' we said.

'I'll make some tea,' I offered our excited friends. Mother had just died. Her coffin would not yet have reached the chapel of rest and here we were serving tea to visitors from Finland as if nothing momentous had happened. They remembered Margaret. They remembered Harry. They remembered the cock crowing at dawn and the donkeys and the goats and the other guests who were in at the same time and what the weather was like and what I'd served for evening meal. They effervesced with goodwill and delight and we were forced to respond for the show must go on and 'the world's a stage and we are the actors'.

At last they rose, reluctantly, to go. 'It's been lovely,' they said, 'Lovely!' They hesitated at the door. If we waste any time in this busy life we lead, it is standing at the door waiting for guests to go. They delay doing so as long as they possibly can.

'It's been so lovely,' they repeated. 'We wish we could stay.' And then it happened.

'Oh, where is your mother?' The tall, blonde girl wanted to know.

I took a deep breath. 'She died just four hours ago,' I said. Perhaps I should have done so earlier. Our welcome had, I knew, lacked the merriment, the spontaneous laughter, the pervading warmth which is our hallmark but they had not seemed to notice and I had not wished to spoil their brief visit.

It was a relief and excused our woodenness, our heaviness and our pallor, but it was cruel to the poor girl and her blonde, handsome husband. I should have spared them but their immediate compassion was sincere and healing. They sat down again and we told them our nearly ninety-four-year-old mother had died with great dignity, had earned her place alongside Father and that we were without guilt or remorse for she had died at home, among family which is where everyone should and that we had done everything we could for her. We talked to the couple from Finland for a while and then they left and the following Christmas we had a card from them saying how sorry they had been to arrive too late to see Mother. We do not know their name or their address. We are ordinary people to whom extraordinary things happen. Perhaps it is normal and other ordinary people have similar encounters and it is arrogance on my behalf to write ours down and think other people will be interested, but it is basic to all conversation that listening to the experience of another excites personal memory. If I can do no more than make the reader remember it is enough. That encounter with the blonde Scandinavian giants will never be forgotten and the memory will always be pleasant and foster a hope that there is an affinity between nations which will eventually bring universal peace.

Less pleasant is the memory of sitting outside the Registrar's office an hour later, waiting to record Mother's death. A young couple were there with a newly born baby which was being pushed up and down the corridor, presumably by Granny. The young, unmarried parents were giggling together and arguing about the surname to use when registering the birth. They were not quarrelling but the girl thought it was important and laughed nervously saying, 'What shall I do if I say it wrong when I get there?' I was born into a much more secure environment. It cannot be that it is increasing age which makes me think things were better then than now. The young people have got it wrong and the generation before them have failed to correct them and dear knows where it will all end.

Just before 5 p.m. the Vicar came. We had met briefly whilst dry-stone walling one day, at the top of the first intake. He had never met Mother having only been appointed two years ago. It was our fault for being too busy making breakfasts when Matins were sung and dinner during Evensong, but Mother used to walk regularly over the fields to church. He arrived just as we began to prepare the evening meal for the remaining twelve guests. Maureen had brought back the lady in a wheel-chair. The day had been so warm they had been able to sit in the garden all afternoon. The stranger, who had felt it his professional duty to come to us in our bereavement, sat at our huge kitchen table and systematically watched us make the meal. He was fascinated. He was still sitting bemused when we served, and did not rise to go until we were collecting the emptied tureens. We were near to screaming point. All we wanted was to be alone, but I repeat, we are actors in life's drama and there is nothing we can do about that!

It is strange what triggers weeping. There is a legitimate end to all life. When premature end is reached then anger and tears are acceptable. We shed gentle tears for Mother but our emotion could not be described as weeping. That sobbing was reserved for the ducks. We had such a flock of them for all fourteen of the large brood had survived and each morning there were twenty-eight of them on the twin ponds. Village children came with bread to feed them and goodwill abounded. Two days after Mother died Margaret shut up the ducks, as she always did, in the boarded off quarter of the garage. The partition did not reach the roof but a heavy, glass fibre sack screened the gap. The Range Rover was a very tight fit and the garage doors had recently suffered a winter gale and would not shut properly. We had been too busy to repair it and had forgotten the vixen would be whelping and the dog fox would be daring to come to the busy farmyard out of necessity. That night he squeezed through the gap the broken door afforded and leapt onto the bonnet of the Range Rover. He pushed his way past the curtain to cause mayhem in the Duck 'Ole. It was high time straw and droppings were taken out. Their depth made it possible for him to get out once more, but he did not leave until he had succeeded in almost a total massacre. We had always believed he killed to live, taking only one of many to feed his cubs. He rarely comes to The Currer at any other time of the year for there is plenty of food elsewhere and to risk our farmyard, alive with lights, dogs, people and cars, is too dangerous. On this occasion he must have been very hungry for he risked it all, squeezed himself into the garage, and caused utter chaos. The quacking frenzy of twenty-eight ducks would drown the barking of the dogs. We, tired as we were with weeks of sleeplessness, heard nothing. Nothing at all.

When Margaret went to the Duck 'Ole door, next morning, to let them out she found only four still alive. Most had been removed completely, some had been dragged into the garage and a few had merely died of shock trying to get out of the door. The four survivors were petrified. Blood and feathers littered the straw. Crying with anger and remorse at our own negligence and the organised killing by the fox, we dragged out those dead birds he had left. We cried as we had not cried for many years, noisily, without control. I suppose we were crying for Mother, for Harry who had lost his main guyline, for ourselves who had too much work, too much trouble, too little money, too much sorrow, for too long a time.

Crying purges so that, when sobs subside, they have done as much good as a heavy shower when drought is becoming serious. But on that tragic morning our tears left us devastated and when the storm of them had passed we still had to remove the mutilated birds and Margaret had to nurse the survivors. The twin ponds were silent and empty for the four ducks would not leave the shed for many days. The Currer was silent and empty, too.

The calm, sunny weather tried to breathe new life into us. Father always allowed us to cry as far as the yard gate, then we had to pull ourselves together and get on with it. 'We'll buy another on Monday,' was his philosophy. We went up onto Baildon Moor, where the caravan-site owner keeps a lot of ducks, and bought three six-week-old ones and later, after the funeral, we went to a small-holding near Haworth and bought eight white ducklings and we repaired the garage doors.

Our greater loss could not be solved by buying another on Monday. We decided to ask Tony Ainley to speak at Mother's funeral. He had known her since he was a small boy living, as he still does, higher up on the moor than we do. He is a Methodist local preacher but he was honoured to be asked to the Church of England to give Mother the accolade she deserved.

The week before the funeral dragged slowly by. The cattle were all out but we had no holiday impending. We had done no decorating but had no intention of starting. Saturday, as always, is a busy day and when we had that behind us we thought we might just steal for ourselves that moment of privacy everyone needs. After Sunday breakfast we intended to down tools and watch the service on television. The bed and breakfast guests were quick to disappear up the Dale and the cottage people were not our responsibility. The washing-up and bed-making could wait while we watched the morning worship coming from the Island of Barra in the Outer Hebrides. All four of us gathered in the sitting-room for this anticipated moment of calm. The sunshine was so bright we had to draw the curtains in the east window lest all four mullions showed on the screen. We relaxed for the first time in a long, long time.

We could not have gone to our own village church for the nature of our work prevents it and in any case it would have been second best to sitting in comfort in our own home, bathed in sunshine, sharing our singing with gaelic friends on one of the most beautiful islands we know. The programme began with splendid views of Castlebay, of deserted white beaches and untrodden sand and of the church on the hill we have seen so many times. Had Mother been spared we would have been on Arran. It was our moment and it was shattered by the fire alarm screeching throughout the building.

In utter disbelief we dashed to the dining-room where the system told us the alarm had been activated in the Loft Cottage. The occupants' car was still parked outside and there was no sign of fire. We hammered on the door. They opened it and we saw the red light on the new smoke alarm and the wife happily frying breakfast bacon. The door from the kitchen to passage was wide open and the delicious smelling fumes had activated the alarm. We heaved a big sigh of relief and hurried downstairs to turn off the ear-splitting noise intending to hurry back to our programme. There was, after the OFF key had been turned, an almost Hebridean Sunday silence at The Currer. Perhaps there always is if we stop and listen. We tend to be so busy only our guests are aware of it.

It only lasted seconds for, almost immediately, we heard the sirens of appliances hurrying along the top road and down the lane. There were three engines and thirty-six helmeted firemen in the yard and a police car. The embarrassment of that occasion, remembered because it was the day before the funeral, was minimal compared with the mortification we now feel after so many false alarms. They have regularly punctuated the years that have followed in spite of a warning notice on the kitchen door, insisting on closure, and a re-location of the smoke detector.

Of course, we missed the morning service on Barra. We had missed the 'moment of calm' Sir Anthony Quayle told us was important. What's new? It's the NFU. Normal For Us! Margaret and Harry went to the washing-up and I to the bed-making.

The spring sunshine seemed here to stay making mother's funeral very lovely. The church was packed. Mother's generation had mostly gone before but the younger people who filled the pews left feeling up-lifted. Tony said all the important things about Mother, the founder member of The Currer family, the successful mother of Harry. He said that Mother's life had been like filling a supermarket trolley with little things. Amazingly it was full to overflowing when she reached the checkout. He said too, that she had chosen a good captain for her ship, that, in the engine room, she had obeyed every order from the bridge, veered neither to port nor to starboard, and come safely into harbour. A few days ago, he said, she'd obeyed the final order which had been, 'Shut down engines.'

A great pride took away sorrow. Without waiting for a cue from the vicar, Harry rose. For a moment he stood alone, bent, in front of the seated congregation. Then he turned and looked at them and in any other venue they would have applauded. Well done, Harry!

We were detained by a multitude of friends outside the church and there was, unexpectedly a college friend, Betty, seen only at Reunion for the past forty years. We arranged for her to come to The Currer and got into the car to go to the cemetery. Hazel and all our friends waved as we left.

The small, unpretentious headstone, the best we could afford for Father eleven years ago, is different from those surrounding it. It does not just say, 'Herbert Henry Brown, died, February 6th 1979'. It also says, 'Of Currer Laithe Farm'. Why not? If The Currer was to be preserved in perpetuity by Covenancy to the National Trust, the memory of those who restored it, and created its atmosphere, bears importance. Their story is part of a social history worth recording. We had only to lift up our eyes from that well kept cemetery to the hillside on the south, and there was The Currer, isolated on the green fields.

What I would most like to know will forever remain a secret. Who the people were, who lived here for four hundred years before Grandfather bought it, may be known by name but never will we know what kind of people they were, what they did, how they farmed, what they believed, what was their philosophy. Were they happy or sad, rich or poor will remain a mystery. Perhaps our successors will like to know those things about us. We have already farmed the land for nearly seventy years, more than any before us, and we have been the people who have determined its future. Yes, I thought, I really ought to do something positive about my laboriously typed manuscript. Life has been far from easy but we have lived it joyously. We still continue in a way which was the norm fifty years ago. We have not thought it the responsibility of society to look after the aged and the handicapped. The status quo has remained at The Currer and on the day of Mother's funeral we were more than happy with our lot.

On the wall of Auntie Chrissie's home on Tiree, the text was, 'Take your troubles to the feet of the Lord and leave them there.' Why not?

So we went home to a way of life where work was still manual and a Creator acknowledged, a life which had once been the way of many and had only recently become the way of the few. Our family was growing smaller but not through disintegration. Lest this way of thinking and behaving and working becomes history, I vowed, I must sort out that record I had been compiling for many years. It seemed important, then, to preserve the knowledge of how Margaret farms, how Joan and I used to teach, how Hazel and I interpreted Guiding, how we cope with disablement and disease, how we entertain, the meals we serve and the philosophy and faith we share. At that moment I really intended to do so but it was a long time before I plucked up enough courage.

Betty came to lunch and I promised to run her home to Wibsey, but Alma came. Alma lives near Manchester and has been a frequent bed and breakfast guest for so long she has become a friend. She came first with her headmaster husband who, following the death of their son in a terrible car accident at Christmas, suffered a stroke which left him virtually helpless. She had hoped to come to Mother's funeral but had not got there in time so we all had lunch together and then Margaret took Betty home. Harry and Auntie Mary went too. They called at our Land Rover specialists for some minor problem and heard, over the garage radio, that an aeroplane had crashed on Northton Hill on the Island of Harris, killing all personnel. In that dirty workshop the three members of my family were transported to the clean, almost arctic beauty of the Hebrides and they speculated on the tragic activity there would be among our

Northton friends whose knowledge of the hill would be needed by the rescue squad.

At home I was unaware of the drama. I was accompanying Alma to see a friend suffering from what had been diagnosed as rheumatoid arthritis. Alma had some experience of homeopathic cure and wanted to share this knowledge and my friend was eager to listen.

We desperately wanted to flee this perpetually active life at The Currer. We cannot shut the door or lift off the phone. If we want to be alone we have to go. So the day after the funeral we went. Miraculously we had no evening meal. It was a perfect May Day. We hurried through the morning chores, washing dishes and making beds whilst Auntie Mary packed the picnic. Then we collected her eighty-eight-year-old friend, Hilda, and we drove to Saltburn. There we left the two old ladies on the promenade and Margaret and I pushed Harry in the wheel-chair, for miles at the water's edge. Backwards and forwards we strolled, barefoot, mourning, remembering, just the three of us, the remains of perhaps the closest of close British families. We had achieved closeness because of spiritual ties. You will argue that The Currer is a material thing but you will be wrong. No way is it just stone walls and window panes. Harry remembers when we first removed here in 1958, after we had rescued it from ruin, Mother, the romantic member of the family, had gone to the music shop to buy a copy of 'Bless this house,' for Margaret, our pianist, to play.

Yet when we most need spiritual consolation we must leave it. It is too demanding. So we found our solitude at the ocean's edge, at Saltburn on the 1st of May 1990. It was the only holiday we had that spring.

Chapter Five

> Let us take life as a challenge, a venture brave and gay
> And with steady hand and with gladness
> we shall tackle each new day.
>
> *Dorothy Clough, Steeton Hall 1900–1981*

If Harry had not been so restless we would have slept away our exhaustion. If we could have got into bed at night, Margaret and I, and stayed there until the day's work had to be begun, we'd have been fine but Harry began to sleep so badly we were up and down like yo-yos.

The National Health inspector who had visited, had suggested Harry might benefit from some physiotherapy. We relayed this to our doctor. He sent a practitioner whose only suggestion was that Harry should be laid on his stomach to straighten his back, a posture which would have killed him. However, a locum we called late at night when Harry's hiatus hernia was giving him great pain, suggested aqua therapy might help. Our doctor made the necessary arrangement for six sessions at the hospital pool which proved only one thing. Harry certainly liked being in the water. When the six weekly half-hours were over he was offered no further treatment so we took the initiative and went to the newly built leisure pool, in town, which had disabled changing facilities and, hey presto, we became regular users and Harry's walking became better and he fell far less frequently.

A relative's wife suggested Harry might benefit from Cemetidine and the doctor agreed to prescribe it. For thirty years Harry had been taking a regular tablet three times a day and it was inevitable that the Cemetidine would be taken at the same time as one of the other pills. We thought the new pill helped his digestion and the baths helped his movement but even so he seemed to be on a downhill slope. We thought the loss of Mother was the real problem. An old lady was heard to say, 'He won't last long after his mother has died.'

We were livid! How dare it even be suggested? We were not going to lose our brother. We got up and down perpetually. We talked to him, took him everywhere with us but the recovery was destined to be long and didn't really come until our Asian chemist had to make up a prescription for both tablets and said, 'You know not to take both of these together, don't you? See that at least an hour lapses between.'

'Say that again!' Margaret said. The pharmacist showed her his machine and pressed the buttons indicating that the two medications must be taken at intervals. From that day on there was an improvement but by then we had had at least two years' struggle.

The information from the Asian Angel was a life-saver. So too is springtime at The Currer. Dr Green, we call it. Just when cattle can stand winter no more, when barns are empty and the feed bill has topped the £15,000 mark, this good friend arrives. The house is desperate for a spring-clean, washing needs bleaching on the line and windows everywhere need opening wide. We are crying for the buds to burst on the trees long before they do and then, suddenly the first sycamore of the year, the one which stands sentinel on the Top Road, outside George's farmhouse, becomes the beautiful thing it is and all others follow suit and become heavy with leaf.

We are seldom without wind, here at a thousand feet. In winter it passes silently through the empty branches of our nearest neighbour but when millions of leaves darken Auntie Mary's bedroom the summer breezes are imprisoned. They play background music and sunshine reflects their movement onto the sitting-room wall.

Below the Five Acre there is a flash of lighter green and we each tell the other, unnecessarily, that the beech on the brow of Jimmy's Wood has leafed. Shortly afterwards, if we check the downhill fences on the boundary and the springs on the Ramblers' Gates, we see that the May blossom on the hawthorn trees is as thick as drifted snow. While the new grass is still short we can see that the wooded dell is nurturing scores of baby trees standing upright among the carpet of bluebells.

Because we use no artificial fertilizer, and because our acres are predominantly pasture, May brings wild flowers galore. The Ministry announced grants for farmers interested in conservation by using no chemicals and mowing late. We applied but the visiting assessor declared we had already achieved what they were attempting and would therefore not qualify!

In May relatives from Glasgow called on their way south and left their ageing mother with us whilst they collected an aunt arriving at Heathrow from Canada. We had to pick ourselves up and entertain. Before leaving, Marion said, 'Anytime you wish to, do use our holiday home at Cove. You'd be welcome.' It was very kind of her.

The bed and breakfast side of our business did well. It provides us with such variety but its incidents are enjoyed then frequently forgotten. Why,

I cannot explain. One remains firmly in my memory maybe because of its animal connection. I cannot place it on the calendar but I know we had the blacksmith here to trim the donkey's hooves. Mr and Mrs Wright had been annual guests for several years. She was in a wheel-chair and as she grew more infirm her daughter and son-in-law accompanied them.

On the morning the blacksmith came all cars had left the yard. He came before I checked bedrooms and attended un-made beds. It wasn't necessary for me to stay throughout his visit so I returned to the empty dining-room and ran upstairs eager to get chores done. I thought I heard a radio somewhere and expected to find one. Descending the steps I heard the voice again. I opened the ground floor bedroom door currently occupied by the Wrights. The good lady was sitting firmly on the the floor, legs out, and she was knitting furiously.

'Goodness, whatever are you doing there?' I gasped. 'I thought you'd gone.'

'I'd to wait for the pill to work,' she said. 'Then I fell off the side of the bed. I couldn't sit here doing nothing. See how much I've knitted.'

Disabled guests are such fun!

'I'll never get you up,' I laughed. 'Wait till I fetch the blacksmith.'

'I've an old lady on the floor and I desperately need help,' I ventured to ask. He was immediately shy. He didn't care how obstreperous the donkeys were but he appeared petrified at being asked to lift an old lady.

Margaret said, 'I'm here.'

'You and I won't manage,' I said. 'This man will have to come.'

Needless to say his embarrassment dissolved into laughter when he saw the old lady knitting away on the floor.

We pick up people from the floor regularly, but not Jack. Most of his three-score years and ten had been in a wheel-chair and he could cope. He came alone because the three-wheeler the State provided for him, could not transport his wife so she stayed with her daughter and he came here, frequently. We never had to do anything for him. Every other day we made him sandwiches and he went to Bolton Abbey and every alternate day he stayed at home. He dearly wanted his wife to come to The Currer to see the paradise he had found for himself so he arranged for his son to bring her. With no transport for her during her stay, the able-bodied partner had to stay home when the disabled one went to Bolton Abbey. One evening Jack came into the kitchen, propelling his wheel-chair, to say we were needed to lift his wife off the floor onto which she had accidentally fallen. He and she were full of laughter for, being stout, the round little lady could not get up again.

We have no problem making the bed and breakfast and evening meal pay. It was the low price of autumn cattle that year which made us wonder if we should neglect to have any holiday at all in 1990. The bank manager cruelly added interest to our overdraft. We had gained my pension but lost Mother's. She had died penniless. Since Father had sold the farm to us in 1962 the only bank account, jointly owned by the three of us, had been in the red. Mother had drawn out the small sum in her name to pay for Father's headstone.

I felt very guilty that Mother died without a penny to her name. She hadn't needed money for she had no bills to pay. Cosseted, cherished, busy, she had lived in luxury. She always had somewhere to go, something to do, someone pleasant to meet and was never lonely. Until Christmas she had never been ill, had taken few pills, had never been to hospital. Until she died she never missed a holiday, an outing or a bath.

Nevertheless I wondered if the Government might think I was holding something back, that we had robbed her of savings or something quite untrue. I asked the undertaker and my accountant. Both said there was no problem but I decided to ask my solicitor. I did. He said, 'Do nothing,' over the phone, and he sent me a bill for giving me that advice!

Always when the product of our labours is at its best, it must go. Just when my eleven-year-old school-children were at their bonniest, healthiest, most active, most confident and capable, they had to transfer to senior school.

As soon as a Guide had become an excellent leader, grown to be such fun, such a source of strength in the Company and had, at almost sixteen, gained a Queen's Guide Award, she had to go to Rangers and when her discussion was becoming mature and her ability to serve the community was almost adult, she left for college.

Similarly when the food we have cooked is just right we must see it eaten. When the soup is perfection, the roast tender and the apple pie is golden they must be served. This is all very good for our ego, no doubt.

The saddest, I'm sure, is that just when the bovine stores are at their most beautiful they must be sold. Our hundred Hereford heifers were very bonny that year. Black is beautiful when it shines and the coats of fit, sturdy Herefords glow as if the black-lead brush has been used to polish them.

We went to Skipton with the first batch. The best of the herd goes first in a belief that, given another fortnight, the rest will have improved. There was a melancholy atmosphere in the newly built Auction Mart. It is out of town and landscaped against a pleasing backcloth of the dale. The price of everything had plummeted. Calves were cheaper than they had been for years and stores were selling badly.

The farmers, on their hard seats round the ring, were bent with work and dejection. Their beautiful animals were barely fetching enough to cover costs. Our overheads were higher than most for we had bought weaned calves and had not made our own hay. Though Margaret got the best of the shockingly low prices for Herefords what she received gave us no profit.

Since beginning bed and breakfast, we had accepted the fact that we couldn't do everything ourselves. Ten years ago a heavy April snowfall had warned us that we couldn't rear calves and feed twenty paying houseguests at the same time. Newly born calves need undivided attention. The small grant we had received from the English Tourist Board was conditional that we stayed open twelve months of a year and provided evening meals every night for ten years. From that obligation there was no escape. The large sum of money we had borrowed from the bank, we knew, could only be repaid by paying guests and we had those in plenty. They came as thickly as the midges in summer and starlings in winter.

Because this trade was more likely to pay off the overdraft than cattle sales, we had opted to buy three-month-old weaned calves but each year the initial cost of them had risen and so had feed and wormers and vaccines and antibiotics. Nevertheless, until 1990, we had managed to make a small profit and when small is multiplied by one hundred, even Mr Micawber would be content.

If the profit was replaced by a loss, however small, when similarly multiplied, it would replace contentment with fear.

We had taken Harry into the buyer's amphitheatre in his wheel-chair, to watch Margaret parade her lovely animals round the ring. We are visibly different from other farmers. We are women, for a start. We sell more cattle at any one sale than anyone else for all are the same age and all must go at the same time, give or take a week or two and the only man in our family is in a wheel-chair. Furthermore, when we go back to our old Range Rover, I immediately begin to pour from our flask and open up a picnic whilst Margaret takes the dogs round the perimeter.

On her return we sat eating in thoughtful silence remembering a sally a dealer had thrown at us as we'd walked to the car park. 'You ought to be buying calves!' he'd said.

'Too true,' Margaret had replied.

The ten year commitment to the ETB was almost over. We are not landladies, really. We do not want to profit in the field in which we do not really belong and lose financially in the profession to which we were born. In farming there is no one way of doing things. When one method, however perfect it seemed, no longer works we have to re-think, change, sometimes do a complete U-turn.

I said, for I thought it should be my idea so that I could never say that the preposterous suggestion hadn't come from me, 'We'll have to buy new-born calves this autumn, Margaret. What d'you think?'

The best Hereford price of the day, the small prize for the best pen of Hereford heifers and yet make a loss, just would not do!! We'd paid £230 for the weaned calves we had bought in June and week-old calves were to be bought for a song.

'Why not,' she replied but she was more apprehensive than I. She knew what it entailed.

It is amazing what the prospect of a new project does to such a crazy family as ours. It did not matter that subsequent sales were equally horrifying or that the year end accounts would be disappointing for we were ready and prepared with a solution. I was the most enthusiastic. Margaret kept a low profile knowing she was leaping into rearing calves in great numbers after a nine year gap. It is a skilled job only to be considered by a professional and she felt a bit rusty. We had no recent experience of calf markets, new milk powders, antibiotics and vaccines. Neither had we any accommodation. The hundred heifer calves, already bought, would fill the two comparatively new sheds and all we had left were two sheds we had built way back in the 'sixties and the dilapidated dutch barn whose rusted corrugated sheets rattled in a breeze and blew off in a gale. We'd need more hay and straw and we'd have to stack it outside but accommodation was first priority. There is no tonic better than a new project.

We began to feel healthier by the hour. Our spirits rose as we struggled to build stone walls on the open-ended north side of each shed, to empty and begin the task of separating them into pens using old tins and bales and splitting railway sleepers for the uprights. We put Yorkshire boarding on the the south and endeavoured to exclude any draught winter might have in store. We worked endlessly whilst our guests were out enjoying the Dales and were secretly ashamed if they came back early and saw the old tools we were using and the old materials at our disposal. We weren't very professional at all but we had neither time nor money. If we were to buy in October we had to make do with what we had.

We shouldn't, of course, have been buying that autumn at all for we were still committed to the Evening Meal not merely to comply with the ETB. We had several, already booked, Guide Association weekends when we would have to feed upwards of thirty. We are impossible, really! If we think of an idea we have to implement it right away and with calf prices at rock bottom we had to snatch our opportunity lest we lose it altogether.

But we needed some period of rest before the fray. We had cancelled the Grange cottage way back in February when Mother was ill. Harry loves new places and we had seen it as opportunity to take him elsewhere. Then, somehow, we had stopped being interested in holidays and had let the matter slide. I think Margaret and I would not have bothered at all if it had not been for Harry on that frightening downhill slope.

Our Glasgow relatives had offered us the use of their second home at Cove, on Loch Long. We began to think it might not be a bad idea. It was Scotland but not too distant, new ground to interest Harry and somewhere for us to sit at a window and lift up our eyes 'unto the hills'. We'd need all the strength we could muster for the winter. So we arranged to go north and Harry began to look forward but all Margaret could think about was where could she buy seventy calves all in one day.

If I took the responsibility for our decision to go back into calf-rearing it was Margaret who made the next suggestion and I, quite unfairly, rejected it. She announced that she wanted to buy at Beeston Castle, in Cheshire, nearly one hundred miles away.

There was a major tug of war and I had weight on my side. The market opened at 8.30 a.m. It would mean leaving before 5 a.m. At least two hours of the four it would take would be in darkness. We didn't know the route, we had an old lady to leave behind and a disabled brother to take with us. I had plenty of ammunition with which to argue but Margaret was determined. She had selected the biggest calf market within a possible radius, on a date convenient for our cattle remover to follow us down and it didn't matter what I said, she dug her heels firmly in and would not budge. She was aiming at as many calves as possible in the one day and she planned to rear seventy, come what may.

We made sixteen pens in the two bullock sheds and the Dutch barn. We bought teats and pipes and buckets and we hammered and sawed until we were whacked, and the stream of guests continued. We made a big stack of straw and covered it with tarpaulin and threw up scores of old tyres from a local garage to weight it down.

At the same time came the blackberry harvest and the plums from Altar Farm. Every year Mary Hamilton sent for us when the wind disturbed the weighted branches and the first fruits fell. We took two clothes baskets and cardboard boxes and brought back thousands to put in the freezer or for Auntie Mary to make into jam. Our neighbours at Marley told us that the Bramleys were ready and we collected three sacks full from the tree and one from the lawn beneath and at the same time we prepared to leave for Cove which was a marathon in itself.

Then, at the last minute, we had an eruption of mushrooms. Never had we seen anything like it before, nor have we since. The pastures were white each morning and we are opportunists, remember! We could not let them go to waste. We gathered them in buckets and had mushroom soup and mushroom delicacies every day. We put pounds and pounds into the freezer and when we left, tardily, for Glasgow we took a bucketful for our cousins.

We drove slowly leaving the A65 just before Kirby Lonsdale and travelling due north through Casterton, Tebay and Orton to the now empty A6 and Penrith. It always takes us ages to get anywhere in the Range Rover for speed is something we can only cope with on two legs. The vehicle is old and its passengers are not as young as they were. The rush to get away from The Currer leaves us gasping for coffee so that we have been known to open our flasks on the first lay-by on the local by-pass. We have frequently had half-an-hour's sleep at Kirby Lonsdale. Our friends are scathing, especially those with new models, no oldies or dogs, and who do not live on a farm and provide bed and breakfast. Those with one or other of the latter partially understand. Those with all are few, but they would understand completely.

We followed the A7 as far as Langholm before turning off onto the beautiful B709, which goes through Castle O'er Forest and approaches Peebles in the most scenic way. We arrived at our cousins' quite late. Their welcome was warm and caring and we were grateful for the lovely meal and the comfortable bed but Harry's indigestion kept him and me awake most of the night. Mother's death had disturbed him badly. She had been his constant companion and we had to learn that role and add it to the numerous ones we already had.

Next morning our relatives took the fast road to Cove, opened up, lit a fire and made our lunch whilst we, their country cousins, took a B road, mentally and physically unable to follow them on the highway. We found the lovely bungalow overlooking the loch and drove up the steep entry and parked outside the kitchen window. They welcomed us into the warmth and showed us the luxurious accommodation and well-appointed kitchen. 'It's lovely,' we said drowsily. 'Just beautiful.' Their benevolence was overflowing. They fed us soup and sandwiches and mercifully did not beg to help us unpack the luggage from the covered rack. We thanked them profusely. At least I hope we did. We were so terribly tired we hardly knew what we were doing.

When their car disappeared from sight we lowered our waving hands and all four of us went inside, drew up armchairs to the log fire and were instantly asleep. We remained so for two hours, Harry sleeping peacefully, too, whilst the logs burned to embers and the sun lowered itself behind the mountains on the opposite shore of the loch.

We refer to this 'bliss', now, as 'doing a Cove.' Rarely do we indulge but frequently long to. What's new about that?

At Cove we meant to sit at a window and rest. We did no such thing. Autumn was richer in colour than anything we had experienced in the Lake District. The Trossachs were so grandly dressed, so aglow with red and orange and yellow we could no more stay in than I ever could with my nature-loving school children. With them I used to spend the afternoon hours collecting the hedgerow harvest and they would frolic in the crisp drift of leaves and chase the airborne seeds and purple their lips with brambles. We could not stay in that very comfortable cottage for the roads were empty and we crawled along in worshipful wonder without being hooted at.

Perhaps we were unwise. Perhaps, considering the wakeful nights, the tragic spring, the burdensome summer and the frantic autumn attempt to prepare to rear calves, we should have hibernated at Cove. Perhaps we should have said, like so many people do, that there is nothing to see in woods and forests, no real reason to climb a hill or skirt a lochan. Perhaps we really should have done a prolonged 'Cove' and slept in the armchairs of 'The Weald' and prepared ourselves for what the winter had in store.

In actual fact we had no choice for to be on holiday with Harry is to be energetic. If the weather permits, at all, he wants to be out. 'What are we going to do? Where are we going? When are we setting off? What did we do yesterday? What will we do tomorrow? How? Why? When? Why? How many? Who? Where? What?' is Harry's eternal interest in his environment. It exhausts us but in spite of activity during the day Harry continued to sleep very badly so that, for Margaret and me, one night in two was a write-off.

* * * * *

We had a busy week following our return. It was half-term and the house was packed with holiday-makers and no moment could be called our own. On Friday more than twenty Guide Commissioners came for their annual conference. We served breakfast, elevenses, lunch, afternoon tea, dinner and bed-time drink and when they left there was a pile of more than forty sheets and pillowcases to wash. Nevertheless we were anticipating a moment of calm, a few days of peace before going to Beeston Castle in the early hours of the following Friday. Four days without stress, we prophesied, and we'll be fine.

Sunday was a cold wet day. Long before dusk the cattle headed home, seeking the warmth and dryness of the shed. Those 'animal lovers' who think cattle should be given their freedom and out-wintered should witness their preference for shelter. Margaret had opened the shed door and she and Jess were rounding up heifers who would have come equally willingly of their own accord but Jess gets too little work. I was watching from the front room window and I noticed an animal, standing in the Five Acre, facing the wrong way. It made no attempt to join the incoming tide of its companions. It seemed incapable of turning round. Indeed it took both Margaret and Jess to motivate it and persuade it to go the last few yards into the shed. It's head hung heavily on it's shoulders as if something inside of it hurt so much it could not bide. Its belly was empty in spite of the fact that its fellows had been able to fend all day.

Next day it was no better so Margaret called the vet who diagnosed meningitis but said he would give it a remedial dose of antibiotic. He was confident of its recovery. On the following day there was a deterioration rather than improvement and Margaret sent for the vet again. He gave it a further dose but saw no reason to separate it from the rest of the herd.

Vets, it must be acknowledged, live a life full of drama and tragedy and a fair old smattering of comedy and a few have done an excellent job of true story telling. On the other hand farmers' experiences are seldom recorded for their hands are too callused to wield a pen, their lives too active and dirty and hours too long for them to relate their traumas and catastrophes. I am often asked when have I time to write and why do I do so? I write before day-break and perhaps I write on behalf of farmers who get more than their fair share of criticism and who live a life few know anything about at all.

Sometime after midnight, Margaret went down to the shed to check the herd and monitor the sick animal. There was absolute chaos. The poor animal was deranged and the whole shed was bedlam. Scores of heifers were charging around the sick one. Margaret came running into the house to rouse me. I clambered out, pulled on something warm and covered my bed to keep it aired in my absence, a routine retained from my camping-with-children-days. I hurried outside. It was raining steadily.

'I don't care what the vet says,' Margaret was muttering defiantly, 'That animal can't stay in with the others. Let's get it into the Donkey 'Ole.' We had left some old calf sheds on the east side of the stack-yard. Taking the tins from the west wall created an enclosed yard for the stack and had left us with a roofed-in corridor and an entrance we called The Donkey 'Ole. It was useful for all sorts of reasons and it was easy to let an animal out of the bullock shed and guide it into solitude. 'It won't take a minute,' Margaret said but when we both reached the shed we found the demented animal had leapt over the feed barrier into the manger. That was a disaster. It meant reversing the animal along the trough and, hind-first, out through the Duck 'Ole. Calves frequently squeezed through the barrier but it was unknown for one to leap. Occasionally one would enter, illegally, through a negligently left-open Duck 'Ole door, and then we had to do this backward drive. It was a tight fit and fraught with anxieties. To attempt it with a frenzied animal was even more difficult.

'Wait,' said Margaret. 'I'll go and get some colic mixture,' a commodity no longer on the market. 'We'll try and make it drowsy.'

She left me, lonely and tense, and went for the medicine which would dope. Every moment I feared the animal would go down kicking and break a leg in the iron bars of the barrier. Margaret returned, grabbed the animal's head and forced down the colic mixture.

'We've not got long,' she said knowing how efficient colic mixture was. 'You force it backwards down the manger and I'll open the Duck 'Ole door.'

We succeeded admirably. The frenzied quacking of a few ducks did not frighten the heifer, so bewildered was she. It was as easy as pie. The animal reversed out, turned and was guided by a suddenly relaxed Margaret, into the darkness of The Donkey 'Ole. I'm sure there would be smiles on our faces proclaiming our success after a difficult animal manoeuvre. See, we can cope, it says, who needs a man?

Pride so often precedes a fall. Margaret should not have relaxed. She could not see into the darkness and I should have followed to bar the entrance. If we had, mind you, we might have been killed for the estranged animal did not like the darkness these women had forced it into. It did a quick u-turn and bolted, without knocking us flying for I am ashamed to say we were not there to stop it. Perhaps it was as well, after all, that we were not in its direct path. It disappeared completely into the darkness.

When there is no mist or fog there is never complete darkness at The Currer. Lights indicate the road along the valley. The dwellings on the opposite hillside, which are well camouflaged in daylight, are illuminated at night. The radiance from the town two miles away, on the floor of the dale, and from the city ten miles down the river, throws a redness into the night sky. If there is no cloud every star in the northern hemisphere spangles the heavens; when there is a moon nothing impedes its celestial journey.

But if there are the mists and fog of late October and November nothing at all can be seen. From the house we cannot see the shed lights and from them we cannot see the lighted kitchen, though the curtains are seldom drawn. Such was the night the sick heifer disappeared down the field, the Dyke Field, riddled with ditches which deepen every time we get a cloudburst or John clears them with his tractor shovel. Seldom have we ever felt so clumsily incompetent. It had all been so nearly in the bag. If we'd just shut the gate quickly enough we would have been able to go to bed. Instead we were in a mess.

I knew Margaret would not rest until she had found the stupid animal. Who, in their right senses, would reject the luxury we offered for a foggy night in the open? 'You go one way and I'll go the other,' Margaret ordered. Increasingly wet and miserable, having been called from my bed, I knew I had absolutely no alternative but to obey. She shoved a torch in my hand and we separated.

We have 170 acres of ungated field pasture and moorland. A healthy animal would not stray far for cattle fear loneliness more than anything and can be depended on to return, almost immediately, to the herd. But this animal could not be relied upon to behave normally. It was very ill, very frightened and, what's more, it had just been given a knock-out dose of colic mixture.

We were almost immediately drenched. The Dyke Field is full of fast flowing streams easily negotiated in daylight but treacherous in fog. We assumed that the heifer would run downhill. I searched the Five Acre and Margaret crossed the Footpath Field but it was the dogs which found her. She had leapt the eastern boundary wall and collapsed on what we call the Rough, the ten acres of heather and bracken which belong to us but which we seldom graze. In fifty years I can only remember once our fields being eaten so bare that we had to turn the cattle onto the Rough. They completely demolished the mass of bilberry bushes and it was several years before the crop was good again. How she managed to leap the wall, which was ancient but solid and quite capable of keeping even high flyers restrained, remains a mystery. There was no way we were going to get the unconscious beast back that night.

Fortunately she was on a relatively dry bed of bracken and she was behind a west wall. Our prevailing wind comes from the Atlantic. It was rising and bringing rain which would, by morning, have cleared the fog. Other than The Donkey 'Ole, there was really no more sheltered place to be. She was doped and would stay there until morning but we needed to cover her. We were shaking with cold. Just to pause and discuss what to do was to cool significantly. We turned, put our heads down into the wind and went back to the barn for bales of straw and to the house for a plastic sheet. Each hoisting a bale, we retraced the quarter of a mile track back to the Rough under sail. The rising wind took the bales broadside and our feet barely touched the ground.

We covered the recumbent animal with the plastic sheet and weighted it down with loosened straw. Margaret arranged its head comfortably and then we went back to turn out lights everywhere. We left our saturated clothing in a pile on the porch floor, towelled ourselves down, filled hot water bottles and crawled into bed. It was approaching morning.

When daylight came the fog had disappeared and, with binoculars and viewing from our bedroom, we could see the black head and the heap of straw. The animal, we felt sure, would be dead but when Margaret went down to investigate she found it still alive. We could not possibly leave it there. If it died John and his fork lift would have to be sent for to bring the dead animal into the yard for the remover to take away. He might as well be sent for now, whilst the animal was still alive. Until a heart stops there's always hope.

A live animal is more difficult to move. The effects of the dope would be wearing off. Margaret phoned first John and then the vet for a more reliable knockout jab. She drove down to town to collect it and was home before John and his wife, Annabel, arrived. Reaching the Rough, prepared to do a well organised job, we were horrified to find the heifer had disappeared.

There are ten acres of bracken, heather, bilberry and scrub and then many acres of silver birched hillside. There is a precipice we call Lovers' Leap, rocks, bog and woodland, all unfenced as far as the Druid's Altar. There was no animal in sight but we calculated it couldn't get very far and all four of us set about finding it, beating the bracken flat with wellingtons and sticks. It took us almost an hour. It had walked a distance but had collapsed and would not rise. We thought it would never walk again. Margaret gave it the necessary injection and we took a leg apiece and dragged the animal over the aptly named Rough. We had to resort to tying ropes all round it and pulling from the front. It was relatively easy on the down slopes but there was as much upward pull and we stopped frequently to take a breath and re-adjust the ropes. An eight-month-old heifer is a sizeable animal! We expected the creature to die any minute but alive or dead we had to get it to the farmyard. Reaching the wall we hastily dismantled about three feet of it, enough for the fork lift to pick up the unconscious beast and the rest was easy. John backed the tractor to the Donkey 'Ole and we rolled the heifer inside into a luxurious depth of straw expecting to say 'Goodbye' to it anytime.

It soon became obvious that the animal had pneumonia as well as meningitis. It was inevitable. A night in the open, like that, is neither good for man nor beast. If we'd had pneumonia, too, we would not have been surprised!

The amazing thing is that the heifer did not die. It did not die at all! It cost Margaret more in time, money, inconvenience, effort and anxiety than perhaps any other animal she has ever brought back from death's door. For a little while it was unconscious. Then the fits began and the lungs started to make an awful noise. Every few hours we turned the poor thing over, day and night. What worried us most was that we had no time to sort out the drama in the Donkey 'Ole before leaving for Beeston Castle early on Friday morning. So Margaret called the vet once more. Until an animal dies she does not give up. I might be a giver-upper. I don't know, for fortunately it is Margaret's responsibility and I am never put to the test.

We had asked Dorothy to come up for the day. 'There won't be anything to do,' we had said. There were no guests other than those in self-catering cottages.' You can walk your dogs all day,' we'd told her. 'It's a doddle of a job. When the calves come, if we buy any, just put them into the pens. We'll sort everything out when we get back. Auntie Mary will be at home. Her friend, eighty-eight-year-old Hilda will be with her. They'll make their own meals.'

So much for the best laid plans of mice and men. Ours went awa'!

For one thing Auntie Mary went unusually quiet. I suppose we were all on edge. We questioned her. 'What's bugging you, Auntie?'

'I don't want to be left behind,' she said.

She has an imagination, our Auntie Mary, and she was going where we were and not being left at The Currer where everything imaginable happened.

'OK, OK.' we said, calming her agitation. 'You can come. Hilda won't want to, but you can come!'

Amazingly Hilda did want to come. It was a trip out, a holiday not to be missed. Harry, too, was excited. Margaret and I felt our cup to be running over with problems. It wasn't a holiday. It wasn't a fun day! We had to leave in Dorothy's care a fitting animal, which needed turning over every few hours. We had to drive the near hundred miles, at dawn, to a place we'd never been to before. Margaret had to brave the auction ring, bid against farmers and dealers and then come home to the task of rearing babies, decidedly out of practice.

There was little we could do but to keep going in the direction we had chosen. We could not change the timing of our survival. We could not postpone our buying. There was one thing we could do and that was to tell the vet that, if Dorothy called him, he must come up.

Incredibly it was Hilda who wakened us all at 4 a.m. She was as bright as a button. She and Auntie Mary had packed up sandwiches. They filled flasks as light-heartedly as if we were going 'where the sun shines brightly and the sea is blue'. I was positively eating the map trying to memorise the discussed route, avoiding all fast roads. Margaret went out to her patient. It was still alive. We turned it over. Margaret and I were not talking.

Long before dawn we all squeezed into the Range Rover and headed south. There was an unmistakable air of excitement we could not join. The driver and navigator were not amused. We had a long way to go in the dark. Somehow we had to negotiate Huddersfield, an industrial town no more than eighteen miles away, to which we had never been. We had to follow a zig-zag trail, a mass of roads whose numbers I had on a lengthy sheet of paper and we had to make not one mistake. We had to be at the cattle market as near to 8.30 a.m. as possible. For this new generation it would have been no problem. For us it was a major operation and we were tense and silent. Others would have headed for the motorway and crossed from east to west and then turned south oblivious of darkness and accustomed to speed. We, who avoid motorways like the plague, scuttled south-west, by small A roads, B roads and short-cuts, following a compass direction by torchlight, accurately but slowly.

Dawn was our best friend, when it came, but it brought a November mist which lay frostily on the fields and delayed its departure over the rivers and the reservoirs. It was very beautiful and our temperatures fell and we relaxed. We changed seats and I took the wheel and Auntie Mary poured out some coffee but we did not stop. We risked spill and choking but we did not stop! Neither did we look for loos all the way. We hadn't asked them to come and we were calling the tune.

It was approaching 9 a.m. when we finally drove into the market car park. Our Guardian Angel had followed us for we had made not one false turning. We had taken a foolish risk. It had been irresponsible of us and presumptuous to think we might get there at all, let alone on time. We had too many dependants on the back seat but, incredibly we had made it within thirty minutes of schedule.

We left the gang. They were desperate for loos but we ignored them. 'I'll be back soon,' I shouted, pulling on wellingtons and running after Margaret.

The auctioneer had been calling for bids for the past half hour but the adjacent shed was full of calves. We squeezed through the male mass of wellingtons, anoraks and red, weather-beaten faces and climbed into the miniature amphitheatre.

Calves were going for a song. Hereford, Aberdeen Angus, Friesian and Charolais. With luck, we had still two hours of the sale. Margaret attracted the auctioneer's attention and the show began. She always seems to be the only women bidding amongst a score of men. I can't quite cope, nonchalantly, with Margaret bidding. Something queer happens inside me and a lump rises in my throat when the auctioneer first knocks a calf down to 'Mrs Brown.' We are both always addressed as 'Mrs Brown' for no-one really believes ladies can farm alone. One child guest, helping Margaret count the one-hundred-and-fifty head of cattle, remarked, 'You're not a proper farmer, are you?'

The 'not a proper farmer' was causing quite a stir in the ring. The first calf bought had been guided to an empty pen on the extremity of the huge shed. Nearer pens had names of recognisable, regular dealers chalked on them but our pen began to fill more quickly than any other. Margaret had come to buy calves and she wasn't going home without them.

Not only was Margaret causing entertainment in the ring. The men who pushed bought-calves to their respective pens, were highly amused and seriously wished they had chosen a nearer pen for us. Such tiny, bewildered babies they were, being pushed on the long journey down the corridor to the quickly filling pen. My job was to open the gate to allow each one in without letting the multitude out.

I could hear 'Mrs Brown' being called frequently in the ring. She was buying anything and everything, beautiful, knock-kneed, splatter-footed, timid, crying babies. Who could harm them? The five foot woman was causing welcome entertainment. Onlookers began drifting to my pen, shaking their heads and asking how many were we buying and how on earth did we expect to feed them all? God only knows, I thought as the last calves were auctioned and the number belonging to us had risen to fifty-seven.

I dashed to the market telephone to ring our cattle haulier, as planned, to tell him we had bought and needed him to come. He said he'd be down in two hours. I did not tell him how long it had taken us!

The ring had emptied and Margaret came back to look at what she had bought. I left her fondling them and reassuring them that they were going to be happy at The Currer.

I took my cheque book to the office and paid. It was 1990 and our calves averaged £24 each. I mention this because it emphasises just how financially suitable was the climate for us to re-enter calf rearing. The time was right for us to opt out of providing the evening meal. The ten years would be up in six weeks' time. According to Grandmother Smith, remember, you can stand on your head for two weeks. We could do better than that. To stand on our heads for six weeks would not kill us and in future years we would drop evening meals during calf-rearing time.

Everything supported our decision, we believed. Beeston Castle had been the right place to come, at the right time, when calves were cheap. We felt great. We rang Dorothy. The vet had been called, had given a long-lasting sedative to the meningitis patient and all was well.

We went back to our oldies and the dogs. We ate and relaxed and waited for our transport to come. Eventually we felt we could wait no longer so we tipped an employee to help Joe when he arrived and we set off home. It was 2 p.m. and, having not enough daylight to do the distance, we decided to take a faster road. It was a mistake. The road was jammed with traffic and we were irritable behind the wheel which is unusual.

Margaret had caused heads to turn in the ring. We wondered if anyone had noticed the passengers in our Ranger Rover, two incredibly old ladies and a disabled brother. That would have certainly set tongues wagging.

If we were apprehensive before, I think we were more so now. It would be late and dark when we got home. We had inadequate lighting in the sheds and a queer conglomeration of pens. We would have fifty-seven crying babies and an untried bucket and teat feeding system.

We got hopelessly lost in Huddersfield. I don't think urban man has any idea of country folk's clumsiness in a town. Even on foot I am a fish out of water. I behave like a fugitive, run, dither crossing the road and do so frightened and agitated. On wheels, in a town, there must always be two of us. If we get in the wrong lane we stay there and go where we do not want to be and sort ourselves out in some back street. In towns and cities we admit to being complete strangers so we avoid them almost totally. Why not? We do not need them! We can now get to Beeston Castle without going through Huddersfield!

Lights were on everywhere when we drove the last quarter of a mile down Currer Lane. The calves had arrived. Dorothy greeted us and we feared we had lost a friend.

'Never again! Never again!' she exploded. 'We nearly lost half of them down the field!'

We were taken by surprise. We admitted it was not easy to unload calves and push them into pens. We'd been doing it for years and hadn't appreciated how much of a shock to the system it would be, to do it for the first time. We showered apologies and thanks and praise. We do not want to lose a good friend. She is now an exceptionally good calf-pusher-into-pens and our relationship remains amicable.

Margaret went down to the Donkey 'Ole and turned over the unconscious member of our herd. The pneumonia was bad. What a mess it had landed itself in! It was quiet and would remain so, so she came in and we all had a cup of coffee.

At 8 p.m. we waved Dorothy away and started the marathon feed. The first drink on arrival is always Ionaid. Each drink must be mixed separately. Those who advocate that calves should be fed en-route are as ignorant as a mother who will over-feed her stressed baby before a car journey. Advocates are invariably those who will, themselves, refuse a cooked breakfast before travelling home. A calf has a very delicate stomach and a vet does not even advise feeding milk after a journey. So we feed Ionaid. Time consuming and expensive.

But that was not the main source of our trouble on that November night. Our feeding system did not work. It consisted of a squared piece of netting and a teat, a tube and a bucket. We managed to feed only three calves in the first half-hour. Three down and fifty-four to go! We felt like novices. It was a shambles and we resorted to lamb teats on pop bottles and resigned ourselves to a late night sitting. Slowly, slowly we went down the line with grim determination.

Around midnight we heard some sort of disturbance up the road. Though unusual we decided it must be teenagers just acting daft at the Moor Gate. Because the view from there is so splendid we had fixed a seat so that elderly villagers could pause and enjoy it. On one or two occasions we have had to pick up empty beer cans left, we presume, by youths late at night. They ought to have something better to do. The yelling continued and we finally heard a word which sounded like 'Help!'

We didn't respond. Silly teenagers! We were not going to be taken in. We had far too much to do. We had hardly fed half of the calves and it was already morning. Margaret went to the house for more buckets of hot water and when she returned she was worried.

'Someone really is shouting for help,' she said so we downed tools and set off up the road. We had only begun to walk out of the farmyard when we identified where the voice was coming from. A dark shadow was sitting on the cattle grid. A guest from the Loft Cottage had been taking his dog for a midnight outing and had fallen awkwardly through the grid. He was in agony with one leg between the bars. He would not let us touch him at all. It was the last straw!

We hastened to tell his wife and to take out blankets, for he was very cold, and we phoned for an ambulance. We left the wife admonishing him for being careless and, I am ashamed to say, we went back to our calf feeding. It wasn't funny. Things which happen to us rarely are so we must be perverted that we find amusement. The same thing never happens twice and I'm sorry if we laugh when we should not but it is the only stimulant we take. I think it was 3 a.m. before we crawled into bed.

It is not unusual for problems to be solved overnight. The solution to our feeding troubles came to me before I got up soon after 6 a.m. I hesitate to make much of this because Margaret says she's never heard the end of it and that I have boasted long enough about our very successful feeding arrangement. I woke with a very positive plan. Years ago we had brought home discarded wooden bomb cases from our Army Surplus yard. We'd used them for mangers and even for furniture in the cottage kitchens. The hard wood was in use all over the place and we still had a few pieces of it on our woodpile. Usually the piece we want from the pile is buried but on this occasion it was on top.

I hardly dare peep in at the calves. Any moment and they would begin to cry for their mother's milk. I brought the wood I wanted to the house and cut two pieces. I joined them with two cross bars and a piece of the squared netting. Then I bored a hole in each end and threaded a brass stair rod across. We had some empty four pint milk containers, each with a carrying handle and I threaded five onto the stair rod so that they dangled. I popped teats through the netting and pipes into the bottles and, lo and behold I had a milk bar. I wakened Margaret. 'See what I've done,' I boasted. She's right, I am big headed.

She was only interested in one thing, 'Did you look at the calves?' she said. She thinks, with some justification that I am more interested in invention than husbandry, but I seldom miss that 6 a.m. check with a torch. It has been mine since the days I used to leave, in darkness, for my village school headship. It is not a job I relish for it is a winter one when the sun is late rising. It is always cold, often wild and wet and sometimes snow-bound but the calves are usually curled up contentedly. If one is stretched out I climb fearfully into the pen, poke it and whisper, 'Are you alright?' Any real problem, Margaret has usually detected on one of her frequent visits in the night.

We tied the new contraption onto the pen and it worked a treat. The trouble was we had to untie it and move it from pen to pen. We had solved the problem only if we had sixteen! I went to and fro from the woodpile all day. I did not have enough stair rods but that was no problem. One was easily carried from pen to pen.

The temporarily empty pens were in the entrance to the Dutch barn, a massive thing we had built ourselves many years ago with second-hand wood and thrice-used corrugated tin. We had roofed and double roofed and it was an eyesore we loved because to build it had been a challenge and it had served us well for countless years. We were increasingly aware that it looked dilapidated at the entrance to our farmyard but replacement of it was not on our agenda.

On the Monday, ten days later, we went to Otley to buy more and again at the next sale until all sixteen pens were full. It is not easy to rear calves. Seventy all at once is a lot. They come from the market with a variety of bugs and a rearer is dependent on the antibiotic, currently on the veterinary market, being effective.

Every farmer knows the importance of good ventilation but attempts, nationwide, to get it right have been largely pathetic. Advice has been to open windows and doors and put in a fan to circulate air. Our accommodation was so primitive we spent all our time keeping the wind out. All three sheds were exposed to the constant prevailing wind from the west and we, however hard we worked, have seldom been so cold. We curtained draughts and piled up bales to make cosy sleeping areas. In the Dutch barn we put glass fibre sheets to roof walls made from bales and still the place was an ice box. There was literally so much air we were fully employed trying to keep it out. The calves, however, thrived like nobody's business.

The bales we piled around the edges of each pen offered themselves as adventure playgrounds for our hyperactive calves. They were stiles into the next pen or into the corridor. We couldn't contain the mischief makers. Trying to control them we were bruised and battered and trodden on and it was all tremendous fun and calf-rearing, we have proved again and again can be anything but. It is a constant worry. It has periods of mourning, of intensive care, of twenty-four hour surveillance. There is always the fear that the bug which has killed might do so again, and again, and again. But the 1990 calves were indestructible. Our makeshift accommodation resulted in strong, healthy animals.

For us, particularly for Margaret, it was all so ridiculously hard. We were climbing gates all the time and we carried so many buckets of milk and of water our arms stretched and our sleeves grew too short. But it is not hard work which depresses and those mischief-making calves only gave us joy. They were so greedy. To see them knocking each other off the teats and stealing from their neighbour sent us into hysterics. I said it was like watching a puppet show of Sooty and Sweep for calves can only bash each other with their heads. They made us laugh again and laughter is normal in our household. Mother had been the generator of it. The calves rekindled the spark.

The season's bookings had to be honoured. We tried, whenever possible, to avoid the evening meal but we had several Guide Association weekends to get through. Breakfast for twenty-five after feeding seventy calves, then lunch and evening meal and all the washing up, came dangerously near to killing us. Having survived a weekend it took us all week to recover before the next invasion.

There came a Saturday at the end of the month when we had snow and gales. The white icy powder found its way through every cranny and piled up in the pens. When darkness fell there were already six inches on the yard and on the roofs of the calf sheds. The hundred heifers were house-bound and it seemed as if we, too, had the five thousand to feed.

Just before the Camp-fire hour we had to call for Guider help. The roof of one shed was lifting. The spring-head nails holding the corrugated tins to heavy beams, were being torn out of the wood. A sheet flew from the north-west corner and the whole roof threatened to go. We tied binder twine round the lesser beams and the Guiders anchored them whilst Margaret, with a rope round her waist, went up on the roof and secured it with six inch nails. The operation was dangerous for the roof slopes to the south and had she begun to slide she would have been entirely dependent on the rope round her waist and on me hanging on. It is always Margaret who goes on the roof!

We had the unpleasant memory of 1981 when an unexpected April fall of snow had brought down that very roof onto a shed full of calves and we had lost nearly thirty. That things can happen twice is always our fear but presumably everything has not happened to us yet for we haven't started on repeats. When we allowed the Guiders to go inside we anchored their ropes to the wall and we did quite a lot of praying for the wind to abate.

Next morning there was so much snow we thought the Guides would not be able to leave, but male friends and relatives of the Company came with a pick-up and a dozen shovels and dug out the road for us.

* * * * *

Towards Christmas we had a letter from our friend, Charlie Maclean, whose Tiree mother had been one of the reasons we couldn't keep away from the island. She was the dearest of friends, a character so well worth knowing we were frequently round her hearth, drinking tea and our Guides were often there singing camp-fire songs. When she died we kept in touch with her sons. Charlie, the youngest, worked for a time a few miles south of us and we saw him reasonably frequently. He was a valuable accordionist who not only played for our island dancing but also came to our Christmas parties at the home of Pat and Alan Roberts. They had an enormous room where we could hold barn dances and cèilidhs and Charlie, when he worked in the area, would play for us.

He was back in Glasgow and we missed our music-man. The letter announcing a proposed visit was welcome indeed. Could he stay with us, he asked, there were friends and work colleagues in the area he wanted to see. Nothing could please us more we replied. We had been using his mother's cottage on Tiree, for our holidays, since the beginning of the 'eighties. We had a lot for which to be grateful to Charlie.

I was alone in the house when he arrived. The others must have been in town or somewhere. He looked so well and, as always, he was so immaculate and courteous. Charlie was a gentleman. I welcomed him in to the warmth of the sitting-room and put on the kettle as his mother had done for us so many scores of times.

We sat by the blazing logs, sipping our strupak and I said, 'How's everything, Charlie?' and he said, 'I've just had an operation for cancer of the colon,' and I was devastated. I mustn't let him see how worried his news had made me.

'And was it successful?' I asked.

'As far as anyone can tell,' he answered.

I said, 'You know Joan Armstrong, my infant teacher, who came to camp with us on Tiree but used Hugh and May's cottage alongside your mother's. She had that operation nearly two years ago. She's fine.'

Charlie said, 'If you're going to have cancer, my doctor told me, the best place to have it is there, so I'm crossing my fingers.' Many survive cancer.

There was most definitely a guardian angel around for not one of the babies died that year. Nor did our friends or the heifer with meningitis or Margaret who had such a struggle. It took the two of us to turn the heifer over every day. After a while she was able to stand provided we helped with a sack underneath her. She eventually stood of her own accord and began to get well. When mobile she had to be restrained and then the five-barred gate, used to do so, had to be climbed and the shed being low, the space above would only allow Margaret if she rolled over. It was easier if I was on hand but Margaret mostly struggled alone.

Since the building of the two big cattle sheds during the 'eighties, feeding cattle had not been difficult for water was piped and the ATV and trailer had ferried the hay and straw daily from the Dutch barn. Now that was impossible. We soon realised that our massive store of bales was on the other side of the four calf pens. So they had to be carried, often through the increasingly boisterous calves, and life became horrendous.

We had water in the big sheds but only taps in two of the calf sheds and nothing in the Dutch barn. We carried thousands of gallons. We took a hose pipe to a bath but the pipe froze and we resorted to carrying from the stream. The taps froze in the sheds and we had to carry water to the other twelve pens. It was a miracle we survived but I think we had to have it tough to resuscitate us after our apathy following Mother's death. Harry, with less stimulation, was finding life difficult and Auntie Mary was indispensable.

Christmas came unusually suddenly. For close on ninety years Mother must have stirred and clouted seasonal puddings. I kept up the tradition a little sadly. Increasingly we needed to make smaller and smaller dumplings as the family shrank. Harry felt it. He was particularly worried because, surrounded as we always were with a kaleidoscope of people, he knew that we were, as we had always been, essentially alone coping with life at The Currer.

Without school-children or Guides, without family or paying guests, Christmas suddenly became very quiet. Too quiet for Harry who does not watch television anytime and finds nothing to tempt him at Christmas. The house was eerily empty. Those spending Christmas in our cottages went to relatives for dinner. Margaret and I ploughed through the feeding of one hundred heifers, seventy calves, two donkeys, three goats and our flock of poultry with as much speed as possible but we realised we had lost, for as much of the future as we could visualise, our freedom to cruise at leisure along the narrow country lanes of our homeland.

Auntie Mary packed the ham sandwiches and we climbed into the old Range Rover to snatch what few hours were left before the evening feed. It was snowing. We drove through the almost deserted Brontë village of Haworth using the Main Street which is forbidden to through traffic. No-one cared. There was no-one there to stop us. With snow falling, the steep, cobbled street, with its illuminated shop windows and pub and its cheerful Christmas tree, looked just like a Christmas card. We began to capture some of the romance which we had felt was lacking. We drove on through Stanbury which was asleep and on to The Old Silent Inn where Bonnie Prince Charlie is reputed to have slept. There were parked cars and, had we gone inside, there would have been a festive atmosphere, but we passed and went over the moors, whitened with snow, to Wycoller. The car park on the brow of the hill was empty but cars for the disabled can go down to the sixteenth century village and park by the ford. We drew up beside the river. There was no other car. We ate our sandwiches in the quiet of the afternoon. I tried to identify myself with the teacher I had been, conducting carol services and building the Crib, organising parties for children and Guides, and entertaining the blind and the disabled. More recently we had cooked an enormous meal for an extended, unrelated family.

Now, incredibly, there were only four of us eating sandwiches in the snow, in ghostly Wycoller. One of the cottage doors opened and a man came to peer into our car. He was obviously suspicious of a farm Range Rover parked by the river when all good men and true should have been behind closed doors trying to recover from having had too much to eat. He was agitated enough to come and ask us what we were doing.

'Just eating our Christmas sandwiches,' said two tired ladies, one disabled brother and one old auntie.

We must have looked harmless for he wished us a Merry Christmas and went back into the warm glow emanating from his cottage.

We drove home, put back our dirty clothes and went out to do the evening chores and the next day we collected Auntie Dorothy. Christmas Day she spent with her daughters but on Boxing Day we brought her back, as we always did, to spend the best part of a week with us. Harry was happy. He liked his 'Auntie Dor'.

At the end of the week we took her home, afraid for her. She was so frail. She had acquired a cuddly Teddy Bear to take with her to bed. She showed it to me when I took her case upstairs to her bedroom. I could think of nothing more pathetic. Less than four weeks later she was rushed into hospital and, when she was well again her daughters took her to an exceptionally luxurious home for the elderly. We visited her almost every week but after the first week Harry would not go in. It was like a four star hotel but Harry called it a prison and would not be talked out of thinking otherwise. I think she was happy there. I hope so.

* * * * *

That winter was hard. As the store of bales depleted the whole of the Dutch barn became a calf playground. Their luxurious accommodation failed to imprison them. They were escapologists extraordinaire and mountaineers to boot. They ruined scores of bales, pulling them about and romping all over the place. We did not lose one of those calves. They ate and drank and cavorted about in excellent health and we were bloated with success. If we needed confirmation that Margaret had not lost her skill we thought we had proof. Our very cheap calves had fared better than we had ever dared to hope. We could think positively about a programme of annual rearing and wean ourselves, in winter, from the drudgery of evening meals.

But we couldn't rear calves again in the same accommodation, primarily because it had been totally vandalised. They had eaten the binder twine holding everything together. As they had grown we had demolished pens to give them space. These could be reconstructed in only one of the used sheds. We doubted if we'd ever be able to clean out one shed. The tractor would never get in it so it would be a fork job of massive proportions. We hadn't thought about that, as we'd seen the depth of manure rise so that the calves were nearly touching the roof. The little blighters had gnawed through planks and had splintered glass fibre sheets and broken down barriers. The weight of the bedding had pushed corrugated tin from uprights so that the walls bulged and nails flew out. Emptied of hay and straw the Dutch barn became vulnerable not only to the winds without but also to the galloping calves within. Just, and only just was the Dutch barn going to 'see the cuckoo'.

Reservations poured in for the summer. We had no weaned calves to buy and my one and only insurance policy was about to mature. We decided we could afford one new cattle shed; a purpose-built one solely for calves. If Margaret could rear them in slums she would certainly be successful in more salubrious surroundings but things do not always turn out as one expects. We began to plan immediately and told John we would need his team of builders just as soon as we could let the calves out. We are always optimistic that everything will go to plan.

On the first beautiful day of spring, which dawns in the latter half of March, we opened the gates of the two old sheds and persuaded the calves to come out. They came tentatively at first, a little insecure in their new found freedom, dragging out manure and bedding on nearly three hundred hooves and messing up the clean yard. It embarrassed us that holiday-makers' cars were dirtied coming to the house. Though people choose to book country holidays they do so in ignorance of what it is really like on a farm. They don't bargain for the dirt and we spare them as much as possible. Neither do they bargain for the sweat and offer only once to take off a load of hay. They watch calves being fed for fifteen minutes and then disappear.

'Thank you, I really enjoyed that little experience,' said one eleven-year-old-boy after one brief visit to the shed at feeding time. Few stay to see a job completed and fewer still come to lend a hand a second time.

No-one was there when those calves eventually realised the farmyard led to the field and that they were allowed to gallop. Suddenly the whole herd of them began to leap, wildly and wonderfully, making strange noises of unbelief. The thundering of hooves, the heavy breathing and the vocal appreciation excited the calves in the Dutch barn. We did not know how to open their gate quickly enough. I have never seen anything to compare. They were delinquent calves. They had never been disciplined and cattle are easy to train. They had belted round the inside of the Dutch barn for months. They had ignored all barriers. We hadn't been able to keep them off the mow and when the gate onto the paddock was opened it was akin to to the starter's gun on the London Marathon. As the barn had emptied of last year's harvest we had brought in calves from the other two sheds lest their growth result in overcrowding. So the score in the Dutch barn had become twice that number but when they dashed out of the shed we were sure they had been reproducing. They flowed out in one solid stream of black and white, red and dun, and when they had galloped all around the paddock they flowed back into the barn for the joy of flowing out again. They played this game over and over until they were exhausted.

We brought them in at night, for a week or two, but we seldom had to let them out in the morning. By fair means or foul they had always managed to break loose and the Dutch barn became a ruin. They threw their weight against the tin until the nails popped out and they had an escape hole. It looked as if it had failed to withstand a siege. We surveyed it with despair. Our only hay barn would never be used for that purpose again. It was deep in three feet of bedding. How could John take it out? It was an aisle barn with upright telegraph poles within it. The modern machinery of this new age would be useless. The job was impossible.

We hadn't planned properly. We had thought only in terms of one year but we weren't going to admit we'd done the wrong thing. Our calves were beautiful and we had no expensive new ones to buy. The money we would have had to borrow to do so would just have to be borrowed to build a new storage barn. A purpose built one into which the haulage contractor could reverse with his full load. Hadn't we reached the age, anyway, when it was ridiculous to have a barn too low for the lorry to enter until the top layers of bales had been unloaded outside and carried in? Our barn had initially been for tractor and trailer not multi-wheeled lorry. That we had also reached the age when we should not have been borrowing from the bank troubled us. To take two steps forward we always have to take one step back.

We argued that to do the two jobs together would save costs in the long term. Each of the two new sheds we would have to build would be able to borrow one wall from the existing cattle sheds. The calf shed could run along the east and the Dutch barn along the south. This was easy for the builders and easier on the purse but posed us future problems which had to be addressed little by little in the years that followed. Only by using them did we discover the faults and only by experience were we able to rectify.

We went once more to the bank manager and told him our dilemma and of the hundred Hereford heifers which would be sold in the summer and the seventy calves we had bought for a song but which had wrecked the place. We showed him the filling chart of holiday bookings and mentioned my maturing insurance policy. 'No problem' he said, sitting comfortably on our Deeds.

We had plenty of problems. We had this awful stack-yard to get rid of with its surrounding corridor of half-demolished calf sheds and the Donkey 'Ole. There was one enormous Dutch barn with its double roof of tin and glass fibre sheets. To put on the roof, many years ago, we had first put in the harvested bales and had had a safe platform on which to work. For Margaret to prise each one off, now with only empty space below, was a different proposition altogether and before the area became a building site we had to do the spring worming of our one-hundred-and-seventy-head herd. We had to finance the whole project by making hundreds of breakfasts and evening meals and we had a brother who could not sleep and too frequently measured his length on the floor.

Oh yes, we had problems galore but the chore of carrying water was over and there is nothing like a new project to stimulate our adrenaline.

Easter came with its regular guests and work was halted for a time. As soon as the weight of them had gone we turned our attention to the stackyard, a formidable job we could ask no-one else to do. The ground was uneven with wet bales, binder twine, tyres and torn plastic. Only we could sort out the mess. Summer heat would have helped solve our problem had we had time to wait but the team of builders was already in the wings.

At last the job was almost complete except for the gale-torn plastic sheeting. It lay in a messy, unfolded heap which we intended to take to the household waste tip. When we disturbed it, to our horror, we saw scores of rats run dazed and frightened into the field. We had known we had vermin, in the Dutch barn in particular, but one cannot put bait anywhere calves might eat it. Margaret had been well aware of their presence but had ignored it. Since turning out the calves not one rat had been seen but when we disturbed that sheeting I think they must have been holding a conference. We saw them scurry across the field following some invisible Pied Piper of Hamlyn and, this is true, we have never seen them again. Where they went and why we have no rats now, remains a mystery.

We left the Donkey 'Ole still standing for it kept the cattle off the concrete foundations the men quickly laid. No longer could we build our own sheds with telegraph poles and second-hand tin. Huge tractors must get in to take out muck which Margaret used to handle with a fork. Laden lorries of hay and straw had to back inside. The Do-it-ourselves era had gone but we were pleased to have lived through it.

The two relatively new bullock sheds had taken John weeks to build but things had changed even in the last decade. Now he had ordered kit sheds. The pieces were delivered by the supplier, cut and holed already and, like Meccano, only needed erection. The men, with tractor shovel, had dug holes for the uprights and poured in a platform of concrete. They bolted the first iron arch and with ropes and tractor power they raised it in position and guyed it to the Donkey 'Ole to the north and to a tractor to the south. By elevenses they had three of the four bays standing and separated by the roof girder at the peak. The structure was standing of its own accord when I returned from town, having taken Auntie Mary for her pre-holiday hair-do. I was amazed. What had taken weeks, not so very long ago, was about to be accomplished in a morning.

Margaret was standing by the window, watching. I brought Harry in, out of the car, and was about to go upstairs to exchange my clean clothes for working-jeans when Margaret, with such a pleased look on her face, delayed me. The shed was growing faster than any mushroom before her eyes.

She said, 'Do you know, I think I've worked all my life to get a shed for my calves like that!'

She is the one who looks like Father but I was the one who felt like him. My smile was like his when we had been able to give Mother something she really wanted. I went upstairs leaving her gloating in front of the window. Suddenly I heard this awful crushing sound. I only heard it. Margaret, still in front of the window, saw the disaster happen. Unable to do anything she saw the whole framework of the shed collapse like a pack of cards. I dashed to the bedroom window still pulling on my jeans. I was too late to see the iron and wood frame fall but I could see it had come to rest on the Donkey 'Ole. Thank God we had left it standing, so that at least the fall had been broken. Men were on the site who could have been killed. That was our first fear. The corrugated tin and beams of the Donkey 'Ole were twisted and crushed every which way but nothing mattered but the safety of the workmen. We found them dazed and unbelieving. One man had been aloft but, miraculously, not one had been hurt. I remember asking John if the men had sworn. He said there had been this incredible silence. No shout or scream or oath. Not one had breathed, let alone spoken. No-one had been hurt. That mattered most.

We walked back to the house unwilling to embarrass the men. We dare not look at the structure fearing it must have been irreparably damaged. We left the men to do what they could without spectators. We knew we could not afford to buy again. We were well and truly heart-sluffed.

The men were undaunted. Thanks to the fact that the framework had to be free standing before any concrete had been put in the holes, and the miracle of the Donkey 'Ole being in exactly the right place at the right time, the men were able to pull the whole thing upright again and this time they roped it to the existing shed and left absolutely nothing to chance. Only the loyal donkey shed was a total write-off. Even so we ran to the bedroom window, first thing next morning, to check the shed was still standing.

After that everything went smoothly. The new Dutch barn posed no problem. Its walls of used railway sleepers gave it a pastoral look. Then before doors could be hung or calf mangers built the silage season began and the men left. We hoped they'd be back to finish before we made the autumn journey to Beeston Castle.

On the anniversary of Mother's death came the news that Eric White had lost his fight with cancer, too. Eric and Valerie and their pretty daughter Marianne, had been coming to spend holidays at The Currer for years. They used The Loft cottage but came down to the dining-room for meals. Our extended family encompasses hundreds of guests who share their problems with us, write to tell us when they are ill and phone immediately someone dies. We are very touched by their dependence.

* * * * *

Leaving everything we went on holiday. the RMS *Lord of the Isles* sailed from Oban at 3 p.m. The slow crossing of the Minch suited us fine. How long was it since we had been to Barra, we pondered from comfortable seats in the observation lounge? Sailing, for us, is akin to writing memoirs. Boats and islands hold so many memories of twenty-five years of taking happy children to camp in The Hebrides. No wonder the box for that purpose is so full it cannot contain the memory of who stayed with us last week, who went home yesterday and what soup we served last night. More important details fill my memory than where I last laid my glasses.

Barra! Barra with Margaret and Janet. Barra with Ann and Shiela. Barra pushing trike carts with the patrol leaders. Barra with the Sea Rangers, Barra by plane on our first journey out to Harris. Tigh-na-Choddie, Whisky Galore, the Campbells at Tangasdale, the gale under canvas in the fir plantation at North Bay, the roar of the Atlantic on Traigh Vais, the touch-down on the Cockle Strand! Barra!

Darkness fell before the *Lord of the Isles* reached Castlebay. The MacNeil castle was but a dark ghost against the backcloth of the harbour lights. We had only a non-elaborate sketch of where our cottage would be at Kentangaval. The others rely on me to find their accommodation. I wondered how on earth we were going to find one cottage, on a dark island, at close on midnight.

Shortly I volunteered, 'This is it,' with more confidence than I actually felt. There was sea, a road, and one or two cottages. The one I suspected was ours was ablaze with light. The others all in darkness. I walked tentatively to the front door. There was no-one there, the door was unlocked. There was no name of the cottage above it, nothing whatsoever to indicate it was ours. I opened the door and went inside. Lights were on in every room. I felt like Goldilocks. All the beds were made and the heating was on.

There was no note, just a warmth emanating from the radiators. The wardrobes and cupboards were empty and the fridge. It must be ours! Must be! This wasn't the mainland! This was Barra! Normal for The Hebrides! I called the others. It was too late to visit neighbours. We unloaded, went inside and shortly to bed. How we love islands and Barra never disappoints.

To say we enjoyed that comfortable, warm cottage would be an understatement. It was not a traditional one being a chalet built with cedar. It had a large lounge window overlooking the bay. It was cosy and convenient and a blessing for many reasons. The Hebridean spring was a little late and our week on Barra was colder than the week on Harris which followed it and the cottage view and the warmth were greatly appreciated. Our memories of Barra were all of tumbling waves and white beaches and nothing had changed. We hadn't noticed in our walking, camping days that the roads did not skirt the shore so that most of the beautiful beaches were not easily reached with a wheel-chair. But we could walk for miles on the Traigh Mhòr, the Cockle Strand where the plane lands and on the white shore at Eoligarry at the north of the island. We had taken two wheel-chairs anticipating that Auntie Mary might like a lift back if we walked too far but hers was never taken off the roof rack. An incident on the Eoligarry beach put paid to that idea early in the week.

We had just had coffee with Margaret Ann, a Tiree girl who had joined our camp years ago. We had pushed Harry, in his chair, across the grass to the firm sand and had lent an arm to Auntie. She, fortunately, is a beach addict too. Her greatest joy is to walk along the line of breakers with open palms to the salty breeze. Health giving ozone enters through the palm, she believes. Everything was in our favour for a long walk on a level hard surface. The day was still and sunny, we were alone, sea birds calling, our lips were salty, we were on Barra. As our Hebridean friends would say, 'It was beautiful, chust!'

Tiree is so flat there is no system of streams. You can walk on a beach from end to end and not have to ford running water. Barra has hills, and rivulets pour from the rocks and eventually cross the beaches into the sea. One such, six foot wide and two inches deep, was flowing sedately into The Minch at Eoligarry. Margaret and I were barefoot and the wheel-chair safely took Harry to the other side.

We unloaded him there and re-crossed the stream with our transport to ferry Auntie Mary likewise. We did not bargain for the fact that Harry's wheel-chair is a tighter fit and she is twice his weight. Mid-stream the wheels began to sink into the softer sand, unevenly at that. Golly, I thought, we are going to give her a soaking, one way or another. The chair came to a virtual standstill and its occupant began to panic and Margaret and I, to our shame, began to laugh. To do so is fatal. Laughter weakens one considerably. We managed to drag the wheels near enough for Auntie Mary to put her feet on firm sand and then we attempted to pull her out of the sinking chair. Her sit-upon was so wedged in the narrow seat that, as she came upright, the chair came too. That was the last straw. 'You won't get me in a wheel-chair again,' she declared. Nor did we, that year.

We missed a frightening experience by just one day. On her next crossing a low flying jet narrowly missed a collision with the *Lord of the Isles*. It missed by a few feet only and caused grave consternation on the island. Caledonian Macbraynes were appalled and it made headline news on national television. There is not a lot one can do if one is on board and an RAF jet ploughs into a mail steamer!

Every evening we left Auntie Mary in the warmth of the cottage and pushed Harry into sleepy Castlebay. The wheel-chair is preferred by Harry to the Range Rover any day, which is saying a great deal indeed. When we are tired and would opt to stay in, he is eager to go out. When the wind is too cold he snuggles into the zip-up bag and doesn't mind cold cheeks. When it is raining and we have to go to the phone he will not be left behind but will suffer waterproofs and hood. Harry will not miss one bit of his holiday.

We left Barra on the evening boat, sailing north to Lochboisdale. We had booked overnight accommodation at a guest house numbered 455, or thereabouts, I cannot quite remember though I could drive there without error. It amused Harry that, in an area of a few houses, one should have such a high number. The accommodation was excellent and when we signed the visitor's book next morning I noticed an entry by a boy I had taught many years ago. Colin Bradford, his name was. I frequently see children I have taught. They visit me at home or detain me in the supermarket but Colin I had never seen since the day he had left. I missed him by a few months at Lochboisdale.

Next morning we drove by road and causeway to Lochmaddy and sailed to Harris. We took spring with us on the boat. A week was too short. The island was utterly, utterly beautiful, carpeted with primroses and peopled by our dearest friends. The sun warmed the hard beaches and Auntie Mary proved she did not need a wheel-chair by walking two miles along the beach at Luskentyre.

We have so many friends on Harris, not least of them Maggie and Calum MacLeod, of Northton. We were visiting them on the last evening of our holiday. Hens were fending all round their house and Margaret remarked that our flock had become too small.

'I only keep a few,' she explained, 'Just enough to teach our guests that hens lay eggs. They never say hen. They call all adult birds chickens. I can't keep enough to serve breakfast but I can tell them that what comes out of an egg is a chicken, nearly always a cock, but what lays the egg is a hen!'

Calum laughed and took her into his barn where a hen and three-day-old chicks were fending. The hen was all business-like, clucking loudly with ruffled feathers and strutting with pride. Calum was hunting for a box. 'See,' he said, 'Chust let me find a box and you can take these three home with you!'

We have long experience of the generosity of the Maggie/Calum family from the days when we were accompanied by fifty children and more. Margaret should have known that she had only to say she was short of hens, or anything else for that matter, and our Northton friends would supply them.

'I can't take a hen and chickens home with us,' she remonstrated. 'No way. It takes us two days to get home and the Range Rover roof rack is full!'

There appears to be no word 'No' in the Gaelic. We have never convinced any of our friends that we do not need another strupack, that we have just eaten and are full to bursting point. We could not convince Angus that we did not want another dog when he was determined to give us Lusky and Calum would not be said, 'Nay,' either. He found a box and caught the birds and Margaret sought desperately for words to tell us that she was taking the chickens home. Of course we were no more successful than she had been in deterring Calum from sending day-old chicks to The Currer. We have brought home some funny things in our time but the journey with chickens was the funniest of all.

We were staying at Airdrie, on our way south, to pick up Mother's elderly cousins, Minnie and David. We had decided not to tell our urban relatives that we had a hen and chickens aboard. Margaret had a legitimate reason for climbing aloft to rescue our overnight bags. In the dark no-one noticed her errand to feed the stowaways. In the morning it was a different matter. Tommy would insist on helping. Our relatives' cases had to be accommodated on the roof rack and the hen aloft was making such a clucking sound Tommy mistook it for some engine fault and our secret was out. Our cousins have every reason to think we are crazy.

There were only seats for five in our old vehicle so one of us had to sit on a cushion on the floor. Years ago when there were seven of us in the station wagon one of us, meaning Margaret or I had to sit in the back. When travelling with guides we sat on a rucksack in the bus aisle. When we have family guests, or friends, or over-book we have to sleep on the floor. It is no problem to us. We find it difficult to understand those who choose bed and breakfast with us in preference to something make-shift at their lonely mother's house. I think one of the most important things I taught my children was to sit and to sleep on the floor.

Chapter Six

> Happy hearts and happy faces,
> Happy play in grassy places
> That was how, in ancient ages,
> Children grew to kings and sages.
>
> *Robert Louis Stevenson 1850–1894*

Considering we love The Currer so much it is illogical to hate coming home. When we have a moment to organise it properly, we always say, we'll make sure we have no guests on our return. That day has not come. To return to a full house and bring an extended family to our own bedroom would be cruel if it were not normal for us and Minnie and David had a lovely holiday.

The contractor came with his huge tractors to take manure out of one calf shed, demolish the roofless Dutch barn and clear the depth of bedding. The mess was embarrassing for a very short time, for tractors work at a tremendous speed and do an excellent job. We finished the landscaping of the floor of the barn, harrowing it level and re-seeding it. An enormous pile of beams appeared against the plantation and we stacked, as neatly as possible, the tins and the glass fibre sheets. Suddenly the farm looked almost as it had when first bought in 1929. We realised what a view stopper the Dutch barn had been and whenever we crossed the yard our eyes strayed to the open aspect and a smile of pleasure crossed our faces. We wanted June to last forever for the thought of July and August filled us with foreboding.

Bernard had announced that he thought Denis was old enough, at fourteen, to come to The Currer, alone, to learn English. Not for one week, or two! 'Denis can come for two months,' Bernard said. We had decided that, should Bernard suggest Denis coming alone we would say, 'No'. We did not expect him to want his son to come for more than two weeks but we considered even that would be impossible. A boy, a teenager who couldn't speak English, in our busiest season, a city boy from Lyon! Impossible! Quite impossible!

But how can you tell a charming Frenchman that you have a handicapped brother and an old lady, that you are a pensioner yourself and your French is fifty years old. How can you say your sister is a farmer, that you have enough problems and that you could not possibly face the summer with a teenager you do not know under your feet? The reason we do not employ summer help, much though we need it, is because we think it would cause more problems than solutions. For years we have been criticised for not getting help but we've been adamant that we would go mad coping with employees. That, we acknowledged, would be child's play compared with coping with a French boy on holiday.

There may be words to hurt a nice man. I have taught that there is always the word 'No' and advocated the use of it but I couldn't say 'Non!' So we spent June trying to forget that July and August would follow. We concentrated on giving Mother's octogenarian cousins a good holiday, on clearing up after the demolition and on feeding the masses.

Married friends were not helpful. Those who had experience of their sons' pen-pals recounted tales of woe. 'The most difficult fortnight we've ever had,' they said. 'He wouldn't eat.' 'He wouldn't talk,' 'He sat in his bedroom all day.'

Suddenly it was July and Bernard rang to say he was bringing Denis at the end of the week and we had only three days to have our nervous breakdown. We saw the Glasgow couple safely onto the coach for Scotland and awaited the dormobile from Lyon.

Bernard is young and charming, his smile is bewitching, his courtesy pleases, he is helpful, capable and wise and we hadn't the heart to comment on the fact that Denis looked much younger than fourteen or say we wished he hadn't brought a bicycle and that we were all petrified because Pennine hills are so steep and all roads are so narrow.

After a few days Bernard asked to reverse his dormobile into the shelter of the newly-built calf shed. It was then we realised that he was nervous too. He left it with the engine running on the slope into the new shed whilst he collected something he had forgotten. He suddenly saw the vehicle begin to move and knew I was behind it, opening the gate. He screamed my name but I had already jumped out of the way. He managed to leap into the van and prevent it coming to any harm. He was trembling and when he and Denis left for the bus station he forgot his wallet. He was travelling home by coach. A connection leaves from London and goes all the way to Lyon. He was returning with the rest of the family in September when they would all go home in the van. Denis came pedalling back to pick up the forgotten wallet and twenty minutes later came pedalling back for a second forgotten article. We decided Bernard was as nervous as we were!

A blue van had reversed several times, purposefully, into our new wall on Altar Lane. This had opened up a new dialogue with the Council. Every so often our phone calls and letters reminded them we were still here and still crying for help but they refused to do more than make promises they never intended to keep.

We'd planned to repair it on that first day of being left alone with Denis. We'd told Bernard to tell him where we would be when he returned from seeing his father onto the London coach and we put our dependants into the Range Rover and went up onto the moor with picnic flasks and sandwiches. It was a beautifully hot, summer's day when we could smell the clover and the neighbour's recently cut hay. There were myriads of butterflies that year, I remember. Nature never ceases to amaze.

We had taken Harry's wheel-chair and Auntie Mary's folding one and we could see our cyclist friend approaching over the turf. He looked very young, in khaki shorts and one-colour tee-shirt. He could have been a Baden Powell Scout. We were two generations older than he, a city boy, and we were farmers handling big and dangerous cattle. We dare not think what lay ahead. We did not know what he was thinking, cycling across the flower studded pasture on that beautiful sunny day. He would see Harry sitting in his wheel-chair, an old lady tatting and two women, with roughened hands, lifting veritable boulders onto the damaged wall. We thought, in our ignorance, that he was probably quite as apprehensive as we. The fact that he, too, might be worried comforted us but, in retrospect, I hazard a guess that he wasn't apprehensive at all, this very composed, laughter-loving teenager whose ability to cope with us and our life-style, our language, our work, our nationality and the quite considerable age difference, was extraordinary. We talked to him all the time. I volunteered not one word of rusty French. His ability to pick up our language fascinated us. He could memorise vocabulary and phrase instantly and correctly and he began to chatter away experimentally without any trace of embarrassment. He created an aura of joy and happiness. It warmed Harry and Auntie Mary and made them feel so at ease they looked eagerly for him to come through the door and talked continually about him when he was not there.

To be with Denis, we soon realised, was to talk about important issues. He did not seem to need the things today's teenagers seem to crave. He was not dependent on television or background music and he had no money problems because he wasn't tempted to spend. He just wanted to communicate with real people. He ate in the dining-room and discussed real problems with mature guests. He sat alternatively listening and being listened to in a way which would have been unusual in one so young even if there had been no language problem. Remarkably quickly that problem was being solved. He ate traditional Yorkshire food with enthusiasm, coming to us after the meal to say, 'That was awful, Jean. I only had three helpings.'

His behaviour was sensitive and quite undemanding. Many teenagers would have been moody, silent and homesick and ruined the congenial atmosphere of our dining-room. Not Denis. His presence added to it to such an extent that guests too began to watch the door for him to enter and to talk about him when he was not there. His command of the English grew by the hour.

His manners were unpretentious. He was not performing. What you saw was real. He never failed to tell us where he was going. He would jog round the boundary every morning and, when darkness fell, he went looking for the fox and the deer. Reports had been coming in that there were deer in the woods which was unknown this century and Margaret had proved the rumour true by disturbing one in the Dyke field. Denis always came to say, 'Good Morning' and 'Good Night' and he would mischievously follow the last with a well pronounced Yorkshire 'Luv!'

A valuable asset we had, was the ATV Yamaha. The boy could maintain it and drive it better than Margaret and I leave it well alone. He began to do a hundred-and-one jobs, finding them and avoiding boredom and loneliness. He was genuinely the happiest boy we had ever come across. Though reared in the city this boy was a natural in the countryside. He soon knew more about what was going on outside than we and he knew far more about the guests than we did. He had an extraordinary compassion, an infectious sense of humour and a judgement of character superior to most adults. He seldom went to town. He spent no money for his parents are not affluent. He was so unexpectedly easy we could not believe our good fortune.

We had heifers to sell later in the month. Sales are monthly on a Friday. I do not know why we went on a Monday at the end of July. There was no reason except to make sure of the date of the monthly sale which we could have done equally well by phone. Harry may know why we went, but I do not. I only know that Denis went round the market attracted to any baby animal whatever species. I heard him muttering in English. 'Just a baby, any baby. A pig. A calf. Just a baby.'

'Hey, you! Stop that!,' I scolded. 'We are not here to buy.'

'Let's see what these calves are making,' Margaret said. She really wanted to show Denis that calves bought in the summer time were quite expensive.

He sat apart from us in the amphitheatre round the ring. I was watching him. He was praying for a baby and I thought we should get out quickly. It was cruel to torment him.

Calves were much more expensive than in November. Then a very small, free-martin Friesian heifer was pushed into the ring which no-one wanted.

'Come on,' bribed the auctioneer. 'Hasn't anyone got a wife with a birthday this week?' Apparently not, 'Come on,' pleaded the salesman. The minute thing had to be sold. 'Who wants a pet? £5?' He looked in Margaret's direction. She looked at Denis, sitting with his chin on his hands and his elbows on his knees praying for a baby.

Margaret nodded her head at the auctioneer hoping someone would outbid her for the last thing we wanted was a calf to rear in the summer. No one did. '£5 I'm bid. £5 ... £5 ... £5'

The calf was knocked down to Margaret and I crossed to Denis and whispered, 'She's just bought you a calf.' He was ecstatic.

Of course you cannot take home just one calf to rear in loneliness so the next cheap calf pushed into the ring, Margaret bought and we bid a hasty retreat. 'You'll have to feed them,' she told Denis, 'We haven't time and you'll have to sit in the back and stop them from flying out of the window when we go round a bend.' Denis didn't mind anything.

Returning we made a pen in the new Dutch barn and Denis bottle fed them for the rest of his two months stay. They were named Denis and Emilie and they entertained the visiting children all summer.

The weeks passed by and we began to dread his leaving date. As he began to speak more fluently we found him to be a mature philosopher whose advice was worth listening to and whose wisdom was profound. Bobby and Kathleen, visiting us from Luskentyre, pronounced him, 'Fourteen going on forty.'

Too true. When he told us he planned to go to Russia for a year, on an American sponsored scheme we were not amazed. Any other boy we would have thought was fantasising. An international group of children would be sent to a Russian school and would, during the year, learn to speak the difficult language. We admired his courage.

If Jess was the first of the good things to happen to Harry, after the death of Mother, and the opportunity to go to the baths was the second, probably the third was Denis.

No-one ever treated Harry quite like Denis did that first of the many years he has been coming to The Currer alone. He made Harry laugh more than Jess, even, for we do not scold our brother often and excite Jess to come to his rescue. Denis made Harry laugh all the time. A unique relationship grew between them for both are jokers. Denis never embarrassed Harry. If laughing made him splutter and sneeze Denis went for a kagoule and found no difficulty if that made Harry splutter so much more that he lost his dentures.

He responded to Harry's too often commands more willingly than we. At lunch it would be Denis who fetched and carried uncomplainingly. 'Warm tea', Harry always wants when we have just seated. 'Salt and pepper', again when we have already seasoned. Pass this, that and carry when all we want to do is collapse, having been on our feet for hours.

We have a rule. Which one of us is standing answers his call, or the phone or the door. Whoever is on her feet puts on the kettle and fetches the logs but, without fail, as soon as both of us are seated Harry wants something else. Denis was a blessing for he would respond without saying, as we do, 'Why don't you ask before I sit down?'

One day Harry laughingly commented at lunch, 'It's nice having a French waiter!' Oh yes, Denis was wonderful for Harry.

July is fruit picking time. We gathered the year's harvest of strawberries at a pick-your-own farm just off the Leeds Ring Road. Perhaps it is as well that we can never leave home early for, between us, we pick too many in the three hours we have. With Denis, Auntie Mary and her friend Hilda we gathered 120 lb. Margaret and I pick as if our lives depend on it, and filling the roof rack with baskets is a penance we must pay. Our brief annual appearance causes amused entertainment for the field staff. We return to base frequently with full baskets and trot off again with empty ones to fill. Students picking for pocket money eye us with distrust as we put them to shame.

Margaret and I can pick fruit better than anyone, I'm sure, and in doing so we forget tomorrow we will be in agony, that our legs will jerk with pain, our backs will give us torment and the spring which helps us rise, when we have succumbed to the temptation to sit, has deserted us completely.

Friends Keith and Edith Whalley came to see us. They were fascinated with Denis's ability to learn English. Their first year's experience of giving hospitality to a now familiar Brigitte, was to send her to camp with us, on Tiree, for a fortnight! 'You ought to pick-your-own at Birstwith,' they said. 'There are gooseberries and currants. All sorts.'

So we went there and Denis sampled the gooseberries and came to us in a panic. 'Don't gather those,' he warned, 'Ugh. They are not nice!' We loved Denis's use of the negative. When his shoes hurt he said his feet were not well. He was not used to fruit being sour. Around Lyon peaches and pears and cherries had made him a sweet tooth.

Auntie Mary, on the other hand, loves gooseberries and rhubarb. In July she has her head continually in her recipe book. A conversation overheard between her and Mother, some years ago, was so amusing I cannot resist recording it.

'Here's a nice recipe for rhubarb,' Auntie Mary said.

'I'm not a rhubarb flan,' replied Mother.

We had already begun to laugh when Auntie Mary's next remark sent us into hysteria.

'I know,' she said. Her eyes jumped to the next recipe whilst her brain stayed with the first. 'But I'll make one, sometime, when I get some bananas'.

Denis's footwear troubled him in the fields and he sat on the grass loosening the laces and muttering he must go the Spastic Shop and find a bigger pair. We laughed all July and August.

Bernard, Yveline and little Emilie were to come by coach in early September taking Denis and the dormobile home. We were afraid the return of his parents might excite him into behaving differently. His two months had been a success story. He could speak fluent English. Perhaps, like most children, he would become an exhibitionist in front of his parents and be embarrassed by their questioning. Would he behave like five-year-old Gillian who was no problem at all in school and yet created abominably in the cloakroom when either of her parents came and who screamed all the way along the churchyard. Or like Alison who had been with me two weeks camping on Harris. I, who had been there, knew just how much she had enjoyed every minute. The ten-year-old had had a wonderful time. She had coped with freedom and island magic and the isolation of the white shore, but could she cope with adoring parents? Not she! She had coped fine with her luggage through busy Glasgow, with a rocking boat and breaking waves. She had slept and eaten in strange places and climbed high hills, been swimming in a strange environment, danced barefoot at the céilidh, listened to the Gaelic, burned the peat. No problem, but she could not cope with the hugs and kisses and questions! She snapped and cried and was generally unlovable for days. Losing patience her mother demanded, 'Didn't you like camp at all?'

'I didn't like it,' the exasperating child declared, 'I loved it!'

We thought that might happen to Denis overwhelmed with French questions when English came more easily but, amazingly he did not behave any differently at all. We were sorry to see him go.

Chapter Seven

Oh a wondrous land is Yorkshire, of hill and lofty dale,
From her windswept limestone cresses to
York's wide spreading vale,
From her wealth of busy cities to the lonely moorland farm,
To the seeing eye she's glorious with an ever varying charm.

Dorothy Clough, Steeton Hall 1900–1981

We had expected, years ago, that most of our guests would be walkers. Our first advertisement promised transport between us and the Pennine Way. In July 1981 we ferried our first serious walkers from Stanbury. They were also our last. A later couple, also attempting the challenge, got as far as us and then gave up, sent for a parent's car and went home. They had set off with a tent but soon found that they were neither walkers nor campers and our farmhouse accommodation was too tempting. To give them their due they did stay a few days, mostly in bed, before they finally decided to give up and phone for Father. Once we entertained two men walking the Alternative Pennine Way which actually follows Altar Lane.

Walkers who stay here, and there are many, are not doing either Pennine journey. They leave each morning and go to High Withens, to Malham or Bolton Abbey or Wycoller and return every evening satisfied with no more than ten miles. We have some surprises. One middle-aged bachelor stayed a week and from Sunday to Thursday visited, by car, all the nearby towns south and east and west and saw not one glimpse of the Yorkshire Dales. His clothes and his town shoes were not suitable for anything else. He was clearly out of his environment, on a hill farm, until Friday morning.

It was his last full day and he came to breakfast dressed the only way he knew how. 'Where are you going today?' I asked wondering which local town he had yet to see.

He replied, 'A man in my office said I hadn't to come home until I'd climbed Ingleborough. I'm going to do that today.'

'Oh, do be careful,' I advised, quietly terrified. He was alone, ill-equipped, without experience or map but he went and he returned and he had been to the top!

'I followed some other climbers all the way,' he said, glowing with achievement.

'Pen-y-ghent, next time,' we commended. Most things can be done by those who try.

For many years the Greenwoods had been coming to The Currer. He had been born, and had lived, locally but had married and was living in Cambridge. He was well into his eighties but both he and his wife were great walkers. They used to drive into the yard about mid-day. They would put on their walking boots and walk to the Altar at once. They would climb the steps over the wall into St Ives and go through the estate to Harden and over the moors and be back in time for their evening meal. They had a routine of walks they followed daily. I had a great admiration for them.

One Sunday I asked where they planned to go that day and was told they hoped to do a walk which ought to be longer but, because of their increasing age they were slicing six miles from it only attempting the shorter track. As most of our guests would fail to complete six miles I was interested to know how many they were actually planning to walk.

They were parking at Haworth, they said, and would walk up to High Withens. Ninety per cent of our holiday makers never make it that far but the Greenwoods were continuing over the moors to Heptonstall and back via Trawden and Wycoller or some such outrageous route.

'The full walk,' Mrs Greenwood said, 'Is twenty-six miles but we'll only do twenty.'

We were all flabbergasted. Those at breakfast table stared open-mouthed. 'You mustn't do that,' I gasped.

'We usually do the whole walk,' they said.

All day we were worried. I normally serve salad on Sunday. It really isn't an easier meal to make but there's not the same heat in the kitchen and the washing up is simpler. At 7 p.m. the couple had not returned. September nights are pulling in and dusk was falling. There was an unusual hush in the dining-room as guests tried to enjoy their meal. We were all terribly worried. The meal was served and eaten and the Greenwoods had not appeared.

I phoned the police but I had really no idea where to send out searchers. I had been so shocked by the twenty miles I had not listened properly to the route. The police suggested I went up to Haworth to see if I could spot their car. I had no idea what their car looked like. Years ago we were able to say, 'The red car belongs to the Smiths and the blue one to the Jones's,' and so on. After years of cars and people we can't anymore. We never know the make of the car and only Harry ever recognises a number plate. Fortunately guests always know these things. They knew the make and the colour and the number plate and one man offered to take me up to Haworth whilst Margaret and Harry dealt, as always, with the washing up.

It was quite dark. We tried to watch oncoming cars but few were abroad at that time of night. On reaching Haworth we scanned the empty car parks knowing full well they may have parked at Stanbury and walked to High Withens from there. Only if we found one abandoned could we really call out a search party. There is a saying, 'There is no fool like an old fool.' I'm afraid I felt more critical than complimentary of two old folks wandering off into the unknown.

We found no car so we phoned home and Margaret had had a call from a remote farmstead to say the walkers were safe and being taken back to their parked car and that Mrs Greenwood was fit to drive home. The relief was enormous. We'd felt in some way to blame for not locking them up for the day and making sure they didn't attempt the distance.

They arrived home safely. The first sixteen miles had been fine. Then the old man had begun to walk sideways and it had taken them four hours to do the last four miles. A hiker had seen their problem and had taken Mr Greenwood's other arm and the three had made slow progress as far as the isolated farm. There they had sought help.

The wanderers sat down to their salad and even the old man ate well and neither seemed any the worse for their ordeal. Next morning I made them promise not to go far and when they returned late afternoon I said, 'Well … ?'

'We didn't go far,' Mrs Greenwood said. 'We only went to Bingley and through Myrtle Park and back to Goit Stock and Harden. No more than seven miles.' No more indeed!

I have a great admiration for the generation preceding my own. They were tough. The ability to be so manifests itself in fewer of the young people today only, I think, because they are deprived of the opportunity to be so. My Guides and Sea Rangers and village school-children were as tough as old rope because of me. At least I like to think so. The long walks we did on far islands, the snow expeditions we undertook, the water we carried, the logs we toted from the shore, the jobs we assumed for the handicapped made us tough, too, like our predecessors. My children were proud to be so but, on the whole, the new generation is too cosseted.

The change from struggle and toil to luxury and warmth has been a quick step, a jolt in the history of mankind. I have fought against, rather than for, the new-age of technology and choice which has brought some laziness, some greed and selfishness.

* * * * *

The summer had been very dry. So dry the duck pond had no water all season. Whilst there is water in it thoughtless children throw stones from the wall to make a splash. What child will not? If we catch them doing so we make them wade in and fish boulders out. During his stay Denis had cleared the dry hollow and put all the stones back on the wall. Shortly after he went the rains came.

The downpour was exactly that, right from the word go. It caught a villager by surprise out on the moor exercising his dog. Heavy rain always brings the cattle home. He saw them hurtling ahead of him and noticed, thank goodness, that one could not follow for its head was stuck in a pylon. Getting wetter by the minute he ran into the farmyard to tell us. We had just served the evening meal and darkness, premature because of the cloud, was falling fast.

We had farmed The Currer for sixty-one years and we could hardly remember a time, if any, when we had not had pylons on the land. We do not like them but there is not one thing we can do to rid ourselves of them until Electricity Board policy is to put wires underground. Over the years they have caused trouble. Men working on them have left bolts and barbed wire and paint cans for inquisitive cattle to lick and eat. There was almost a tragedy when a bullock lifted a coil of wire over its head and was very nearly strangled. A gate left open by pylon workers let out cattle and one poisoned itself with St John's wort and, on one occasion a suckler cow broke its leg in a newly felled pylon when old ones were being replaced.

Never before had we had a heifer with its head stuck between the bars. Told it may happen we would have said it was impossible.

We left our guests finishing their meal and we accompanied the villager up onto the moor and sure enough the heifer was standing, trapped, resigned to a long wait for the angel of death. The downpour wasn't helping. Even in waterproofs we were feeling wet. We tried to get it out but it is common knowledge that human and animal heads push through bars easily one way but ears are in the way when either want to reverse. Children are notorious for getting heads stuck in railings but one can turn them upside down and allow the head to come out the same way it went in. Even a small calf cannot be handled so, let alone a hefty heifer.

Margaret is very careful to fit a piece of wood in any gate or feed barrier where heads might get stuck. Never in a million years would we have expected an animal to find a pylon suitable for experiment. All our efforts to get the animal out were doomed to failure and the rain was torrential.

'Run and phone for the Fire Brigade,' Margaret said. So I hared off downhill to ring for help.

'You'll need permission from the YEB to cut the bar,' I said.

I got the Range Rover out and put Harry and one of the guest's children into it and drove up onto Altar Lane to await the fire engines. The sirens could be heard approaching. Firemen are only little boys enjoying a bit of fun if there is no danger to human life. Two appliances came with at least a dozen-yellow-coated, helmeted men. They parked behind the Range Rover and the men dashed across the pasture to the patient animal.

The release was easy for, with masculine strength and a huge spanner, one bolt could be unscrewed. That was all that was necessary!

* * * * *

We were on schedule. All August we had ordered load after load of hay and straw until the Dutch barn was full. We thought it was easy unloading. The lorries were able to draw in, right up to the stacked bales. Our guests thought it was torture but, they had not had the old Dutch barn experience.

We still had no doors on either of the new sheds. The men, leaving the job for the silage, had never returned. We couldn't keep the village children off the hay. They ruin the bales, do children, and building sheds had left us with a bigger overdraft. We could ill afford to have waste when we had managed to rid ourselves of the stack-yard. We feared fire and chased the children out all holiday and every weekend.

The men thought it would be OK so long as we had doors for our return with calves and we tried to hurry them but women fail to be able to persuade men to do anything quickly.

'The door and mangers'll be ready before you come 'ome from Beeston Castle,' John said and we believed him but we wanted them done before we went on holiday. We did not succeed.

Years have passed since then and we still have this problem. By now contractors, builders, electricians, plumbers should all know that we take an autumn holiday. We have been doing so since long before Father died in 1979. What we call the final date for readiness is the day before we leave. These otherwise reasonable young men are completely unreasonable about our peace of mind and annually put us through the trauma of leaving with a job not yet done, not even started. We have to phone from holiday to instruct and cajole and demand when we should be relaxing.

We had one new, full barn, open to marauding children and one completely empty calf shed needing concrete mangers and doors. Sometimes I think the rest of society, the whole of the animal kingdom, our few mechanical appliances and even the elements themselves gang up on us before we go on holiday. We knew exactly how we wanted the mangers to be constructed and the milk-bar grids fastened but we had to leave, in disgust, before the job was even begun. We'd bought numerous sheep pen fences to make partitions. We were well organised, but they could not be assembled until the mangers were done. I think we were quite angry, really, but men are bigger and a different species. 'Stop worriting,' they said and we had to believe them.

We were anxious but we were not going far. A friend had died in East Witton, in Wensleydale, a few short years ago and her daughter had kept the lovely bungalow for holiday purposes. She had recently been made Head of Haworth Middle School and was more than willing to rent it to us for a fortnight, mid-October. We wanted to explore home territory without responsibility. Swaledale, Teesdale and the North Yorkshire Moors are accessible from The Currer if you are a guest and have no washing up, or beds to make and if you have no animals, or oldies and need not hasten back to prepare evening meals galore.

We left reasonably confident on one account. The recent downpour had filled the ruts in the Altar Lane and opened up, again, our fight with The Council. A near miracle had happened. Bradford Met had acquired a lady solicitor. She came suitably dressed in waterproofs and wellingtons and walked the mile of mud and lake, gullies and ruts and declared that not closing it immediately was utter and inexcusable nonsense. What men had failed to do in twenty-seven years she promised to do in an instant. And the apple harvest was exceptional.

We left, too, knowing we would be only an hour-and-a-half away should we have to get back in a hurry. We left at 5 p.m. I remember distinctly having tea from picnic flasks on Baildon Moor only twenty minutes later. For most people it would have been logic to eat it at home, but for us, the longer we stay at home the more we find to do.

The Paddock, at East Witton, was just right for us except that we were tempted to do more Yorkshire miles than are ever possible on an island. We drove slowly and joyously, as tourists in Herriot Country should. We visited Beamish Museum and found Hannah Hauxwell's High Birk Hatt Farm in Baldersdale. We left Harry and Auntie Mary in the car and walked the dogs along the Pennine Way. We are Daleswomen of the same age but Hannah's experience of being alone has not been ours. We have known poverty, isolation, manual struggle to survive, we have fought the elements, disease and handicap. We have lost and mourned and wept but almost all the time we have had fun. Enormous, adventure-loving fun, laughter-loving fun, joyously-singing fun. We have seen the comedy in life as well as the sheer beauty of the Universe and we have recognised success as narrowly managing not to fail. We do not think it has been through personal greatness but through the greatness of our good fortune and an ability always to be able to trace a rainbow through the rain.

From East Witton we went to Saltburn and Whitby and Staithes and we caught the North Yorkshire Railway at Pickering. Harry wanted to see the highest pub in England so we went to Tan Hill. We asked for tea and were told they only had coffee. 'There's nowt so queer as Yorkshire fowk.' We strolled round open markets and went to the mart at Hawes. We forgot about unfinished sheds and the looming visit to Beeston Castle.

Each evening we pushed Harry round the village green at East Witton, in darkness, chatting about the day's events and peeping through the lighted windows of The Blue Lion.

Lusky didn't like East Witton very much for a gun went off every few minutes to scare the birds. He preferred the old fashioned Worzel Gummidge to modern methods and would not get out of the Range Rover except up the dale. It was a pity for there was a spacious garden at The Paddock.

There was little traffic on the roads for the season was late though the weather was fine. We spent a day in the vicinity of Kirby Stephen and and stopped, on our return, at the head of Wensleydale to fill the rear of the Range Rover with wood for the open fire in the sitting-room. It was dusk as we drove gently down the valley, through Hawes and West Witton. Milking herds were hovering at the gateways waiting for their call. Their ability to tell the time is perfect.

Happy, fulfilled, perhaps a little hungry, I was driving at about 35 mph for the road is narrow and meanders. Unlike Scotland there are few passing places. West of Bainbridge a police car appeared in my rear window.

'Make sure you are below 30!' Margaret advised.

Leaving the village, I suppose I had collected quite a line of traffic but no suitable place presented itself for me to let them pass. If they wanted to travel down the dale, at speed, at dusk, I was willing to pull in and let them go but there wasn't anywhere, and a police car was on my heels. So I just kept my average speed, in my farm vehicle, laden with wood and two collie dogs looking out of the rear window.

At a left hand bend in the road it widened slightly and the police car risked any oncoming traffic and drew out to pass me. There was no car travelling in the opposite direction, fortunately, but there was a milking herd of about sixty cattle heading towards us. The policeman's reflexes were good even if he had been foolhardy enough to pass on a bend. He braked in front of me and the line of traffic came to a halt behind him.

The bovine ladies, with their full udders, ignored the lot of us and headed home to the mistal. When they had passed I thought the policeman would be on his way quick sharp, with his tail between his legs, embarrassed at having overtaken on a left hand bend. Not he. He flashed his light for me to wait, got out of the car and came to poke his nose through my window. I struggled to think what I had done wrong.

'Is everything alright?' he asked with the nonchalance of youth. 'Yer driving rather slowly.'

'It's milking time,' I replied. 'We might meet a herd of cows.'

He saw the twinkle in my eye and became quite confused. 'I know,' he blushed. 'I'm sorry.' Then pulling himself together and assuming official status afforded by his uniform and flashing roof-light he said, 'I have to ask anyone I stop, to show me their licence, etc.'

Mine were all at The Paddock. 'Follow me to East Witton,' I suggested. He declined for my pace was too slow for him. 'We are on holiday,' I said.

'Take it to Leyburn, or into the police station when you get home. That'll do,' he advised.

Huh! No way was I going to take my licence locally and say I'd been driving too slowly. I'd probably have taught the policeman, or delivered his granny's milk or taken his mother to camp.

So when next in Leyburn I took it there, along with the MOT and insurance papers, and said I'd been told to do so by a young policeman who thought I was driving too slowly, at dusk, down Wensleydale. The senior officer wasn't as amused as I was.

'These bloody youngsters,' he growled. 'They think everyone should drive fast.'

Until relatively recently there was no opportunity for urban man to explore the countryside except on foot. That, of course, is still the best way. To see only and not to hear and feel and smell and lick one's lips in the great outdoors is to be deprived.

The invention of the car has given to a city dweller the means, at least, to see great beauty but he has thrown away the privilege by driving at speed and scorning those whose slower pace enables them to admire the vista round every bend in the Yorkshire Dales. What ails modern man seems to be infectious but we, thank God, are immune.

The Paddock phone rang unexpectedly one evening and we received the welcome news from home that the lady solicitor had phoned to say the wheels had finally been set in motion for the closure of Altar Lane. She had managed what several generations had failed to do. We could not believe our ears. 'It takes a lady!' we said. 'Men! Men never do anything'.

* * * * *

Our holiday had been timed for us to leave East Witton on the Thursday and for us to go straight to Beeston Castle to buy calves. We had booked overnight accommodation at the Travelodge at Crewe. We had taken too many risks the previous year, with Harry and the two old ladies. That we had been successful once only emphasised the fact that there was too much at stake, too far to go, too early in the morning. We had taken one risk not knowing the enormity of it. Knowing the pitfalls we were not prepared to be foolhardy again, so we left East Witton on Thursday, in the teeth of a frighteningly strong gale.

Having promised to pick up Auntie Mary's friend and take her again, for the outing, we called at home on the way south and were horrified to find there were still no doors on the two sheds. The mangers were complete and we showed Dorothy how to fix the sheep pen partitions and told her she would have to build a bale wall against the open north wall.

Margaret rang John to say she was on her way to the cattle market and he said again, as men always do, 'Stop worritting.'

So we left and picked up Hilda and drove against the gale all the way to Crewe. It was a doddle. Rising refreshed next morning, with no guest-house lady to insist we had breakfast, we had a cuppa and went out into the darkness. We were armed with flasks and hot water bottles and plenty of wherewithall to make sandwiches.

We were at the market at 8.30 a.m. and, would you believe it, everyone knew us. A woman buyer is not necessarily unusual, but a lady buyer not five feet tall, buying as near seventy as possible is not often found at the ring side. That she has an ex-headmistress sister somewhere in the rear-collecting-shed, shoving calves continually into a pen marked Mrs Brown, again distances her from other buyers. When the passengers in the Range Rover become known, then Margaret becomes unique.

We had been to the market only once and twelve months had elapsed since then, but as sister and I crossed to the buildings the auctioneer was doing likewise and immediately recognised her.

'We must stand out like a sore thumb,' I said.

So we bought a new batch of almost sixty calves and they filled the new shed to capacity. The average price had risen by just eleven pounds and we were happy with that.

We soon realised that the new shed was so completely sheltered from the prevailing wind by the existing sheds, that there was no movement of air in it at all. The south gable was all Yorkshire boarding but the rising hill, in front of it, was steep and even a south wind was pretty useless.

John had fixed the two heavy doors on the north face but we never closed them all winter. Dorothy's half wall of bales remained and the doors were left open. What's more we began mutilating the east wall and the Yorkshire boarding by putting in windows, five of them. It was one of the most difficult jobs I have ever done. Auntie Mary had asked what we wanted for Christmas and we had said, 'An Alligator saw, please.' Women farmers make these strange requests. We bought the Christmas gift prematurely because we needed it and, as the local Do-it-Yourself store gave pensioners plastic cards and allowed 10% discount on a Wednesday, we bought this heavy piece of male electrical equipment mid-week and ignored the funny looks of the salesman.

I stood on bales, precariously handling this heavy apparatus with its dangerous jaw and its electric, dangling wire which all the calves wanted to eat and I cut out two windows from the Yorkshire boarding. We realised the openings would have to be closed when the wind veered east so we had to make and hang shutters. We were fed up.

The new shed was full and we hadn't got our final contingent of calves. We hastily carried gates into the bullock shed and made room for the ten more we planned to buy at Otley on the next possible Monday.

John and Mamie came down from Horgabost, on Harris, to stay with Otley friends, Arthur and Olga and they came to tea at The Currer and to watch Margaret buy calves at Otley. A few days later we had given the babies their afternoon feed and they were all curled up in the straw as happy as Larry. We'd lost three calves at the onset. They were puny ones, already scouring but the panic seemed to be over and all were doing well. We went indoors and there was a sudden, most violent storm. One burst of thunder followed closely after brilliant lightning and nearly split our eardrums.

When the storm had passed and we'd had tea, we went outside and milk-fed the babies in the new shed. Having done so we went into the bullock shed and continued the nightly round. Rearing calves is not easy but it is the least difficult of all tasks to spot an ill one. Initially it does not get up for food. If it is chewing its cud, all is well. It's as simple as that. We have said frequently we would never rear animals that did not ruminate.

One calf, after that awful storm, did not get up nor was it ruminating. Perhaps it was just asleep. It was curled up in luxurious comfort and was after all relatively newly-born. Margaret hopped over the gate saying, 'Come on. It's supper time.' The little fellow was against the iron gate and he was curled up quite dead.

It was a shock! We had had no warning. We looked to the roof of the high shed and saw the gate on which it was leaning was directly under the ventilation. Could it have been struck by lightning? We looked for burn marks but there were none. Even so we could find no other explanation.

Deeply depressed we took the four stiffening bodies in the Range Rover to Jerusalem Farm which is our local knackers yard.

'There's nothing for it. We'll have to go to Otley again,' Margaret said when Monday came around again and that is how we bought Samson.

Chapter Eight

> God grant me the serenity
> To accept the things I cannot change
> The courage to change things I can,
> And wisdom to know the difference.
>
> *Rev. Dr Reinhold Niebuhr*

You shouldn't buy an animal, let alone a strong Hereford bull, if you know darn well he is blind. The vendor was partially honest. He followed the little fellow round the ring with an envelope with which he tried to make the calf blink. ''E's just short sighted,' the selling farmer said. ''E can see where the bucket is alright,' he assured the bystanders as the little fellow bumped his way around the ring. He continued wafting, hoping the calf would blink and only Margaret bid. She bought him for a song but he was totally blind.

He wasn't the only calf she bought that morning but we do not remember anything of the others except that they all lived. Because the calf was blind we called him Samson and, because in his confusion he used to turn in circles, on the spot, there was born the phrase 'doing a Samson'.

We 'do a Samson' frequently, in the kitchen, when we don't know which way to turn. I have often said that there are three stages from sanity to abnormality and we go through all of them every summer, beginning the season capable and ending it round the bend. The healthy way to behave is to go to a cupboard knowing what you want. After several weeks of evening-meal-making you begin to go to the right cupboard but do not know what you want when you get there. By October you begin to go to the wrong cupboard for something but you can't remember why. Then you need to start worrying. There is no cause to worry about 'doing a Samson'. We all need to 'do one' to get our compass direction right.

Samson had no sight but he had hearing and smell and touch and could find his way to food. He was a very worried little man, holding his head awkwardly and trying to understand his environment. How Margaret loved him, but from a distance because he didn't snuggle or want cuddles like the sighted calves do. He didn't lick fingers or have head on fights or anything. He hung back trying to fathom what was going on around him.

Fortunately he never ailed a thing. We have never been able to afford expensive week-old calves. When the bidding gets high Margaret drops out. Samson was a thoroughbred she would never have bought had he not been blind. He was an expensive calf, quality written all over him but, sold to a softee like Margaret for a knock-down price, he was nearly as cheap as Emilie.

We soon decided we might lose more calves if we didn't get them out of the bullock shed so we reconstructed pens in the yard shed whose accommodation we had scorned. We sought every way to let air into the new shed and every way of keeping it out of the old one and after the first few weeks all were thriving.

At the first lull after buying we phoned the lady solicitor from the Council to shower our thanks. We were told she had left, had moved into North Yorkshire and had been replaced by a man. He assured us that closure would be advertised publicly so that objectors could come forward. Only a little unease germinated. Things were moving steadily in the right direction.

We were too busy anyway, and we judged it to be just a matter of time so we did not get too uptight when fifty or more hippy caravans, buses and furniture vans moved onto Altar Lane. 'They'll just excite the Council to move more quickly,' we hoped.

Although half-a-mile away, the noise of their music was awful. For a little while, feeding calves in a strange new shed, not yards from quite a noisy stream, we did not know which was torrent and which was torment. The latter it was. It drummed incessantly until we thought we would go mad.

They, of course, effectively closed Altar Lane. Now no off-road Land Rovers could make matters worse. The New Age Travellers did what we would have been prosecuted for doing. We felt secure in the knowledge that legal closure was imminent. Indeed notices had been nailed to telegraph poles informing the general public. We hoped any objector would be overruled.

One weekend at the beginning of December we had the house full of trainee Guiders. Shortly after feeding twenty-five of them, and the calves, Margaret thought she ought to check the yearlings who hadn't yet decided they wanted to come in for the winter. She went off happily enough and came home in a frenzy. Counting cattle on the move is always difficult. In summer the herd will be lying down and the job is easy and by the late autumn Margaret is looking forward to the herd coming back to the sheds every night so that they can be counted in. Only two, less hardy, had succumbed and sought shelter. We had not lost one of the first batch of Beeston Castle calves but two, no three, had done less well. Two had been brought in and, every now and then, Margaret put the third in so that it could fill its belly even though it preferred to stay out.

Margaret reached the pasturing herd and counted them. I find counting a kaleidoscope of moving cattle impossible but Margaret does not give up until she has counted correctly, twice. Each time on this remembered Sunday morning she counted one short.

She was suspicious of the Travellers, their dogs, their lifestyle and their reputation but she did not wish to antagonise them. However, after a thorough search of the hillside and wooded area and every extremity she could think of, she decided to tell them she had lost a bullock. Their dogs were running all over the place but she thought there were fewer parked vans. The atmosphere suggested they might be moving on. They had not seen our missing bullock they said.

There was nothing for it but to hurry home and ring a few neighbours to see if, by any chance, the missing bullock had jumped a wall. Passing by the muck heap, left by John to mature and spread in the spring, Margaret picked up a new jaw bone. Meat was still adhering to it.

'They've killed and eaten my bullock!' she panicked, picking up the remains. She was crying when she arrived home. She picked up the phone and dialled the police.

They were unhelpful. It was Sunday and they were short staffed and human life was not in danger. Farmers said hippies stole animals but it was difficult to prove. It was not possible to identify a jaw bone that had never been to the dentist. They'd send someone up later, if someone was available. Margaret was incensed. If the police didn't come quickly the thieves would be gone. She rang Dolores, who lives on a farm nearer the entrance to the lane, who offered to monitor the moving off of the Travellers. She saw that indeed they were going and she came down to tell us. I rang the police again but failed to motivate them. I can see us all now, sitting fuming round the big kitchen table, Dolores, Guiders, us, all angry, all ready to annihilate hippies, police, the Council. We hadn't a good word for anyone.

Margaret reacts scornfully if I even suggest that she might have made a mistake. It was only one bullock missing so I thought my query was ridiculous and hesitated before saying, 'You did account for the two in the shed, didn't you?'

'Of course I did ... ,' her mouth did not close and her eyes widened in unbelief. 'There aren't two in the shed. There are three ... !' she remembered too late. We didn't know whether to laugh or cry. The relief was incomparable.

We did neither. We all sat silently round the table accusing no-one. Just breathing again. Eventually Margaret pulled herself together and picked up the phone and told the policeman the bullock had been found. If he had shown any interest in the first place she'd have admitted where, but she was still hurting. She put the phone down and perhaps we laughed then. I can't quite remember what we did, we were in so much shock. Margaret went out unnecessarily to verify that the animal was indeed there but she knew that it was all right.

* * * * *

I particularly like to tell this next story though it is not in chronological order. It happened to a Guider but not to one who was here when the hippies were. We had a group of Guiders who came from Birmingham for a pleasant, get-together weekend. They went to York on the Saturday and one of the elderly members of the party saw a picture she wanted in the window of an art shop in the city. There was no price on it so she went inside and asked to look properly at the three dimensional picture of a very attractive cat. Motif upon motif had given the picture depth and it really was a skilled piece of work. But it cost more than the Guider had in her purse. I think it was selling at £25. The old lady, in uniform, asked that it be stored for her. She promised to send the £25, plus postage, as soon as she got home. 'When you get my cheque, please post it to me,' she said.

'Oh, take the picture now,' the York shopkeeper had said. 'It may break in the post. Just send a cheque when you get home.'

The Guider found it difficult to tell this story without tears. 'It wouldn't have happened in the south,' she said.

On another occasion a male guest rose early to drive up to the village post office to buy a morning paper. He had enough small change in his pocket but when he got to the main road he noticed his petrol was very low. The nearest filling station is only half-a-mile away so he decided to kill two birds with one stone. Having filled up his tank at the self-service pump he remembered he had not brought his wallet. He was horrified. He apologised and was dumbfounded to hear the Yorkshire voice say, 'It's alright, lad. Drop it in when yer pass.' Nothing like that had ever happened to him before,

Southerners repeatedly tell us how much more friendly we are, who live in the north. 'You come from the south and you are friendly,' we say with justification. They say, like the German lady, that it is an echo.

It gives us great pleasure when our guests receive courtesy and generosity from those who live in our locality and come back in the evening to tell us. 'There is the wherewithall in Kirby Malham church for visitors to make themselves a cup of tea!' they report with wide-eyed appreciation.

'The man at the petrol station is my friend,' says our Lyon guest, Bernard. 'He always remembers me.'

'We were talking to an old lady in Ilkley,' two pleasant guests said with amazement, 'And she's asked us to call for a cup of tea tomorrow!'

* * * * *

Less pleasant to remember was the mess the New Age Travellers left when they moved away from the entrance to Altar Lane. Litter stretched for a quarter of a mile. They had closed the road but the Council did not. They had erred. The notice to close had not been advertised nationally. The meeting scheduled for mid-December was cancelled and the whole thing fell through.

We insisted that the new, male solicitor walk the mile length of the boundary we share with the road, to see for himself the state of the ancient way, the litter left by the Travellers, the crumbling estate wall, the smoke blackened patches where joy-riders had burnt out cars, so that he could fully understand our problem. He came, young, inadequately dressed for the country. He had smart polished shoes and an umbrella. He slithered along beside us deeply envious, I am sure, of our wellingtons and kagoules.

The road couldn't have looked worse. Top stones of the newly-built wall had been removed in several places, to end up thrown into the ruts. December was on our side making a string of miniature lakes en route.

The delay caused by the advertising error had given objectors time to send in letters against the proposal. The solicitor told us any effort to close would fail and that we might all be better employed. To try would be a waste of time. We threatened to close it ourselves. All it needed was a telegraph pole barrier. We would be committing a crime, he said.

'The hippies have had it closed for a month,' we argued. He shrugged his shoulders.

We expounded on the vulnerability of our expensive new wall. We'd only dared to use that sort of money because Council officials had promised closure. We did not need to exaggerate the danger to the public if our two hundred cattle stampeded round the Nature Trail in the estate or galloped down the road to join the A650 two miles away.

'I don't think we are ever going to get a closure on this road. It's an ancient highway,' the young man said.

I was sorely tempted to stumble, walking alongside him along the banking of a road which was a river. The sidewalk was two inches deep in wet mud. I could have used my age as an excuse for stumbling. I could have steadied myself by grabbing the young man and making sure he and his umbrella went sprawling in the mire. It was tempting but we have been taught to clean up our own mess and I couldn't face scraping the mud from his immaculate suit.

'There is one other method we could maybe try,' the young man volunteered, 'but it would take at least twelve months.'

It was December 1991. We knew full well he was only procrastinating. Our file, if indeed they had kept my many letters, would be put aside and forgotten. I had kept copies of all mine. I sent them to the Country Landowners' Association, to which we belong, and to the National Trust and let them eat away at officialdom for us. We had other things to think about.

* * * * *

Margaret was lucky to survive the winter of 1991/92 for umpteen reasons. The new calf shed, which should have taken away problems only added to them. We could not get a circulation of air. All winds insisted on coming from the west.

We went to Burnsall for our Christmas ham sandwiches eating them in a deserted car park beside the river, missing Mother, the village schoolchildren and all the Christmas activity they and the Guide Company seasonally brought. The Range Rover seemed empty and the house was empty for only the holiday cottages were occupied. Life is such a contradiction. We were eager to relieve ourselves of the burden of bed and breakfast guests over Christmas yet the dark and empty dining-room depressed us and the Snug with its beams and inglenook reprimanded us.

There were compensations. A winter without evening meals meant a return to the traditional cold weather food we had been reared on. Though we were renowned for our traditional table we served affluent men's fare. We were hungry for a poor man's diet and we had tripe and onions frequently and lots of broth and dumplings. I really don't think there is anything more delicious than tripe and onions liberally sprinkled with home-made chips. On a cold day its comfort is incomparable. And dumplings lifted from the broth and eaten first with lashings of sugar? There is nothing more delicious. We doted on egg and chips, on cold beef, mashed potatoes and ladlesful of onion sauce. We had rice puddings baked in the oven until the skin rose and crisped on top and we ate egg custards by the dozen. We made bread-and-butter puddings and apple dumplings and scalded jellies. We bought liver and lamb chops and for tea we toasted crumpets and thickened them with butter. We lived on prunes and dried apricots and boiled pears and we baked buns and cakes and Auntie Mary had a go at baking bread. We had pan pies and savoury ducks. In spring Auntie Mary made dock pudding with passion docks and nettles.

We had had more than ten years of eating what was left of the evening meal served to the guests. This return to a peasant's menu was truly welcome.

A heavy downpour raised the water level in the hill behind The Currer and the stream which had perpetually trickled or flowed, according to season, through the laithe and into the yard, reappeared in the Snug. The carpet was saturated overnight and, barefoot next morning, we were immediately aware. We piled up the furniture and rolled back the carpet and tried to stem the flow with towels but, until it decided to stop, our efforts were futile.

Resident builders, erecting office blocks in the valley, put flagstones along the north wall and pointed every cranny they could find. The flow eventually stopped, we dried the carpet and put back the furniture.

Then five things happened very quickly.

We had barely coped with the flood when one disaster after another threatened to defeat us. It was still only January. We were having trouble starting the car. Occasionally, when the ignition key turned it failed to start the engine and just made a quite appalling noise.

I had taken Auntie Mary to a 'do' at the Chapel and she had to be collected at four. We are habitually late, for something always detains us at the last minute. Realising it was long past the hour I dashed out to the car just as Margaret was letting in the herd of seventy hefty yearlings. She had opened the gate on the black and white multitude when I pressed the unpredictable starter and the awful screech stampeded the herd. Margaret was knocked down and flattened to the ground and I don't know why she was not killed.

She was aware of scores of feet galloping over her as the herd catapulted into the shed. I knew immediately what I had done. I leapt out of the Range Rover and dashed to what I believed to be a tragedy.

It has always been understood that, at all costs, we must not cause a stampede. We never approach penned cattle in the dark without talking to them and warning them we are coming. We do not make strange noises and if one is unavoidable we make sure we are well out of the way.

We are always aware that other people might make an unexpected noise and put our lives in danger. We know some idiot might blow a horn in the yard, or the phone will go, or a heavily sweating jogger may appear from nowhere. We always face an oncoming herd, if it be approaching speedily, and we position ourselves against a wall if animals are leaving or entering the shed. A rising lark or a spooky owl in the moonlight, a sudden zigzagging of a hare or a low flying helicopter can set the thundering of eight hundred feet. We have seen the whole multitude plunge down the hillside in daylight or heard it in darkness. We have seen the black torrent plunge towards the fence round Jimmy's Wood curving and swirling like a whirlpool and we have held our breath and prayed. When the frenzied running has stopped we have followed the trail, beaten by heavy hooves, to check no animal went down and none had hurtled over the fence and rolled down the wooded hollow.

We know all this and yet I pressed the noisy starter and whatever had happened it was all my fault. But miracles, little ones and big ones are happening all around us all the time. When the animals stampeded Margaret had no escape. It all happened far too quickly. She lay there expecting to be dead and not one animal trod on her. By the time I reached her it was all over. She was lying in four inches of cow muck, plastered from head to foot, hands and face and clothes, but unharmed.

I cannot remember what we did with the clothes she took off. Maybe we threw them all away. All I remember is that when I set off for Auntie Mary she was already at the foot of the lane, only fifty yards from home and equally angry.

Even though we had the starter repaired I never press it without looking to make sure no cattle are being driven. I was at the garage having it mended when Margaret was nearly killed the second time.

We had a fifteen-month-old Friesian bullock which had not been properly castrated. It is a job the vet does annually when calves are about two months old. We had first ignored his aggressive look, his more masculine bellow and the hefty shoulders he was developing but later had recalled the vet to repeat his performance. Even then we were still suspicious of the bullock and concerned that the castration was still ineffective. Whenever handling him, we were always wary. Whilst I was at the Land Rover specialist, getting the starter fettled, Margaret let the cattle out. It was a sunny day albeit still winter.

Where the gate leaves the yard into the field the builders had rolled a big beam of concrete. Mud was deep everywhere. Margaret was near this concrete block, close to the fence, when this dubious bullock turned on her, rolled her over and angrily helped her over the barbed wire fence. It was the first time, ever, that any animal had wilfully tried to harm Margaret and, taking her completely by surprise, he could well have done so had the block and the fence not been near. She was understandably shaken and re-opening the shed door she drove him and one or two more, back into the shed. When I returned she was very positive in her determination never to let him out again. Supposing he attacked a walker on the moor? Friesian bulls are dangerous animals indeed!

Actually he seemed to have a one-track mind and was only interested in maiming Margaret. He looked past me and went straight for her. It is not often we are intimidated but we decided we had too much responsibility, too many footpaths and too many holiday-makers to take any risks. We prepared to sell him at the next market.

We had an hilarious few days before the auction. Feeding him was safe enough for the feed barrier was between him and us. Putting down the bedding straw was a problem. Every time he caught sight of Margaret, he pawed the ground and bawled. We put an escape ladder against the wall dividing the two sheds and when the beast began to charge I, the look-out man, yelled, 'Run,' and she was over that wall like nobody's business.

This was a state of affairs never previously known and, dangerous though it was, we did see the funny side of our predicament. Every time she was sitting safely on the wall, we laughed. But it was no joke for that animal really had it in for Margaret.

We sent for the cattle haulier and fortunately John and an employee were working here so they helped load the animal on to the wagon. Margaret showed him in the ring and told the auctioneer that the animal was dangerous, at least as far as she was concerned but the errant animal did not throw her over the ring fence and the watching men probably thought the little woman was imagining things. The buyer had been warned and there was no come-back so we presumed it was only Margaret the fellow was after. We have learned, in the years that have followed, that inefficient castrating is a recurring problem. This was the first but not the last animal to cause us trouble. We longed for Herefords, whose males are benign, but which have almost disappeared from the calf markets.

They say things happen in threes and we were therefore due for a break but such was not for us. Margaret was doing her last, long after midnight, check of the herd when she heard a bellow of alarm. We are convinced animals have a multi-syllabled language, hens certainly have. Just as they understand the odd word or two of English, so we know a word or two of theirs. We know a bellow of alarm when we hear it, even half-a-mile away. It has sent us running in its direction scores of times.

This time the agonising cry came from within the top bullock shed. Margaret had not got as far as turning out lights so she saw at once what had happened. The accident should not have been for we were aware of potential danger, and for years had been 'getting round to' doing something about the iron partition which had separated the sixty Hereford heifers into two thirties, in the days when we had a hundred of them. The animals we were now rearing were much, much bigger and the iron partition, which was hinged, was potentially dangerous. Instead of its three bars being parallel, the lower one sloped up causing an acute angle. Sooner or later we expected our quite enormous animals just to walk over it. It was a new experience to have bullocks so big they dwarfed my sister.

Unfortunately one does not always let anticipation of danger lead one to do something about it. At that after midnight hour cattle were marlacking, maybe mounting a bulling heifer. Whatever the reason Margaret found a Hereford had rolled over the gate and caught its leg in the acute angle between the middle and the bottom bar. It was a heavy animal and it was upside down in a bonny pickle.

Seeing she could do nothing alone she came to get me out of bed and together we struggled for as long as it took to realise there was no way we were going to free that animal without sawing through the iron bar. We made a feeble attempt with a hack saw and our strongest crow bar and then sent for the Fire Brigade. Two engines and over a dozen helmeted men came rushing to our assistance once again. It did not take one of them many minutes to saw through the bar and Margaret and I rolled the animal over onto its belly and hammered on its haunches to force it to its feet. The firemen insisted Margaret send for the vet and waited until he came. The poor man had to get out of bed and give an injection which Margaret was quite capable of giving herself at much less cost. But the firemen had no faith in her and insisted on getting the tired vet from his blankets. It was great, safe entertainment for them!

Needless to say the firemen were instructed to lift off the offending gate and we eventually re-hung it under the sycamore.

A few days later, or was it a few days before? Who knows in this eventful life we lead in which there is no time to keep a diary. Somewhere in late January or early February thieves stole our ATV Yamaha. To our knowledge no-one had ever stolen from us before. Margaret had been using it all winter to move straw and hay from the barn, up the calf shed passage to the bullock sheds.

We had bed and breakfast guests in the house and their cars were in the yard and there was a movement of traffic up the road. Outside lights and shed lights were on everywhere. Margaret came in at 6 p.m. for her tea and brought in the dogs because cars were going out and our dogs give chase. They barked from the house a few times. Jess jumps up onto the window seat and Lusky who cannot because of his arthritic hind legs barks loudly enough every time a car goes out. We expect this and don't fly to the window.

An hour later we went out to feed the calves and the ATV had gone. We could not believe it. We searched everywhere expecting to find it abandoned by thieves disturbed by guests or intimidated by barking. They must have had a pick-up for Currer Lane is too steep to push even a light Yamaha. We never saw it again. To steal it someone had to push it many yards through a lighted shed in front of fifty inquisitive calves, through the perpetually open door of the winter of 1991/92, up the incline and through the busy yard. They had a nerve. I'll say that for them. A car is on view all the quarter-of-a-mile of our road and our binoculars are strong. Cars following our road by mistake do not often get away undetected.

We were indignant. The police could find no trace. The insurance replaced it with a second-hand one which Denis uses all summer and we seldom use all winter. Margaret reckons she can do most things better on legs than on wheels. She reasons that, having lifted a bale it is easier to carry it to its destination rather than stack it on a trailer and lift it a second time from the ATV. She is possibly right. As she does almost all of the lifting of bales the choice must be hers.

The early weeks of the new year were just a succession of accidents. We began to think they would never stop. On the last Sunday of the month Auntie Mary's friend Hilda came, as she frequently does, at the weekend. Harry had not fallen for ages. When he does his arms are no use to him at all so that every time his head thuds onto the floor. Being bent he loses his balance and falls forwards, never backwards. Sometimes he rolls sideways as he did on one occasion whilst washing at the bathroom sink. He had wobbled to the left and toppled into the bath and somehow his feet were almost in the wash-basin. On this particularly Sunday he measured his length on the rug in front of the Aga. He went down with a frightening thud. Had he dropped dead he could not have gone down more heavily. He crashed his head on the hard floor and lost consciousness long enough to really frighten us. We decided he must go to Casualty to check for any fracture. The noise of his skull hitting the floor worried us but an X-ray showed all was well.

February was a quieter month. An earlier dawn and a later sunset made our spirits rise. Our nearest neighbour, a few years our senior, decided to sell his milking herd and retire. Along with many others, scenting MAFF changes, struggling with milk quotas and all the impositions placed on farmers, George decided sensibly to almost quit. He kept just the followers, a few heifer calves and yearlings and let out his land. We bid for the acres which are adjacent to us and we have rented eight small fields ever since. Folk think we should be following his lead and retiring ourselves. No way can we do that, nor do we wish to do so. Our decision was wise for the small fields gave us a temporary holding place for calves and bullocks we want to worm next day, or sell. What's more we saw, in them, two fields of our own hay again. Aren't we idiots, opting for more work rather than less?

* * * * *

The Yorkshire and Humberside Tourist Board began sending warning notices to members that the Government was stepping up its rules of hygiene and that all catering landladies should enrol for instruction. Scare stories began to circulate and a farmhouse seemed to meet none of the requirements expected by authority. Everyone loves a farmhouse kitchen and finds every excuse to enter it and loiter. Everybody returns to a farmhouse for the food, not for the animals or the work experience. Our guests, I am sure, return again and again for the farm kitchen hospitality and the traditional fare.

Rumours were that no-one may enter the kitchen where food was prepared, no animal must come anywhere near, no food must be eaten within it and there must be no washing of clothes or linen. There must be no wood surfaces. Tables must be stainless steel. There must be no carpet or mat or open window and no washing of hands in the sink. Rumours were that an unexpected inspection could close a business at once.

I decided I could not stand the stress. We work harder than anyone in the vicinity but stress is not a frequent visitor to The Currer. I was not prepared to open a door and let it in. Quite obviously farming and catering were incompatible. One would have to go, for our farmhouse kitchen, loved by so many, met none of these requirements. We lived and ate in it. There was a dog under the table and a cat in front of the Aga. We fed workmen and corn travellers, the coalman came in for payment, we'd a sickly chicken in a box and sprinklings of straw on the carpet.

We will never stop farming so the obvious choice was to rid ourselves of the catering. 'I will not leave myself open to such worry,' I vowed and I told Margaret we would have to find an alternative. Any ordinary farmhouse providing bed and breakfast could just have brought in the sign and stopped. We couldn't do that. Apart from the fact that we had a full chart of already booked holiday-makers, our tourist accommodation was quite separate from our house. It was almost, except for the kitchen, a self-contained unit of six bedrooms, two bathrooms, an enormous dining-room and a sixteenth century sitting-room. It could not just disappear. We had to do something with it.

But I was adamant. The overdraft was still high and we needed a second income to clear it and rebuild our dry-stone walls. I was sixty-two years old and all my contemporaries were retired. I totally refused to be harassed by authority.

How the solution was born I have no idea. Sometimes you wake up in a morning and your problem is solved. Sometimes you debate all sorts of alternatives. The first idea seemed to be to turn one of the ground floor bedrooms into a kitchen and the whole unit into a self-catering, five-bedroom establishment. It didn't really seem the right thing to do. We were tormented by visions of all the handicapped people we had catered for whose disappointment would be profound. We had created something of infinite value to a great many people. Could we really just ignore that, throw it all away overnight?

The solution was born before we did, thank goodness. Some voice from above must have whispered, 'Freezer 'Ole.'

It was an 'Ole indeed! Three freezers resided in the untidy store which had once been a smokehouse for curing hams centuries ago. Since our occupation of The Currer it had first been a ruin without a window or roof. Below it was a cellar. We had closed that because Mother had feared there would be rats and, besides, the well in it flooded the crypt with a foot of water every winter. We had re-roofed in order to keep our camp equipment there. Later we had transferred tents to a field store and filled the area with coke. Then, when we opened up the cellar, and controlled the well water with a pump, we used the smokehouse for freezers and paint cans and all the paraphernalia of a farm store.

Our Freezer 'Ole was a mess we only tidied before going on a holiday. Its roughly plastered walls had been lime-washed but cobwebs frequently appeared. There was a hole where one of the gnarled oak beams sat on the outside wall and a squirrel knew its way in.

The idea that this could be turned into a spotless, stainless steel kitchen was a Brown one. We saw that we would need to break through the thick wall in two places, one into the passage connecting the ground floor bed and breakfast toilet with the bed and breakfast sitting-room and the other into our own wash kitchen so that we could enter it from our own accommodation. It would have to be plastered and underdrawn. We are visionaries, of course, experienced at seeing order where there is currently chaos.

We sent at once for Brian, our extraordinary builder, and I applied for a hygiene course at York. We were as happy as proverbial kings. We had a project and a solution. We did not need to close. We need not say to all those long standing guests, 'That was it! You'll have to go elsewhere!' The relief was tremendous.

York brought back memories of my second School Practice during the summer term of my first year in college. Every so often, during my teaching career, I had taken children to visit the walled city. It was an opportunity, this essential visit to York, to travel alone, by train, a pleasure I hadn't had since University days. For such an active person, I do like just sitting watching the world from a train window, being unsociable, not talking to anyone. I've done far too much talking in my lifetime.

In the lecture theatre I sat with caterers, not teachers but the atmosphere was familiar. I quite enjoyed it. The lecturer was reassuring, confirming our belief that food properly cooked and served immediately is not going to kill anyone. She also confirmed the rumours concerning stainless steel and colour coded knives and cutting boards. There were others who served Farmhouse bed and breakfast. To us she said, 'Your worst problem is your dog!'

'What nonsense!' Margaret said on my return. 'Our worst problem is us. You and me!' We are the farmers at our house. We have no husband to do the dirty work whilst we prepare the food.

'The dog doesn't touch the food. We do!' Margaret laughed. Boy, were we glad we were getting a purpose-built kitchen!

* * * * *

The work was well underway before we went on holiday at the beginning of May. All our pre-holiday, last minute capers are dramatic and leaving with workmen in the kitchen area was no more traumatic than usual. We try to leave everything as clean as possible for Dorothy and George but that year we left dust sheets on the kitchen floor and workmen creating a constant cloud. Work in the Freezer 'Ole disrupted everything. On the morning we were to leave one of the workmen severed the electric wire working the central heating pump and we had to send for the Aga man.

Dorothy had called before breakfast with the contents of her fridge and I had not even defrosted ours. I was ashamed and tackled the job immediately she had gone. We called warnings after her not to come early. Like as not it would be well into the afternoon before we left. For the first time we did not want to go on holiday at all. There was such a mess. As I swept the last accumulation of things we had assembled to take with us into the final holdall, I glowered at the men sitting in my kitchen calmly eating smelly fish and chips from a newspaper. I could have screamed!

'Don't you dare leave anything on this table,' I said ungraciously for we were going on holiday and they, poor souls, were staying at home to work.

The postman brought a cheque for just over £1,000 from the ATV insurance and I knew our first call would have to be at the bank. Realising that no matter how long we stayed we were not going to make anything look cleaner, we climbed into the Range Rover, a very old friend, and tardily left for Scotland.

Once on the road we lost momentum and a speed of a steady 40 mph was never exceeded. We ignore hooting motorists, laughing friends and sarcastic relatives. Their derision makes no difference to the route we choose. Nothing diverted us from the road to Peebles through Castle O'er Forest and Ettrick Bridge. Peebleshire appeals to us with its rolling hills and variety of suckled herds. We had no wish to drive quickly through its unparalleled beauty. We only met five cars in fifty miles. Where the vista recommended we ate our sandwiches and opened our flasks and where a country park or picnic area offered toilets we pulled in.

Inevitably darkness came before we expected it and we muttered, 'We won't let it happen again. We really will set off sooner next time.' Unfortunately we turned left at a crooked crossroads and did not notice our mistake. We lost valuable time and arrived at our Airdrie cousins late and entered their lovely home very sheepishly. We had every excuse for being late but no-one ever believes us. All our friends and relatives appear to think we are incompetent. Once one bought us a poster which said, 'Tomorrow I will get organised!' As Margaret tells me, 'Yer gotta be meek.'

Increasingly we wished we knew a route which avoided even the outskirts of Glasgow. We were sure it wasn't that we were growing old. Driving conditions and road systems were leaving us behind. Lorries were bigger and their drivers more aggressive. There were more and more cars and we, in our backwater, had not kept pace.

We left Airdrie soon after breakfast and headed for Oban via Fintry, Port of Menteith and Callander. We had booked a cottage at Tobermory on the Isle of Mull and sailed on the ferry by that name, to Craignure. It is all very modern. Forty years ago we had to leave the ferry in the middle of the Sound and go by small motor-boat to Craignure. There was no pier in those days.

Our week on Mull seemed the wettest we have ever spent among the islands. It may have been just an illusion for we are used to treeless islands where rocks and sand dry immediately the rain stops. When it did on Mull, water still dripped from the trees and the long grass, unlike the short machair of the barer islands, retained the deluge until the next downpour. We wore wellingtons most of the time and the dogs were always wet. Our cottage was only yards from the bay and the garden was a sponge. The picnic table and benches were unusable. To walk along the path round the headland, with the dogs, was to slither and slide quite dangerously.

There was one quite hilarious coincidence the like of which is a several million to one chance of ever being repeated. There was no television in the cottage and Harry was missing the weather forecast. Over a 10 a.m. breakfast he insisted Auntie Mary tried to get one on her radio. We said there would be no weather forecast at that time of day.

'It will be Story-time or some such,' we said.

We were both right. A man was reading a story. The very first words we heard were, 'Drip, drip, drip ... drip, drip, drip.'

The explosion of laughter from all of us completely drowned the rest of the story. Life is such fun!

We had loved Mull previously when we were walking and cycling and moving to a new bivouac every night. Now the roads were too narrow for our wide vehicle and distances on that relatively large island were too great for us. We clocked up nearly one hundred miles going to the Iona Ferry and back. Reaching Fionnphort we found coaches galore, empty on the pier.

Presumably the island would be sinking under the weight of tourists so we decided to keep to our memories of sailing to Iona on a small boat, sitting in the bow overlooking a sea so transparent one could see the sandy bed and watch the sea life and believe one was on the coral reef in Australia. Things had happened in forty years we did not entirely approve of!

Nevertheless we were happy on Mull. We are always happy on holiday! We decided to call at the Kinloch Hotel on Loch Scridain, for tea, on our way back to Tobermory. We had stayed there in 1951. They were not serving Afternoon Teas we were told.

'It's a pity,' we said. 'We stayed here forty years ago.'
'Come in,' the lady said. 'We'll make an exception.'

So we went in and enjoyed the newly-baked scones and good fresh tea. Fresh tea anywhere is sheer luxury. If we make tea it is usually stewed and cold before we get round to drinking it. The Kinloch Hotel made lovely tea and we were not called away to the phone or by an arriving guest or a barking dog or cattle in the yard or whatever. We felt privileged.

We left the wet weather behind when we went to Tiree for our second, all too short week. Never again, we said, would we go to Tiree for one week only. It was all hello and goodbye. Having re-visited Barra and Mull we had satisfied a yearning and were content.

Tiree is very different from Mull. It is just a fertile sand dune alive, still, with memories of the magic of our camping days and our hundreds of bare-footed, brown-legged children. To go to Tiree you must love sand and sea and to appreciate it properly you must have some instinct for farming. It is heavily stocked with beautiful animals. Every inch of it has potential and every square yard is farmed by crofters, like Hughie MacInnes, whose cattle fetch top prices anywhere. It is firmly fenced and the crofters are busy every hour of the day, just like mainland farmers. We have always admired and respected them for squeezing a good standard of living out of their small acreage. The climate is blander than ours on the Pennine Ridge and they can out-winter but the cost of living and freight charges are so high, they have to be good at their profession.

We are at home on Tiree.

The morning of the departure was quite wild. Any wind on Tiree comes unimpeded from the great Atlantic Ocean. There is no shelter from it anywhere and trees just cannot survive. It wasn't what one might term a gale. We loaded the roof rack without Margaret being blown off and pulled on the cover without it becoming airborne. Perhaps the wind speed was accelerating faster than we realised. When gales sweep across Tiree, driving to Scarinish is comparable to crossing an exposed bridge over a wide river, but there are no signposts on the single track roads to warn you against cross-winds. You feel them immediately but are careless of them for there is not a precipitous road on the island and the sea is shallow for a long way out.

Strangely, I cannot remember the wind feeling very strong that morning. It was only when we reached the pier and joined the queue waiting for the *Lord of the Isles*, and I opened the car door, that I felt it torn out of my hand.

It was then that we began to wonder if we were good sailors for we could feel the wind hitting our car broadside and foot passengers walking down the pier were doing so with enormous difficulty. A tail wind was helping them along more quickly than was safe and they were hiding behind the buildings lest they be thrown overboard.

'It's windy,' Margaret said.

Nevertheless the *Lord of the Isles* appeared on the horizon and made a steady passage over the bar, into the bay and on course for the pier. She came within feet of it and the same wind which threatened to throw people overboard met her broadside and held her off, listing her to starboard. The crew set about their ritual with capstan and hawsers but thrown ropes were blown seaward. The boat was listing so alarmingly we, sitting watching in the queue, thought she was going to be blown right over. The men on the pier and the crew on board persisted for half-an-hour but the wall of wind defied them. We saw the big mail boat draw away from the vicinity of the quay and pull back out to sea. We expected every minute that we would be told the boat was unable to dock and that we had best go away and come back on Monday. No-one moved. No cars pulled out of line and the pier staff disappeared without instructing us to go so we sat tight, waiting and wondering if we were not crazy anyway, to contemplate sailing with Harry and Auntie Mary and the dogs. Margaret hates leaving the dogs on the car deck. I always have the feeling, should the boat begin to sink, that I would be left with the oldies and 'St Francis' would be rescuing the dogs.

Eventually the boat, lingering on the horizon, began once more to cross the bar and head for the land. The pier gang was blown again to the end of the jetty and once more there was the performance of trying to get ropes from the boat to the men ashore waiting to catch them. Once again the *Lord of the Isles* listed some 45 degrees to starboard and could not come within yards of the pier. We began to think the boat was going to roll right over but we continued to sit there, glued to our seats.

Then the boat was pulling away again and we were watching the gang return and we were sure this time we were going to be told to go back to Salum and re-open the cottage. The boat was disappearing behind the headland. That's that, we decided. We would have to phone the cousins in Airdrie and Dorothy to say we were island prisoners.

There is a strange, quiet acceptance of a situation, a tolerance only to be found in The Hebrides. No-one showed impatience, no-one was in a hurry, there was no anger. When God made the wind he also made time and, so help me, there is plenty of both. We were quite amazed at ourselves willing to sit there, in such weather conditions and even contemplate crossing the Minch ourselves.

Nobody moved. We were not told to go. Every car remained in the line. No driver wanted to know 'what was going on'. It was quite an experience. We poured a cup of coffee from our flasks. The Range Rover was warm. There was nowhere to go so we watched the tossing ocean and the buffeted seagulls.

Then, two hours late, the boat appeared once more from behind the headland, crossed the bar into Gott Bay and approached the pier. This time someone had acquired a small boat with an outboard motor. When the ropes were hurled unsuccessfully toward the pier the little boatman watched where they fell in the sea and, because he was sheltered from the wind by the higher pier, he was able to pick up the ropes, one at a time, and bring them ashore. The rest was easy.

The ropes tightened slowly and the boat was pulled to bump against the quay and the slow, unhurried ritual began of offloading passengers and stores and mail-bags. No-one appeared one bit concerned that there had been delay. There was no excitement. Passengers dispersed and, yes, we followed the line of cars onto the mailboat like lemmings following a leader. We all went to our fate without a backward glance. The human being is a funny animal, too.

We had, of course, a tail wind all the way across the Minch. The *Lord of the Isles* makes some queer noises even in calm water when one gets the impression she has grounded or is breaking up, but apart from that she is stable and we did not get the tossing we fully expected. We did not attempt the loos or the cafeteria until we entered the Sound of Mull but otherwise we had no complaints.

We were embarrassingly late at Airdrie. We are unhurried travellers at the best of times. After a fortnight on islands we possibly drive more slowly than ever. There is an attitude out there which is infectious and we are less immune to it than most.

We were taking the ninety-year-old Glasgow cousins back with us as we had the previous year. As before, one of us sat on a cushion on the floor. It was a hot, uncomfortable perch and for some reason I found it more trying than ever. I believed it was the heat making me itch so. Even when I was driving the discomfort remained. Undressing that night, I discovered some sort of heat rash all across my back, following the elastic of my jeans round to the front and down my sciatic nerve to scatter spots all over my thigh.

I went to the doctor hoping it was a heat rash and not something infectious which could cause problems in the kitchen. He assured me that I was not a danger to all our numerous paying guests but said I had rather an overdose of shingles. He gave me pills and a tube of ointment, no longer or fatter than my little finger.

'Squeeze that over the rash, five times a day,' he said, 'and if you need more let me know.'

My rash was so extensive the little tube could not possibly last even one day, I thought. I used it very sparingly indeed and rang for another bigger tube. The small one was quite ridiculous. I went to collect my prescription and checked it for a larger tube but the receptionist said that was the largest tube the pharmacy supplied. 'And,' she admitted, 'each tube costs twenty-five pounds!'

'Good heavens,' I gasped. 'I'll have to put up with the shingles.'

'You'll do no such thing,' she said. 'Have this and another if you need it. You've never cost the National Health Service anything before.' Nor had I and I used three tubes and that, along with the pills, cleared up the shingles a treat. We did not tell the guests or the cousins.

* * * * *

Dorothy first noticed that Samson had a friend. Perhaps he stepped into the breach when we went to Scotland. Dorothy said on our return, 'There's one calf which stays behind Samson all the time.' She was right.

We had wondered how he would cope with the great out-of-doors, this totally blind calf of ours. All winter he stationed himself at the gate, taking deep breaths of fresh air. Handling distressed him. In the black world in which he lived he needed space. We talked constantly to him but he did not want to be touched. He seemed to have perfected some radar system which told him when solid matter was ahead. Having made contact with wall or fence he could then find the opening. He needed infinite care but went to no-one for it. He did not follow the herd and in the first days of being let out, in March, he had been perpetually lost. He, not knowing how to follow, had to be driven. When first let out we'd feared we would not be able to allow him the full freedom afforded all other animals in summertime. For a month or more, after the first fine day in March, we allow the calves to come home at night and sleep in the shed. After only a few days they have acquired the routine and come home of their own accord, disappointing Jess who likes to round them up herself.

The first fine day, like the first swallow, does not make a summer and there are days after that first wonderful taste of freedom when Margaret does not let calves out at all, when they would not go even if she opened the door.

Samson never came home of his own accord. At dusk, when the infant herd wended its way home he became stressed and started going round in circles, 'doing a frantic Samson'. To go for the blind member of the family was frequently my evening job, for Margaret and Harry do the washing up of dishes and cooking pots.

'You need a dishwasher', everyone says. With sinks like ours and a team like Harry and Margaret, who needs one of those? What's more, to deprive Harry of one job he can do well would be criminal.

So it was my job to bring Samson home with the extraordinary help of Lusky. The virtue of compassion is not often attributed to animals but we saw so many instances of it, in the two-and-a-half years we had Samson we are sure it exists.

Lusky was never a brilliant cow-dog, by any means. He herded everything but had his own peculiar idea of where to take his quarry. He had no difficulty in moving the herd forward for he could nip when necessary and providing the cattle knew where they were going Lusky was a treasure. If we were doing something different and cattle were being forced in a strange direction he was a menace.

All we knew about Lusky would have made him quite unsuitable for the job of slowly bringing home an animal which did not know where it was going. He was nowhere near gentle enough. Sighted animals eyed him with wary suspicion. In front of them he forced them to turn and behind he could not resist a nip. He was also tempted to nip over-amorous guests who leapt out of cars with arms ready to embrace us.

With Samson, Lusky was quite different. As sure as I live that dog knew he was blind and had compassion. Instead of drover he became companion, a presence, just there, usually on the up-slope guiding Samson downhill towards the steadings. He did not excite or stress the animal nor did he force the pace or attempt to nip. It was one of those recurring miracles of nature which contradicts any belief man might have that he is superior and adds weight to a realisation that members of the animal kingdom, too, have an awareness of disability and suffering and maybe bear witness to God themselves. I dunno. I just know that Lusky and the bullock we eventually called Friend, were as caring of Samson as we were and never did Margaret, or I, ever see any member of the herd hipe him out of the way.

The only distress Samson knew was loneliness and he did not know how to prevent it so I brought him home every evening and Margaret kept watch all day, which was a much more time-consuming task. Anyone who thinks farmers keep animals for money and do not love them should have come to The Currer whilst we had Samson.

It took me half-an-hour, sometimes more, every evening to guide him home and Margaret went to his rescue several times a day. She was frequently seen with binoculars. If a lone black dot in a field proved to be Samson she and Lusky would go and take him to the herd, or the water. If no Samson was in sight she would go and look for him. Together we spent hours of every day looking after that healthy, beautiful Hereford whose eyes had turned into their sockets so that they were both completely white.

If he could feel infinite space Samson would gallop bravely and joyfully into it. One day the four foot fence circling the coppice we call Jimmy's Wood failed to stop his cavorting down the Five Acre and he was found on the other side of it teetering on the brink of the quarry. We had worried about leaving him with Dorothy whilst we went on holiday but, as Father used to predict, something turned up. Two things in fact. The renting of George's fields meant we could prevent Samson from straying out of sight. Secondly, as Dorothy had noticed, Friend adopted Samson so that for two years the caring of Samson could be left entirely to him.

This extraordinary Friesian bullock grazed behind Samson all the time and followed him even though the rest of the herd went elsewhere. We have always known that cattle form friendships, often connected with breed and colour. The Charolais will be together, predominantly white ones often pair. Those calves bought from the same farm, with almost identical eartags, follow each other through the crush. Margaret takes pains to sell those she knows are dependent on each other, in the same batch, at the auction.

'I want those two to go into the ring together,' she will instruct the drovers.

The friendship enjoyed by the blind animal was more than a normal one for it depended totally on the sighted one. Though the herd went a mile away, Friend stayed with Samson and was more help to us than any human employee would have been. Animals are equally as interesting as people.

Chapter Nine

There's nowt s'queer as fowk.

Yorkshire proverb

Our guests fall into four categories. The outstanding, the normal, the dippy and the dishonest. We reckon 85% are normal, happy people who add to the friendly atmosphere. They come and go like ships passing in the night. They like it here, they like traditional food, they keep bedrooms tidy and make their own beds which is an enormous help. I have never understood why these considerate people have never trained the generation that followed them to be similarly courteous in other people's houses. Margaret remarked one day, 'I think we'll advertise "No children between the ages of eighteen and thirty".' I am often appalled at the mess in which some leave their bedrooms.

Whilst they are with us, these 85% of normal people are part of our family and when they go, they go. They are happy encounters, here today and gone tomorrow. Only if they return again and again do their faces become familiar and their Christian names make any impression. We are eternally grateful to them for being so nice. Those, and there are many, who return expect us to know which room they used last time, what the weather was like and who were the other guests at the time. They also expect us to remember the extraordinary incident which happened during their stay and they think they know what sort of people we are and the kind of life we lead and, as we say in Yorkshire, 'They know nowt o' t' sort!' They think the major part of our lives is entertaining guests but my scribblings prove otherwise.

A handful of people, 14.5%, make such a lasting impression they only need to come once and if they come more than that it is a bonus. Into this category fall most of the handicapped people. It is not possible to forget Christine who sends a card from Australia having travelled there alone and who is wheel-chair bound, or Joe whose wheel-chair has been on many marathons, who sends a postcard from Japan and, when asked by a press photographer what his problem was could only say, 'I've been made redundant!'

Exceptions are not confined to the disabled by any means. A passing American who described England as 'One big park' came only once. A German lady who told us a human response was only an echo, a group of Chinese gentlemen who bowed before us all the time, our Belgian Mr Van whose wisdom and charm captivated us for many years and the jovial vicar whose one week stay made the rest of the house rock with laughter and fun.

It is sad but inevitable that we remember the 0.49% of dippy people. Because so many people come it is to be expected that a certain number will be 'round the bend'. One doesn't book accommodation at a farmhouse and leave because there is a dog, or because the house is not in a village, and be 100% normal. Life they say is stranger than fiction! We had an exorcist who believed we had a ghost, a clairvoyant who had been talking to someone locally who had been murdered and a girl who believed she was the wife of the singer David Essex, using his name with the Social Services. We were so afraid, during her stay, that Margaret slept in a sleeping bag in the dining-room in case she set fire to the house. There was a 'healer' in the Loft Cottage who was found sleeping under his duvet in front of the inglenook fireplace because of the power he felt there. A man looking uncannily like Dirk Bogarde sat at the dining-room table planning his route on an outspread map down which he had drawn a line over which he dare not travel. All add to one's experience of the diversity of the human race and all seem to find The Currer a welcoming place.

The 0.01% that are dishonest we don't forget either. We had one man who came for a few days and disappeared leaving his case. The police tried to trace him for his wife was looking for him but they would not relieve us of the case which, they said, was not lost. The man knew where it was! So we kept it a year having taken it to the Five Acre and made sure there was no bomb inside it.

It is impossible to detect the dishonest. They are all so plausible. One young workman, a shopfitter he described himself, became very popular with the current guests who took him to the pub at night and treated him to a beer. We were less attracted to the extrovert but did not detect a criminal. We remember most that he climbed over a wall into a bed of nettles and went berserk. We were gathering dock leaves in hysterics.

He walked home from work at the end of the week saying he had had all his expensive tools stolen. He said he had been to the police and reported the theft and he won everyone's sympathy. Then he did not come back at nightfall and was brought to collect his luggage next morning by the police who told us we had been housing the biggest con-man in the area. We were told to make sure we had our cheque books and credit cards. As far as we could make out he had stolen nothing from us or our guests and he had paid me at least some of what he owed!

Occasionally, some pay nothing at all. Early in our adventure into holiday cottages a man occupied one for the three months of winter and never paid a penny. He said he was a freelance artist working for the local Metropolitan Council on a new guide-book and was awaiting payment from them and from the publisher for whom he had done artwork. He was embarrassed at not being able to pay until he was reimbursed. He frequently phoned 'a publisher' asking for his fee to be paid because he could not pay his rent. He acted as if he really was frantic about it and I was careful not to press for payment fearing he might have a nervous breakdown or commit suicide or something. Now many years later I think we were just taken in. He left when Easter came, signing an IOU and leaving an address. I wrote once and he answered saying he would pay shortly but he never did.

After long experience we are no more skilled at detecting the dishonest. We put into this category also those who book and never turn up and those who book and cancel at the last minute and will not pay. Like the wedding party of nine who phoned at 9 p.m. to say their car had broken down and they were still at home and we were left with nine empty beds.

Most of the people who do not pay owe very little. There was the workman who booked one night and was so profuse in his commendation of our service he said he would stay another night and sample the evening meal. It was all 'my eye and Peggy Martin'. He never came back. It is our practice to present the bill when guests finally leave.

Apart from the freelance artist who owed us a considerable sum after three months, non-payments have mostly been for small amounts until the June of my return from holiday with the shingles. That really took the biscuit.

An upholsterer, working in a unit in an industrial complex, asked if we could accommodate his small work-force for three weeks. It consisted of his wife and five other nice Geordie ladies, in pinafores. No problem! They would want an evening meal after which they might occasionally return to work overtime. Poor souls, they were working as hard as we were.

The employer begged that the three men, who were sleeping in the workshop, could have evening meals too. We don't do that normally, but it was an exceptionally good booking and I could not refuse. He paid by cheque at the end of the week and all the ladies came back on the Monday. At the end of the second week the man was in Manchester, his wife said, picking up more rolls of material and he would pay on his return.

'That's alright,' I said. 'Have a rest this weekend. You all work too hard!'

Next morning, Saturday, their cheque was returned from the bank. I did not worry. They were coming back for their final week. And so they did, and the wife was apologetic. Their cheque book had been stolen from the office and they had advised the bank to stop all cheques.

'We'll pay you by cash,' the lady said. 'I'm going back north to the bank to sort it out and we'll pay cash.'

She returned next day and said all was well and we would be paid on Thursday night before they left on Friday. Her husband phoned 'from Manchester' to say he was delayed and would come up on Friday morning to pay me the £800 plus he owed.

He didn't, so I phoned the unit and his wife reassured me he was already on his way from Manchester and would come as soon as he arrived, but he didn't and when I phoned on Saturday only the daughter was there and she said it was nothing to do with her!

On Monday I was beginning to be suspicious. I phoned the owner of the industrial complex only to be told the team of upholsterers had done a moonlight flit and owed him far more than they owed me.

As I put down the phone a van pulled into the yard driven by an irate young man demanding to see those he believed were staying with us. He had been working for them for six weeks and had gone to work that morning and found an empty shell.

'They've taken all the hired equipment,' he stormed. 'They've gone off in a hired van and they haven't paid any of their local employees. I haven't had a penny for six weeks.'

The police said non-payment of debts was a civil offence and they could do nothing about it. We just had to forget the whole experience and accept that such might happen again. We even laughed. What else could we do?

Anger must not last long for there is always something pleasant just around the corner. Denis was coming!

I have not accounted for the sum total of our guests if I neglect to say that Denis is in a category on his own and is an important part of the whole. Anger at being cheated was over-shadowed by the joy of his arrival. He stayed for three weeks prior to his year in Russia. He was only fifteen and we hoped the American organisation which was sponsoring him, and accepting responsibility, was reliable.

No-one could describe Denis as a normal teenager. His wisdom and mature conversation put him well out of any category except maybe the 14.5% who are different but we deem him even more unique than that simply because of his age. He is definitely not normal, certainly far from being round the bend and he seems to be uninterested in money.

There are very rare occasions when we overbook, when Harry has to use our room and, less frequently still when all three of us have to camp out in the front room. In the early days of our venture into tourism, when the overdraft was not only red, it was positively luminous, we snatched any opportunity to overbook, but Mother's illness and Harry's sleeplessness made us more cautious. One night during Denis's stay we had to vacate both Harry's and our bedroom and Denis was already behind the curtain in the sitting-room. We dare not tell him we had nowhere to sleep so, when everyone had gone to bed, we put up a camp bed in the new kitchen. Both its doors could be locked and privacy and secrecy were assured. Margaret and I slept on the unfinished floor and the smell of the new wood and paint would have kept anyone other than us awake. The tendency to laugh threatened us, at one point, but we conquered it and even Harry managed to sleep.

Denis finished painting the new kitchen and helped bring in the hay we made in George's meadows. We hadn't made hay since Father died in 1979. It was scorchingly hot and we felt as if the old days were back. The Lathams were here on holiday and Denis got on like a house on fire with the near octogenarian. He called Ron the 'moustache man' and it was several years before Denis admitted why. 'I could have called him Right but I couldn't call him Wrong, I couldn't.'

We led the hay into the new barn on the trailer of the ATV, Denis driving. We were apprehensive about his year in Russia and he, too was all of a doo-dah. He searched the Spastic shops for a navy sweater to keep him warm in the winter but could not find one, so Auntie Mary said she would knit one and Grace Latham and a guest we always call Mrs Weak Tea, borrowed knitting needles and clicked away. I even did some myself.

Denis went to Leeds to get stamps for the foreign collection of the boy with whom he would be staying in Russia and he came home with a plastic tool box and a selection of essential screwdrivers, pincers and etc. He was was so pleased with the genuine pleasure we showed when accepting his gift. 'Now I know what to buy for Currer Ladies. Not chocolates, or flowers or wine. Just tools.' He had us weighed up all right!

Whilst he had been in Leeds I had tacked the jumper together, anxious to see if it had any hope of fitting the boy. So many people had had a hand in the making of it. I confessed my fears and Denis said they were a mere nothing to his nervousness. Whatever would he have done if the garment had been too small or too large?

He caught Margaret and me arguing one day, about the crazy, lunatic way we have to start each holiday. We were not even in the same room, Margaret being in the kitchen and I dusting the sitting-room.

'What's wrong?' we were asking ourselves. 'Do we have to have this charade every time we go?' Gone are the days when the journey did not start in such a panic.

Denis thought we were arguing and asked if anything was wrong.

'Not at all,' we said. 'We're just trying to solve a problem so far away from each other we are having to shout.'

Denis said, 'My father and grandfather and my uncles were all shouting in my uncle's garage and I asked my father if there was something wrong. He said 'Denis, when there is silence, then worry'.'

Too true!

* * * * *

It is surprising how many people come year after year and do not know we are predominantly farmers. We have hundreds of square yards of stock-rearing sheds and hay barn and yet many never see our cattle. One-hundred-and-fifty are penned every winter. Few go to see them. Because they are all well fed and content they do not make their presence known in the same way as donkeys do and so guests think we have just a few farmyard animals: goats, donkeys and poultry.

In the summer they scan the horizon for Jasper and Chocolate and bring apples and carrots and leave them with us because the donkeys keep their distance. To feed them and the geese and goats is comfortable for they are pets.

The herd which grazes the pastures and gardens the landscape so beautifully, embarrasses them because they think we keep cattle for food and profit. They do not know that as calves they were born so that their mothers would give milk, and butter and cheese and yogurt. We nurse them and love them and give them two happy years on our hill. I'm sure many of our guests feel it morally safe to ignore them.

Coming home from Scotland one year, pausing at Inverary we acknowledged a man's smile as he and we leaned over the promenade railings.

'It's a beautiful world,' said he for the sun was lighting up the surrounding hillside and sparkling on a dancing sea. Then, as an afterthought he said, 'And isn't man a funny animal?'

Whatever our problems other people seem to have more. A booked-long-in-advance cottage, on several occasions, has been cancelled because of today's malady, separation and divorce. A couple were having problems on holiday and the wife walked out on her husband one mid-week morning, taking baby and daughter only to return sheepishly at nightfall. The man looked relieved but on the Friday she disappeared and did not come home. Because of her earlier action we presumed she had caught a train home and left him with the luggage and children to travel back next day. He seemed to think this would be the case, but at 6 a.m. next morning a taxi drew into the yard and a very wet and frightened lady came to borrow payment.

'Wherever have you been?' I wanted to know. She was soaking and blue with cold.

It transpired she had walked away from an argument. It is not easy to resolve your differences in a happy holiday atmosphere. She had thought a walk in the fresh air would do her good. In her agitation she had walked blindly and when dusk was falling she had lost her way. She had panicked on the moor, stumbled and fallen again and again and when it was too dark she had curled up in the damp heather and waited for the early dawn. She had been petrified.

As soon as she could see she had struggled to civilisation, found a call box and hailed a taxi. Fortunately she did not need to tell anyone. She was warm and comforted by the time the other guests got up for breakfast. Had she broken her leg in stumbling, we would not have called a search party for her husband would have expected either to find her at home on Saturday evening or with parents or friends. None of us would have searched on the moor.

Nevertheless the summer passed smoothly and we had nothing but gratitude. Our Guardian Angel 'on high with nothing to do' had looked after us well and snatched us out of danger whenever it dogged our heels. Every near disaster had been followed by a miracle so we claimed success, for such after all, is only the ability to avoid failure.

* * * * *

We scoured the Yellow Pages for a firm selling stainless steel canteen equipment and found one at Thorpe Arch, Wetherby, that sold reconditioned sinks and cookers. We took a day off and found just what we wanted for the new kitchen. We bought twin canteen sinks, an enormous cooker, a grill and ordered a made-to-measure steel table top. We were incredibly pleased with ourselves.

Some years earlier, someone had dumped stone sills and lintels on Altar Lane. They had been lying there so long they had become partially buried in mud and grass. We had kept our eye on them saying that, one day, we would get Brian to replace the modern window in our wash kitchen with mullions. The double-glazed expanse of glass had looked wrong from the minute we had it installed. Brian found mullions in a builder's yard and we dug out those sills on Altar Lane.

We had them installed whilst the men were working on the new kitchen. When that was finished and decorated we proudly showed it to the guests. We thought it was beautiful with its gleaming silver surfaces but our guests curled their noses up in disgust and preferred the homely farmhouse one. To please the Government we had spent £7,000 and we had disappointed all our guests who come solely to enjoy the hospitality of our comfortable kitchen.

'You can come in the old one, now,' we explained. 'If we hadn't built the new one we'd have had to keep you all out of here!'

'You wouldn't have succeeded,' they all laughed, continuing to let in our dogs, put carrots on our dresser for the donkeys, take washing from the line and strew it on the table and hang wet clothes over the Aga.

So everyone was satisfied and really that new kitchen was the best spent money, ever, even though it was Mr Barclay's. The cattle belonged to him too. What a pity!

* * * * *

At the end of August it was Harry's sixty-fifth birthday. How proud Father would have been! He had died from cancer fourteen years earlier and, until his terminal illness had prevented him four years before he died, Father had shaved and bathed Harry. In his hospital bed, after the first operation, he had worried about what would happen to his son. Margaret had said, 'I can shave him, Dad,' and has done so almost every day since.

I had said, 'I can cope. He won't have to miss his bath,' and Father had cried. If he is the Guardian Angel, as Margaret believes, who is responsible for our well-being, he would be proudly hovering over Harry on his sixty-fifth birthday.

I wish the two grandmothers could have lived to see Harry become a pensioner. Grandmother Brown was the midwife who assisted the family doctor with the difficult birth. The baby was so weak, so blue with lack of oxygen, she had cradled it at the back of the house, alone, waiting for it to die. But the determined human baby was not going to do so. Mother and her mother, Grandmother Smith, had slept in the big double bed with the weakling child on a pillow between them, praying for its survival and Father had had to sleep in the spare room. One night Grandmother Smith had said to Mother, 'I'll keep watch. You go to Herbert.' Mother had crept in beside him and Father had cried.

The strong, pillar of the family that he was, tears came when he was moved. Struggle, work, weather, difficulty, poverty, illness, none of these made Father weep but his family could bring tears of happiness.

Well, Harry's had been a success story! He had been born too soon to have the benefit of state aid, education and speech therapy but had had a far more exciting life than most able-bodied people because of parents, The Currer and his two crazy sisters.

Those born now with cerebral palsy have outside help and access to new inventions. When Harry was born even the wheel-chair was a clumsy innovation but what he lacked was minimal compared to the advantages the average family unit of the 1920s and 30s provided.

A local genius currently makes walking frames for brain-damaged children who cannot walk and has had a phenomenal success training them to do so. Parents bringing children for assessment, and later to collect their specially built frames, frequently stay overnight at The Currer and nothing gives us more pleasure than to introduce Harry and relate his success story. It gives them hope that their child will walk and talk and live a long and full life, too.

Tommy and Veronica McGinty were staying with us the first time we unexpectedly housed children with appointments at David Hart's. Tommy had been born with a far from perfect pair of legs, powerful arms and shoulders, a brain that could interpret anything and a wit which so outstrips my own I dare not compete. In spite of the extreme difficulty to do so Tommy labours upstairs, scrambles over the rocks at The Altar, crawls under and repairs any vehicle, manipulates any physical operation and has even been known to climb a ladder and repair a roof. Distance defeats him so, for that, he uses a wheel-chair or an electric buggy.

He and his wife Veronica, occupied a double room the week the first children came to keep an appointment with David Hart. The house was full to capacity and all were at the breakfast table when a minibus pulled into the yard with three hopeful mothers and three handicapped children. The driver came to the door and said she had driven from the south for their afternoon appointments, after which they would be very tired and would need a night's accommodation before the long journey home.

Guests left their breakfast table to join the unexpected conversation. I said we were absolutely full, that Margaret and I were already sleeping on the floor. Tommy and Veronica said they would vacate their bedroom. If Denis could sleep behind the curtain, they declared, so could they. Their double room had a single in it. 'Please,' said the mothers, 'just put a mattress on the floor for the children.' The atmosphere in the house that night was wonderful and the story of the McGintys behind the curtain is a legend.

Mother had no frame to teach Harry to walk but she had what mothers do not always have today. She had a husband, other children, four grandparents living near and very supportive sisters. It is hard if you are a working, single mother. Children do not learn to talk by watching television or to walk by riding in a car. Harry was walking by the time he was five and has remained on his feet ever since.

Coming at the end of August, Harry's birthday is cause for great celebration among the guests who buy him presents and cards and generally make the fuss of him he has not been used to, having no friends of his own outside his immediate family. The bed and breakfast has done this for Harry. Certainly he was picking up after Mother's death. The lovely birthday, the swimming, Jess, Denis, the sorting out of his medication all resulted in better sleeping habits and less measuring of his length on the floor.

If these things were destined to take Harry happily into retirement the New Kitchen was our best hope yet of a possible old age. Self-catering in the bed and breakfast was now a possibility. The Guide Association could cater for themselves. So could church groups, universities and National Trust volunteers. Suddenly we felt we could cope with old age which will come in the dim and distant future.

The cattle sales were exceptionally good. We were selling cattle we had bought as calves costing £24 compared with the previous £230. Of course we had had to rear them and milk powder is not cheap and they eat an enormous amount of corn in the first three months, but the signs were all that we had made a profit and reduced the overdraft.

* * * * *

Pat and Alf were here in September when the phone rang at 11 p.m. A neighbour who lives on one of the two farms at the Altar had been returning up the passable half of Altar Lane between the farms and Bingley, a mile away, and he had seen our two donkeys. The stoney track leads straight into the A650 so we had no alternative but to go and attempt to drive them back. There was no guarantee that we could do so. Our donkeys are not broken to the halter and are a law unto themselves. However, if they could get out so could the hundred-and-forty-strong herd and that would be a disaster.

Donkeys are nocturnal. We know this for they make an infernal din in the middle of the night and waken our guests. As further proof that the man-in-the-street is tolerant of donkeys, guests never grumble but should cattle bellow in the middle of the night someone put it in the Visitors' Book: 'Please give your your cows sleeping pills.' Cattle are not nocturnal and only bellow if they have just been bought at market and feel strange in their new environment. They will not wander in darkness. They lie together, close enough to touch each other, until daylight. Then they will stray if the opportunity presents itself. At all costs we had to ascertain whether or not they were already abroad and, if not, we had to find the breach in the boundary and close it. If we could get back the two donkeys that would be a blessing. If not, it would not be a disaster.

Pat and Alf insisted on coming up onto Altar lane with us. It was impossibly dangerous underfoot. We took the Range Rover as far as the quagmire on the road. Then we left it with Harry securely wrapped in a blanket and the four of us attempted to traverse the lakes and gulleys into which I had so wanted to push the immaculate young solicitor. Didn't I just wish I had him now, with an emergency on our hands. Suppose the cattle were out? One-hundred-and-forty of them would gallop up this morass and we had guests who might well be trampled under a stampede. God help us, I thought, sliding all over the place. We knew the road! Pat and Alf didn't!

It was with relief we looked over the wall and saw our recumbent herd, heads raised to the stars, placidly ruminating. Now we only had to find and repair the breach in our new, expensive wall. That, too, was easier than expected. Someone using a motor-cycle had lifted off the new Ramblers' gate and put it on one side to cycle round the pasture, out of sight of the farmhouse. Having trespassed, he had not replaced the gate. His wheel marks were clearly visible. We decided to leave our friends there guarding the gap whilst we went in search of the donkeys but we were only a short way down the road when we heard Jasper bray. Chocolate was never so competent vocally. We stood transfixed listening to the unmistakable, beautiful sound from below. The donkeys were at home. Right now I cannot think of any other animal which actually tells rescuers where it is. Certainly we have never had a cow or bullock which would stray and then go home and announce it had done so!

The donkeys must have been turned in their midnight tracks by the headlights of our neighbour's car. They must have walked to the open gate and, I believe they must have galloped downhill, for Jasper, apparently out of breath, only brayed once. It was enough!

We lifted the gate back on its hinges resolving to come next day and fix it so that no-one, not even us, would be able to lift it off again. Also, on the following day we had another shouting match with The Council. Letters were ploughing back and forth between The Country Landowners'

Association and The Council and we had not interfered all summer. We were hopping mad about the motor-cyclist and the gate and broke our silence. It did not have any effect.

* * * * *

It had occurred to us that we might kill two birds with one stone if we took a holiday in Wales and called at Beeston Castle on our way home. What could be simpler? If we went to the North Wales coast we might be able to leave a holiday cottage early and get to the market by 8.30 a.m., without undue stress. So we booked a terrace house on the West Shore of Llandudno and had a quiet, restful holiday there, partly because we had no wish to use or cross the fast A55. Therefore we stayed seaward of it most of the time and only ventured once into Snowdonia.

The West Shore was deserted at the end of October and we found ourselves pushing the wheel-chair back and forth along the promenade every morning, and driving towards Colwyn Bay every afternoon. It was not our normal kind of holiday at all but it was more restful, driving-wise, than our autumn holidays had previously been. The town house we rented was warm and comfortable and even though the steps were steep and the toilet upstairs we saw no real reason why we might not return the next year if getting to the market in time was possible.

A phone call informed us that someone had pulled down, to the ground, about five metres of our new wall and paved an area of the morass. Finding it, Dorothy and George had been horrified and had struggled to the boundary with fencing posts and barbed wire. Were we fed up of that road and of The Council that owned it!

In a little second-hand bookshop in Conwy I bought a paperback, an autobiography of a woman who had bought a Welsh hill farm. It was a pleasant, readable book. I enjoyed it and we even found where the lady lived on our one venture into the hills.

There are many of our friends and guests who think we fall into the category of smallholders. I do not really care what status people confer on me but Margaret is a qualified, professional farmer competing comfortably with others in the Dale. Dairy farmers, milking a hundred head, come here on holiday from Cornwall and Wales and Cumbria and consider Margaret as an equal. The older vets in the practice were in their twenties when she was and recognise that she has had experience and knows what she is talking about.

Probationer vets are far too cocky. One came in the spring to an ailing calf, one of seventy robust ones. In the fields bordering the lane down which he must have driven blindly, were close on a hundred sturdy yearlings.

He administered the antibiotic and left Margaret enough for another two doses. Walking beside me, back to his car, he had the arrogance to say, 'Make sure she gives all three injections even if the calf appears better tomorrow.'

'Margaret knows that. She doesn't cut costs where sick calves are concerned,' I retorted.

'Probably not,' he said. 'She is more of a hobby farmer.'

Reading that interesting book in the little town house I again thought I ought to do something positive about the seven hundred typed pages of my own autobiography.

But it was calf buying time.

We rose very early on the last Friday morning and drove in darkness nearly all the way to Beeston Castle on that fast highway. Fortunately it was fine. Lorries passed us all the way and we were very tense. We were just as screwed up as we had been the first time we drove down to the market in 1990. We only needed a small mishap, a puncture, a wrong turning, or a traffic hold-up and we were scuppered.

The fast road was already busy. Margaret was driving. We were all praying for the dawn. It came but not before we were nearly there and we were nervous wrecks. We sincerely wished we had done the sixty mile journey the night before and stayed in a Travelodge and agreed that next time we would, however great the temptation was to cut costs. Harry and Auntie Mary were sleepy. 'They are too old for this lark,' we thought.

The buying went well. Inflation in market prices meant we had to pay an average of £76; twice the amount of the previous year. There had been no better time than 1990 to re-enter calf rearing madness. We still had problems with the new shed. All was reasonably well providing the doors and east windows could be left wide open. The prevailing wind is from the west and cannot enter. Faced with problems one is entirely dependent on clean air coming in as well as an efficient antibiotic.

Periodically a new one has to be found because the bugs become immune and it can be very serious if the remedy does not work. We had a good antibiotic that year and did not lose one of the calves though temperatures were common and one or two lungs were damaged with pneumonia. We had one calf we called Poorly. After a struggle we began to call it Poorly-getting-better, and the title stuck. In summer, focusing binoculars on the slopes, Margaret would say with pleasure, 'That's Poorly-getting-better eating away like nobody's business.'

* * * * *

Immediately on our return from Llandudno we reported the desecration of the wall and took photographs. We sent for the press and the evening paper made a feature of it and took a snap of me in dirty wellingtons and all the stones from the gap, paving the mire. Nothing positive came of that little episode either!

We had a Guider weekend and the opportunity for them to self-cater. It was wonderful. Then, shortly after breakfast on the Sunday morning there was an electricity cut. The Guiders could cope. The Aga is a blessing but we had new people coming into the Loft Cottage which is all electric.

When the Guiders left I kept their dining-room log fire blazing and when the new family arrived I entertained them in the warmth. We had to light candles and I made them tea and provided crayons and paper for the children. 'As soon as the power comes we'll turn on every heater we have in the cottage,' I said.

The husband carried cases into the Loft and when at 5 p.m. the lights came on, he and I lit all the cottage heaters. 'Don't go up until it's really warm,' I advised, switching on the television in the dining-room.

Those occupying the Loft, park their car behind the house. We did not see the family the next day. When no light came on at dusk we presumed they were out enjoying their holiday. We changed our deduction to 'they must be visiting family' when eight o'clock came and there was still no light in the cottage. Quite late at night, when everything was still in darkness I went and let myself in. It was not only empty, it had never been used. The man must have walked straight out with the luggage. The two quite boisterous children had not even creased a bedspread or moved a cushion. We could not understand why we had not noticed it was in darkness the night before until we realised the meter would not have emptied and turned out heaters and lights until late.

They must have decided a Yorkshire farmhouse in November was not the place for them. Nor for a young Asian workman who booked bed and breakfast for a period and arrived, in the dark, in a cold sweat.

'Where are all the street lights?' he gasped. 'I can't stay here. I'd be petrified!' The yard light was on and the sheds were illuminated and all our lighted rooms uncurtained. We could not believe what he was saying. He was a charming man. He decided to stay one night because, I suspect, he dare not drive up the road again in the dark.

Next morning he thanked us profusely, paid for his night's accommodation and fled. It is difficult for us to relate to the city dweller who finds our semi-isolation unbearable. It is because we are only two miles from town that guests take a risk, book in and then panic because our farm road is a quarter-of-a-mile long. That we cannot see the houses of our neighbours is surely the attraction and that we can see the ribbon development across the valley is a pity. Amazingly there are many whose eyes open in fear.

The production team for a television series, Roland Rat, booked accommodation whilst they filmed at East Riddlesden Hall, our nearest National Trust property. It was winter and they found us almost impossible to find. After several phone calls and wrong interpretations of our directions they finally arrived.

To enter our dining-room in winter is to be greeted with a blazing log fire and steaming cups of tea or coffee. I am sure our hospitality left nothing to be desired. They were ill-tempered with the frustration of not being able to find us but I assumed our graciousness and the log fire would mellow them.

I re-lit the fire early next morning and was putting out the bread and the perishables on the table when the leader of the group came to cancel the booking.

'It's far too bleak here,' she said.

Weeks and months and years pass with no dissatisfied guests and only a few like it here but offend us so much we say to ourselves, 'Never again book in those.'

Cambridge University booked the Loft Cottage for a fortnight in mid-December. They were hang-gliding on Baildon Moor and caused us no trouble at all. The ground floor cottage was not to be occupied until Christmas so I had washed all the counterpanes and chair covers and everything was ready for the festive season. Then students from a different college phoned to ask for a week's use of it. I was reluctant to take the booking but the students in the Loft were well behaved and, after all, we have to keep the bank manager at bay. I stressed I had just washed everything for Christmas and let them come on condition that it was returned to me in the same condition when they left, a few days before Christmas.

The student caller was adamant that there would be no problem for most of the party were Christians! They knew a deaconess up the Dale! One of them carried the cross in church!

About twelve students arrived and I must say they were far from noisy. I presumed they were tired after a term of hard academic work. They did not draw back their curtains until noon and they did not go out until Thursday afternoon. The leader called pleasantly to say they were going to see the deaconess and I believed them. They were nice young people. That was not the problem.

In their absence I went into the cottage for something I had to check. The electric, I think. I was so shocked by what I found there, I probably forgot why I had originally entered. The chaos within was a hundred times worse than anything I had seen. Ever! In twelve years we had never seen our self-catering cottages used so disgustingly. Sleeping bags, sheets, blankets and personal underwear littered the floor of every room, including the sitting-room. It was impossible to walk to the meter without stepping on the confusion. The bathroom was wet and dropped towels were soaking up the moisture. The state of the kitchen was unbelievable. Everything had been left used and unwashed. Unscraped plates and cutlery were piled precariously. Spaghetti floated on the cold water in the sink bowl. A 4 lb tub of soft margarine lay half empty, messily used and full of crumbs. A sliced loaf had scattered itself on the table and a cold joint of beef sat amongst the clutter.

They are university students and not hippies, I thought, and I am a headmistress not a dustbinman and it is time they knew.

I gave them five minutes on their return and then I knocked smartly on the door and asked if I could come in. I did not admit to having been in before. What had Bernard told Denis? When there is silence, then you worry. I think my silence worried them. I walked dramatically round every room, my eyes widening, my feet gingerly stepping over the rubble. One girl thought it wise to disappear into the bathroom but I forestalled her entering first, picking up the soaking towels and dropping them into the sink.

I'm sure my silence worried them but I don't think it prepared them for my final, loud outburst when it came.

'How dare you?' I shouted. 'How dare you?' and I followed that with the biggest dressing down I'll warrant they had ever had.

'We were going to tidy up before we left,' they said.

'Huh,' I exploded in disgust. 'Everything here will have to be re-washed for Christmas!' I stormed out announcing I would be back at seven and I expected them to have cleaned up the mess.

At 6.30 p.m. they came to say they were ready for me. Order had been wrought out of chaos. Everyone, every child, has a fundamental right to be disciplined. To deprive is cruel.

To re-establish good relations I took them a newly baked apple pie for their late tea. They were sullen and received it less than graciously and I said to Margaret, 'That's the last time we'll take a booking that shade of blue.'

* * * * *

We had a lovely, different Christmas made possible by the new kitchen. The Wilson family, three generations of them living as far apart as Kent and Dumfries, met together to spend a proper Christmas. It made our experience easy but traditional. There was a warmth and friendliness everywhere. Hilda, sleeping Christmas Eve in the spare bedroom, could hear the goings on in the Snug next morning and reported that Santa had come down the inglenook chimney. I think it may have been his first opportunity for centuries. The Wilson family were having a wonderful time and the responsibility was not ours.

Feeding cattle was, but we completed the chore as quickly as we could before driving out to Malham for our Christmas picnic. It was a day glistening with hoar frost. Some angel had taken a sack of glitter and sprinkled it all the way up Airedale from The Currer to the Cove.

The car park at Malham was almost empty. Christmas morning is a good time to see the Dales for very few drive on the B roads. Those going somewhere make haste to the nearest highway but we, who are not going anywhere in particular, meander along. Increasingly incapable songwise, we continue to carol for, though the way we celebrate is less traditional, the anniversary remains the same.

Chapter Ten

Its heather-clad bens delight my soul,
I love its dear shores where the waves always roll,
Its sands are of silver, its grass is of green,
An island like Harris I never have seen.

Harris Song

It is necessary to mention that the first three months of 1993 were not the harbingers of the nine that followed. Try as might I cannot recall any rocking of the boat when the new year came in which is why we should have worried. Margaret is worried of calm but Harris was on the agenda for spring. We were relaxed.

On the first Saturday in March, as always, we held our Guide Reunion. The event fell on a weekend when the house was full of Brownies and old and young gathered in the Snug for camp-fire.

Leading used to be my job. In spite of my vocal deficiency my repertoire was inexhaustible. However, Sandra was home on a visit from Canada where she must have lived these past twenty years. She was one of my Guides and a Sea Ranger and she is a Guider in Canada, so she led the camp-fire. It was lovely. We had such a wonderful evening. The Brownies were fascinated by all these grown women who knew all the camp-fire songs and how to enjoy themselves so well. We had a lot of weekend Guides and Brownies before their own adventure centre, WYNCHES, was built. We still have some but not as many. Though we supported the building of the County Centre we knew we would be the losers. It gives us great joy to hear grace being sung in the dining-room. No other group sings before food. If we have a church group someone mumbles, 'For what we are about to receive,' but Guides sing, 'Oh the, Lord is good to me,' or 'Oh God, the giver, take the thanks we give,' or 'God has created a new day, silver, green and gold. Live that the sunset may find you worthy of His gifts to hold.'

Instead of Brownies we began to get Op Skivvy Off. Many years ago Guiders, grateful for the continuing use of Steeton Hall for Camps and Winter Trainings, offered help with the spring-cleaning of the ancestral home. They named themselves Operation Skivvy and for scores of years the same group came together, twice a year, to clean the Hall after the training season. They became friends and had an annual Weekend Off. It included the owner of the Hall, Miss Dorothy Clough, and had become known as Op Skivvy Off. Miss Clough died in 1981 and the Hall was sold for an hotel but the, now old, members of Op Skivvy still have an Off together. We loved their weekends here. Their graces were older and I remember them nostalgically from my long association with Steeton Hall and its owner, Cloughie. Op Skivvy Off sing the 'Wayfarer's Grace' and, 'If we have earned the right to eat this bread, happy indeed are we. But if unmerited Thou givest to us, may we more faithful be.' I like it.

All nostalgia gives me pleasure. I must have had a lovely life! It was spring and Harris was calling us urgently and May would not come quickly enough. Our peace was shattered before it did for in March the new EEC regulations came into being and farmers received their IACS form to fill in.

For decades cattle have carried eartags in the right ear. Each has a long and complicated number on both sides of the metal clip. It has long been necessary to record these numbers in a Herd Book. As each animal grows older it becomes increasingly difficult to read the letters and digits for wax and dirt obliterate them.

The spring of 1993 was alive with well founded rumours that all male cattle would be given an identification document and a CID, bearing a fourteen digit number and the eartag number. It would have to accompany male animals to the market and would qualify the animal for an EEC subsidy. When such subsidies had been given, many years ago, steers and heifers had been eligible and ears had been punched. It had all been simple.

Rumours were that the new CID procedure would be horrendous. We would have to return the initial Green CID to the Ministry at eight months old. The Authority would retain the document forcing farmers to keep each animal until a Blue CID was sent two months later. All animals dying during that retention period must be reported. Ministry inspectors finding irregularities would disqualify the farmer from receiving a percentage of the total subsidy. At twenty-one months the Blue CID must go to be replaced by a Red one, after another two months retention. Our bullocks would have to be twenty-three months before sale. It gave us just one month between selling and re-buying calves and we had to squeeze our holiday into that. It was possible, providing we sold all our bullocks on one day and were able to identify each eartag so that we could hand over the appropriate CID That was not possible unless we had a secondary, simpler identification and we thought it would be easiest to freeze brand each bullock on its backside.

We bought all the necessary DIY equipment to do this and read the instructions. We were told to practice on one. In a few weeks, it said, we would know whether the application had been strong enough. We hadn't time. It was March and we were letting cattle out for the summer. Secondary marking had to be done at once or not at all.

We decided to have a go and trapped the first large animal in the crush. He was a quiet fellow. We do not have many so benign. The instructions said we had to shave the area, then apply the template, then spray the freeze. It meant every animal would have to be in the crush several minutes and most would object. The one in the crush was a perfect angel. We did every thing just as it said but nothing happened that we could see and there was not a glimmer of a brand weeks later so it was as well we abandoned that idea and bought large coloured eartags and simply marked them with as big a 1.2.3. as possible. We recorded the matching eartag and bravely put one after another seventeen-month-old bullock into the crush. Battered and trampled on we were glad to know we would, in future, secondary eartag only the calves. Margaret could barely move her arms and hands after using the clippers all morning.

A new CID department was set up at the Regional MAFF Headquarters and we have it on record that a neighbour rang and asked for 'The Joke Department' and the telephonist put him through without question. It must also go on record that the staff, at the northern office at least, are the most courteous and helpful people we have to ring and they must be far more bewildered than we are.

Incredibly we received notice that, at the March meeting of the Council Transport Committee, the state of Altar Lane was to be discussed. The Country Landowners' letters were piling up on a Council desk and had to be acknowledged. I arranged to go to the meeting but, at the last minute it was postponed. Here we go again, we thought.

In the meantime we had to do something about this important IACS form. To fill it in was not only obligatory, there was a deadline after which, whatever the excuse, eligibility for subsidies was forfeited. The form had to be filled in 100% correctly and arrive before the 15th of May. Incorrectly submitted forms would also be disqualified. Farmers were advised to have their forms in early and checked well in advance so that there was time for correction.

Maps and Field data were needed. Arable farmers had to buy expensive Ordnance Survey maps which were in short supply. We were able to get our information from the Planning Office but our land was on three maps, two held by one Borough and one by another. Area was in acreage and had to be translated into hectares by decimal multiplication and farmers flocked to Agricultural Valuers and Estate Agents for professional help. We coped alone but it was all very complicated and someone in the *Farmers Weekly* prophesied that farmers would be jumping over the cliff like lemmings.

It was required that every small detail was accurate. A lot of it concerned milk quotas and arable information and there were pages galore. It was easy to miss relevant points in the mass of questions and farmers nearer to the Area Office went in to have their forms checked. Being fifty miles away we decided to take a risk. A small one, we thought, because we sent the form in at the beginning of April. We thought we had plenty of time even though we were leaving for our spring holiday at the end of the month and not returning until 15th May, deadline day.

The end of the month drew near and we had received no acknowledgement of receipt. I phoned. All forms were apparently piled on receipt and all acknowledgments were being sent together, later. The poor demented office girl had no idea whether ours was in the pile. She could, guarantee, however, that we would be contacted early in May.

'That won't do,' I almost wept. 'It's Monday and we go on holiday on Friday and don't come back until the fifteenth!'

The poor girl was sorry. They had thousands of forms dropping through the post each day. Anticipating this chaos we had been warned in the farming press to take photostat copies. The risk of loss was great. We decided the only thing we could do was to go to the area office ourselves. The girl agreed that was the best solution.

Losing one whole day in the week before holiday was a major disaster. We arranged to take Hilda. If we were going fifty miles to Northallerton she might as well have a day out. The postman came at 7 a.m. and brought the awaited postcard telling us our form had been received. It did not say it had been checked and as everything was dependent on it being correct we decided to keep our appointment.

At 9 a.m. the phone rang. It was the over-worked girl in Northallerton to say that they had found our form, because of the urgency caused by our holiday plans, but they had subsequently mislaid it. The postcard had been posted but the form, taken from the pile, had been lost again in the utter chaos the Government directive had caused. The poor girl was distraught. I have only admiration for the office staff. How they kept their sanity I do not know. The form had been on her desk, she said and then it had gone! She'd spent all afternoon looking for it.

'Oh dear,' I said and remembered to follow it with, 'Never mind.' We were all ready to leave anyway and I had a photostat copy. The bedmaking and washing-up could wait. It was a sunny day. We might as well all enjoy an outing. Everyone, and the dogs, piled into the increasingly old Range Rover and we threw in the picnic basket in holiday mood.

Reaching MAFF we noticed a long queue of seemingly tolerant farmers enjoying the opportunity, seldom afforded, of talking to their fellow professionals. We had, I believed, an appointment, so we went to the desk. The girl was sorry but their were no appointments. Everyone must take his turn in the queue. As no-one else was grumbling, how could we?

'How long are we likely to be?' we asked.

'About an hour-and-a-half,' she said.

We told the family to picnic without us and prepared to wait our turn. It's not the waiting. That we can do quite happily. It's the standing. Only then do you discover how many back pains you have. The farmers were talking about anything and everything except the I.A.C.S. form. No-one was complaining. Amazing!

Two hours dragged by as we slowly drew nearer the enormous workshop, littered with desks, where farmers were receiving individual help to check their forms. Having received and lost ours a helpful assistant brought me a new form and watched as I copied from my photostat. He did not hurry me.

I signed and dated it and he took it away to photostat the various pages. The family in the Range Rover had been waiting three hours but the coming and going of the farmers had kept them amused. I need to add that several months later a phone call came from the Ministry.

'Is that Mrs Brown?' the caller asked. We always say, 'Yes.'

'We seem to have two IACS forms from you!' I'm quite sure we did not get twice the subsidy.

The Tuesday fiasco put us terribly late with our preparations for holiday. We had the car serviced but the mechanic omitted to top up the battery with distilled water. Do they normally? I have no idea! Men know these things better than ladies.

* * * * *

We were going to Harris on a Caledonian MacBrayne Hopscotch ticket. We left Ardrossan for Arran on the mid-afternoon boat believing all our troubles were behind us. Our first bed and breakfast was at Pirnmill, on the west coast. We followed the road round the island pausing at Corrie to look at the cottage we had cancelled the year Mother had died.

We found, with several misgivings, that the bed and breakfast we had booked for the night was perched precariously on a hill. The rough track to it was almost vertical and the parking space no more than a postage stamp and barely adequate for our 'tank'. But the accommodation was fine. We liked the lady and we liked the view and in the evening we went down to Flag Point on Machrie Bay and had a nostalgic hour remembering the three weeks we had camped there with Guides and Sea Rangers in 1962, when gales had devastated the camp-site.

It must be terrible to lose one's memory. Its capacity is incredible. It continues to record information ad infinitum. Wouldn't we just have loved, Margaret and I, to walk along the coast to Blackwaterfoot and see if we could find again the field which had been so full of mushrooms. We had picked them knowing there was ample for our fifty-strong company.

We wondered if the lady who had accosted us and taken our mushrooms saying they belonged to the vicar was still there.

The evening was too short and the time at the small guest-house too limited. Next morning we had to catch the 9.30 a.m. ferry from Lochranza to Cloanaig, at the neck of the Mull of Kintyre. Our hostess came to the Range Rover to wave us goodbye. Margaret removed the stone we had put under the wheel for the parking space was almost as vertical as the track. We could not stop chatting to the lady. She was equally as talkative as we were. Suddenly Auntie Mary pointed to the car clock which indicated we had barely enough time to get to the ferry.

Shocked into activity Margaret pressed the starter and got no response. She did so again and again and there was no response at all. The battery was dead. 'Never mind,' she called, engaging third gear. 'Cross your fingers everyone!'

We could not have been in a position of more advantage. Perched on a pinnacle, facing downhill, no traffic, and a good distance before the road junction. Thank God for the house on the hill. Halfway down the track Margaret let in the clutch and the engine responded. Everyone let out inspired breath and we got to the ferry just in time.

We spent the short sail worrying in case the engine would not restart on the other side but it did and because it was such a beautiful day we took an enormous risk and went down to Skipness where we had had two memorable holidays whilst the family was still complete. Smithy Cottage, on the shore, had been the first of a score of holiday cottages we have used.

Infected with nostalgia we continued down memory lane and called on our good friend Vi MacAllister with whom we had first stayed at Salen, on Mull and who now lives at Ardrishaig. We had not seen her since 1958 when she and her family had stayed at The Currer before we had a road and electricity. Her bounteous table was set for our arrival.

'Remember,' we said, 'When you put a couple of mattresses on the living room floor to sleep seven of us when we were back-packing among the islands?'

Growing older has its compensations for there is such a wealth of happiness to remember. So much has happened to us in an ordinary sort of way. Surely other people, too, must have similar pleasure!

We had sensibly parked on a hill in case we had battery problems but we seemed destined to get at least to Oban. There never was a more beautiful day. We arrived at about 5 p.m. and parked outside the toilets and the telephone kiosk. The family used the one and I phoned the AA from the other. 'Help will come in about one hour,' the receptionist said.

Auntie Mary returned to the car ashen. She got into the front seat without speaking. 'Is there something wrong?' we asked. She had been so happy on the journey. Now she looked positively ill. We were so many miles from home if an old lady was poorly. Had we got to Harris we would not have worried so but we had a long isolated route to Mallaig to do that evening. We had a night in berths before sailing to Skye. We had to traverse the island and cross the Minch before we were among friends. She couldn't be ill! She just couldn't!

'I've had a do,' she said. 'In the toilets. I've had a funny shaking do. The floor seemed to be moving.' She was very frightened and so were we but we said, 'It's just the travelling. The "Ladies" was dark after the lovely sunshine.'

Margaret said, 'I've got to go myself. I bet it's dark in there.' She did not need to go inside before she learned the truth. There was a plaque on the adjacent building which said, 'Connel Pumping Station,' and even while she was inside the pump started up again and the public convenience began to shake. It was laughter which shook the Range Rover on her return and put Auntie Mary's fears to rest and it increased immediately for the AA pulled up in front of the kiosk.

'We were told you'd be an hour,' we smiled in pleasure.

'Have you sent for me?' he said. 'My mobile phone won't work.' Amazing!

He discovered the dry battery, put some distilled water in and said we would be wise to buy a new one on Harris. I suppose it was negligence on our part not to check. It is not the big jobs which will eventually kill us for we have them pretty well organised. It's the thousand-and-one little tasks which wear us down. Checking our battery must now be on the list. The AA surely ranks with the dog as being man's best friend.

We drove up to Fort William and took the empty road to Mallaig. We have years of experience of the isolation of the 'Road to the Isles' from the open carriage window of the West Highland Line. That evening we drove, alone, through that wilderness, that indescribably beautiful outback. There was just the ribbon of empty road before us and behind. Just the trees and the mountains, the lochs and somewhere hidden, the red deer.

When light began to fade we had reached Morar and then the darkness was not total for we were beside the sea. Soon we were among the lights of Mallaig and parking on the pier, ready to drive on the next day.

We guided our oldies up the gangway and the purser took us below to our cabins. We were desperate for a cup of tea but our flasks were empty and the cafeteria was closed. I think a lady stewardess recognised us. Once seen with our invasion of children, Margaret and I were not forgotten. Equally they recognise Harry who has been going to the Hebrides for thirty years.

'I'll make you a cup of tea,' she offered and she sat chatting with us in the empty buffet.

I am aware that our adventures are small and insignificant compared with those of some other people but I am writing for the ordinary people who will relate to our life which is made up of millions of funny little episodes. We have a particularly sensitive fuse wire and a rare appreciation of little kindnesses. We do not need television comedy or drama for our lives provide enough.

We slept well on the anchored Skye ferry and awakened to two inches of snow on the boat deck, a white Mallaig and a glistening Skye horizon for the sun was brilliant. Margaret went ashore, early, to start up the engine and drive the Range Rover on board. Watching from the rail I had a few moments of anxiety lest it would not start, but it did.

We had not expected snow after the almost perfect day we had just experienced, from Arran all the way to Mallaig. I had not seen the Highlands blanketed so. Margaret had on her Easter wanderings north, before Father died, but I had not and I was grateful. The Cuillins were sparkling white, and Benn Na Caillach, and the Red Hills. I do not ever want to be deprived of snow. It did not last long, of course, for the brilliant sun soon stole it and only the mountains kept a tablecloth.

The sun was our companion all holiday. We were staying at Scarista Mhòr for the first time having booked too late to secure the Seilebost Ceul-a-Mara. The new cottage was further away from Luskentyre where our roots will always cling, but the wee pied-à-terre belonging to Mary Ann and Donald MacSween proved to be an idyllic place, for a host of reasons. The view, of course is our most basic need. The cottage sitting-room has a large window and we gathered in front of it to appreciate the breathtaking panorama of beach and sea. That, in itself, would have been enough for

Margaret and me but the view also included the road and the MacSween guest house, more crofts and Scarista House itself where it would cost us as much to stay for one day as it did for two weeks in 'The Cloisters'.

Auntie Mary was responsible for christening the otherwise unnamed three-bedroomed dream cottage. She thus named it because in our Dales dialect 'All t' bedrooms were clois t' bathroom an' it was clois t' sitting-room an' t' kitchen an' orl'. So we called our holiday home The Cloisters and, from the big window, Harry missed nothing. The road was full of interesting activity. He saw the school bus and the cars going to church and the tourists of modest means going to MacSween's and those with executive salaries going to Scarista House. He saw Donald moving his sheep, John MacKay's wandering cattle, the post van and the shop and the golfers on the course on the machair.

Our fatigue manifests itself in many different ways. We were so tired when we reached Scarista I, for one, had no fight-back to stand up for myself when an unusual refrigerator defeated me. A very spacious model resides in the kitchen of 'The Cloisters' and I emptied the contents of our cool box into it gratefully. During the week before every holiday, I cook joints of beef and pork and ham. Sometimes I also cook a turkey breast. Originally I kept the cooked meats in the freezer until the moment of our departure when I transferred them to the cool box and we put them on the roof rack. During the two day journey to the Outer Hebrides they gradually defrosted but, each year, they kept fresh in the fridge and lasted almost all our holiday. At 'The Cloisters' I put them into the spacious lower half of the fridge/freezer along with a six pint carton of milk and several blocks of butter and cheese.

Next morning, I found to my horror, that £25 worth of cooked meat had been re-frozen, the butter was as hard as bricks and the milk was solid. The family cruelly pointed out that any fool should have queried the fact that the lower half of the upright unit was all baskets and noticed that it was the smaller, top space which was shelved. It was obvious, they said. You never have baskets in a fridge! Baskets are in freezers! It was inexcusable. Whatever had I been thinking of? They would never have put defrosted things into baskets! They would have known better!

They would not shut up. 'Were they quite defrosted when you put them in?' Margaret wanted to know.

All I could say was that they had been rock solid when we left home two days ago. The cool box was not a freezer!

'Were they properly defrosted?' Margaret persisted. 'It's important. Think!' I couldn't. 'Try to remember.' She would go on at me and all I could think was that you should not re-freeze defrosted food, that I had probably thrown away £25 worth of meat and given myself the job of cooking fresh meat all holiday, a job I avoid like the plague. Fresh meat is not easy to get on the islands, the shop is not near, tinned is not nice. It was a disaster! Cooking was time consuming. Perhaps there was a timer on the cooker! What of that? I had no idea how to use such a modern device. I had only ever used the Aga.

I was too tired to think, too tired to answer back so I let crocodile tears fall. I had a rotten family, ready to kick me when I was down. The tears brought no sympathy but they ignited, rather than dampened that pilot light to humour, we thankfully have never lost. I blew my nose and we all laughed. We asked Kathleen Morrison, the district nurse at Luskentyre, if it was safe to eat re-frozen cooked meat. She did not know but promised to put something suitable on our death certificates. We ate the meat and did not die.

Compared with the corridors and the distances between the rooms in our vast 'mansion' at home, the cottage at Scarista Mhòr was so easy to run and to clean. When Harry was in the bath we could call, 'Are you alright? Are you ready?' from an armchair. At home we have to walk miles and climb stairs every call.

All fortnight we had the most wonderful weather. Each morning, with the dogs, I walked across the machair to the deserted white beach and the turquoise sea and sauntered, barefoot, along the water line. Lusky, having recovered from a heart attack, was able to go everywhere we went. Harris is his homeland, his birthplace and he is a beach dog willing to play ball all day, mischievously dropping it in the sea whenever he can dodge past Margaret. She walked miles with the dogs during the fortnight. Auntie Mary watched the sun go down over Chaipaval and the open sea whilst we pushed Harry in his chariot south towards Northton or north towards Borve. As I have said, 'It's Land's End to John O'Groats as soon as we have nothing else to do.'

* * * * *

Amongst our Hebridean friends we are beginning to feel to be country cousins. Since electricity, the motor car and television, they have caught up with the modern world more so than we have. The EEC is taking them grandly toward the millennium. We love them dearly and the appeal of their humour and their character and their outlook remains the same but we are glad we knew them forty years ago in the same way as we are glad to have grown up in the first half of the century before technology took hold and everything became easy and convenient, shop bought and computerised.

Perhaps because we had not so far to leap as they, we have hardly leapt at all. While they have gone from griddle to microwave, from home-made to pre-packed, from céilidh to video, from legs and hands to petrol and plane we have suffered less of a sea change. Friends, nostalgic for the environment of their childhood, think we have not changed at all, that we are in some sort of time warp. Nothing could be farther from the truth for active people do not stand still but Margaret and I still walk rather than get out the car, we still carry rather than use the ATV. We are addicted to home made food and why not? It is infinitely superior.

A young mother, appreciative of my apple pie whilst eating the dessert at a full table asked, 'Jean, do you make your own pastry?'

'Well, yes,' I replied, amazed to be asked the question.

Another guest at the table said, 'I never make my own. I always buy it frozen from the supermarket.'

A third lady said, rather ashamedly, 'I always buy an apple pie.'

Sadly many Hebrideans now buy apple pie for the novelty is tempting and the supermarket in Tarbert or An Clachan in Leverburgh can provide a much more varied menu than once they enjoyed. When we first started going to the island in 1951 pancakes and floury griddle scones were a staple, delicious diet. They were buttered with fresh butter and piled high with crowdie and cream. Sadly, now only Angus and Katie have a cow in Luskentyre and UHT milk is stacked on the shop shelves. I do not wish to deprive them of the excitement of their new lifestyle but I am glad I knew them before. When something is gained something is always lost.

The bringing of electricity to the islands has been a good thing and we at The Currer, who knew life without it, know the advantages of it more than most people but, when its convenience was brought to us, we lost something of infinite value. In a small way we lost a little of the 'Feel Good Factor' absent in the nineties.

The whole population is seeking it. Government is trying to create it and all are looking for it in terms of wealth and comfort. It is no such thing. A 'Feel Good Factor' comes from being active. More especially from being active in securing one's own, and one's family's survival. Society has lost the opportunity to be responsible for growing and preparing food, for building and providing shelter basically because the purse has become heavier. When it is light, and you cope, you have a 'good feeling' indeed. When it is heavy the knowledge of how to feel good, really good, is lost and then when the purse becomes not heavy enough, there is downright depression.

We learned the 'Feel Good Factor' building our own sheds, our own extensions to the house, doing our own decorating, growing our own vegetables, milking our own cow and churning our own butter. Because our purse has never grown heavy we have not lost the art. What fun it was to buy material from Lancashire Arthur's stall in the market to make our own tents or to cut bracken and gather branches in Jimmy's Wood to build an overnight bivouac, to cook on an open fire out of doors, and to make furniture with green sticks and strong string. It is a 'Feel Good' activity to climb a mountain, to canoe upstream, to swim in cold sea, battle against wind and breathe fresh air.

That wonderful thing which has been lost is being sought in the wrong places. It has to do, not with money or even security but with the ability to cope with the shortage of either. Society, the human being, has always known it until now and I fear for the next generation who might not 'See the cuckoo'.

So, we do not have a video or one of those camcorder things everyone talks about. We do not have a microwave or a dishwasher or a kitchen full of gadgets. We do not have a computer or closed circuit television or a fax machine. We have no answer-phone or mobile. We have no word processor and our Range Rover is very old but we have the 'Feel Good' philosophy all right and are happier than most. Did I teach it to all those hundreds of children I took to the Hebrides when water had to be carried and cooking was over peat? I sincerely hope so. Certainly they all look happy enough when they come to see me.

Things have gone wrong with education since my retirement in 1981. 'Aren't you glad to be out of it?' visiting teachers ask. No way! I would just love to be there showing them how to put things right. But I am a landlady now, and a farmer, and what I cannot change I must accept.

We had such a great feeling of well-being on Harris in 1993.

I am glad I really do not know what is round the corner waiting to frustrate and anger us.

* * * * *

The holiday was perfect. We joined the New Age islanders and went to Rheinigidale by the new road. It is a wonderful feat of highway construction over what can only be described as moonscape but the experience of driving along it compares negatively with the joy of walking the track. The feel good reward for doing that is incomparable. I had walked the precipitous path several times with children and once with Margaret and, given choice, that is the way we would still go and ignore the newly-built road. We would have preferred to climb the hill from the Kyles Scalpay road and to descend The Scriob, that tortuous route, fearsome at the best of times and doubly so with a huge party of nimble, eager children in front and clumsy ones too far behind.

Fortunately, for the few people who live in the ten or so houses in the village of Rheinigidale, the age of mechanical monsters has not left them in their centuries-old isolation but the road which now reaches out to them may well tempt them away. It is a miracle of modern construction. It crosses uncrossable terrain and I must admit that it does not impose itself on the landscape which is all rocks anyway. It leaves the Stornoway road at Maurig and does not follow the route of the green track at all so it would be possible to forget it exists until you reach the village.

The road, like the track, frequently looks as if it will leap into the sea and must have been a very expensive project for so few people. There are, of course, no public conveniences which does not matter on a track with miles of privacy if you have an agility to leave it when necessary. Arriving at the village in a Range Rover, with our kind of passengers, we had occasion to sample those provided by the Youth Hostel, a small butt-and-ben perched precariously on the hillside. Its only path was a trail of stepping stones up the banking and it needed one of us in front to pull and the other one behind to push, confirming Margaret's persistent observation that, in almost everything, it takes two.

At the end of our stay we sailed to Lochmaddy and wended our way south down the Long Island. Nowhere is the sea so turquoise as it is on the Uists, due no doubt to the shallowness of the water around the archipelago. Its insignificant depth has made possible the ribbon of causeways which link the Uists with Benbecula and Grimsay. We ate an evening meal, cooked on our camping gaz stove, on the sunlit shore at Ludaig. There is rock there similar to the one in Copenhagen harbour and we have a photograph, taken years ago, of Mother and her two sisters sitting like the Little Mermaid, dangling bare feet into the water. Our association with these islands now encompasses many decades. Somewhere in one of the small townships west of the main road, there is a bus. An island lady has covered it with shells gathered over a lifetime of living by the shore. Strangely, when new sewage works were being developed in our home town, a man driving a JCB lodged with us for many a week and we learned he was the son of the shell-loving lady.

Similarly we once provided weekend accommodation for two shopfitters working to complete a job whilst the shop was closed. They were brothers and they rose at dawn and did not return until late for the job had to be finished for re-opening. After breakfast, ready to pay their bill and leave, they came into the kitchen and one took out his wallet. They had booked accommodation at the last minute and we had seen nothing of them all weekend. I had not even taken their name, which was remiss of me. I sat at the kitchen table with pen poised to make out their bill.

'What was your name?' I belatedly said.

'MacVicar,' the elder replied.

We had had no conversation with them so I volunteered that the only MacVicars we knew lived on Harris. I could see sudden recognition dawn on the brother's faces. 'We work in Glasgow,' one said, 'But we come from Tarbert.' They had grown up on Harris during the years of my invasion with children, had been at our camp-fires and céilidhs and had danced with my girls at the hall in Leverburgh. A relation of theirs had driven our bus and Elsie MacKie, our friend, had taught them at school. We are privileged to have these unexpected encounters.

Again I have digressed.

We boarded the *Lord of the Isles* at Lochboisdale for our overnight sail to Oban. I wish the boat would not creak and groan so! It kept Auntie Mary awake all night so that she slept much of the journey south, through Inverary and down to Dunoon. We had enjoyed a taste of summer for a fortnight and it was a surprise to find six inches of snow in Dumfrieshire. The crossing of the Clyde to Gourock opened up a route for us which avoids the outskirts of Glasgow. 'This is the future way for us,' we declared undeterred by the snowfall.

The overnight sail had shortened our driving miles by many and we would be able to get home by nightfall. We were as sunburned as if we had been to the South Pacific. We progressed slowly through the white splendour of a county always beautiful and reaching a green Gretna we joined the M6 for what we always accepted as a compulsory evil till we could escape onto the empty A6.

Margaret was driving. We were running late and she suddenly said that, if she felt at ease on the motorway, she would hold her course south to the Kirby Lonsdale turn-off. Dorothy, she said, might be in a pickle. Perhaps we should get home as soon as possible.

We had a few misgivings about the status quo at home. They had not had our weather. Torrential rain had found a roof entry and leaked into the kitchen and a cold tap had needed a washer and had sprayed the lino with water before the mains could be stopped. Little things had blighted the fortnight for our caretakers. A hen had died.

'What must I do with it?' Dorothy had wanted to know.

'Give it to the fox,' Margaret had told her over the phone. 'Take it to the bottom of the Five Acre and throw it into the wood.'

Dorothy had picked up the dead hen, gingerly, and taken it to the fence. Then, not wanting to witness the impact as the dead hen hit the earth, she had closed her eyes and hurled it into space. There was no thud. When she opened her eyes she saw the hen had been thrown upwards and had been caught in the branches of a sycamore. It looked alive, perched there, out of reach of the longest stick. And there it stayed, weeks, until the next gale.

'We don't want Dorothy to lose her nerve,' Margaret said, silencing my protests. I fear motorways nearly as much as we both fear the return. We know how accident-prone all farms are and one day there might be a real emergency for Dorothy! We pray there never is.

Harry was asleep in the back and Auntie Mary does not mind the main highways because she is not the driver. Margaret said she felt competent. The A6 junction was only a quarter of a mile away. She was not going to take it. We were going forward, in spite of heavy lorries continually passing us.

Suddenly the Range Rover began to bump along and I said, unnecessarily, 'We've got a puncture!'

Margaret pulled onto the hard shoulder and we both got out. One tyre was absolutely flat. It was the right-hand front tyre and bending to inspect it our backends were dangerously close to the passing traffic. We had recently bought a trolley jack before which Margaret had had to crawl underneath the vehicle. Since buying this professional piece of equipment we had never had occasion to use it. We lifted it from the back of the van and read the instructions amid the roar of traffic and choking exhaust fumes. It said, 'Familiarise yourself with this equipment before you have to use it.'

'Oh, beggar it!' we said. 'Let's send for the AA!'

We rang from the phone in the lay-by. Motorways have their advantages. I refrained from saying, 'I told you so,' but apparently Margaret believed that someone had issued a warning. Leaving the telephone and returning to wait in the car, she lifted her eyes to heaven and said, 'OK Dad, I heard,' and when the tyre was changed she pulled off onto the safer A6. Everything was fine when we got home. Our fears had been unfounded. Presumably our Guardian Angel had known.

* * * *

The next meeting of the Transport Committee was not cancelled. I met a college friend in the city and we went to the meeting together. Thanks to the pressure from the Country Landowners' and the National Trust I was allowed to put my case and show my photographs. There were protesters. Loudest amongst them were the motor-bike enthusiasts who wanted the right to use a closed road even as horse riders would. We were as tormented by motor-cyclists as by stranded car drivers but those in authority would only believe that the trouble was being caused by the off-roaders.

'They create the mess which makes the road impassable to ordinary cars,' I conceded, 'But they can keep going. They do not breech our boundary. It isn't they who pull the wall down!'

No-one listened!

Then came an incredible proposition. The representative from the motor-bike club offered to help repair the road in return for being allowed to use it. The chairman said the Council was prepared to spend up to £15,000 on repair and closure if agreement could be reached.

I was flabbergasted. All that was needed was a telegraph pole horizontally placed at each end of the road. It was all that had been needed for thirty years. Had that been done we would never have had the expense of fencing and we would never have incurred £12,000 in a new wall.

Agreement had to be reached somehow and the Committee accepted the man's offer to employ his members to barrow hard core, costing thousands of pounds, into one mile of mire.

I asked to be allowed to question, 'Have you done anything like that before?' The man admitted he had not. 'I'll tell you this,' I said, 'I'll be sitting on the wall to watch.'

The offer, ridiculous as it was, swayed the Committee to vote for the expenditure of £15,000. Horse riders and motor-cyclists would have access. It was not the end of the story but it was a start.

Chapter Eleven

Take care to wonder as you wander, never hurry by an open door
For we live in a universe full of miracles galore.

Source unknown

The summer passed relatively peacefully. No day can be without its comedy and its drama and the little miracle which prevents insanity. Denis came for the summer and laughter was again on the menu and Harry was having the good time Denis always provides. We spent long hours sitting at the kitchen table listening to his story of twelve months in Russia. He was a sixteen-year-old French boy who could tell us, in English, of a year in a strange land where he had spoken only Russian. We were full of admiration. We neglected things and it was Denis who cut the thistles with the rotary cutter behind the ATV saving Margaret long, uncomfortable afternoons on the saddle and if the dust wasn't taken from the farmhouse sideboard there were other, better things to do.

We never had any vacant bed spaces but we did have one not very welcome visitor. Mouse turds appeared in the cupboard in which we keep flour and sugar. It might not seem a disaster in a farmhouse where there are cats capable of coping but to have a mouse when you have bed and breakfast guests is not acceptable. We could not remember when we had last had a mouse in the house. We hoped against hope that it would go out the way it had come in. We removed the flour and sugar and put a piece of bread and every morning we found it had been nibbled.

We knew the only thing to do was to take the Welsh dressers from the wall and the cupboard used for the flour, and see if there was a mouse hole in the washboarding. It was not a job we relished with guests in the house. We knew where the mouse was. Any disturbance was dangerous for we could lose the mouse in the kitchen. It could get under the staircase door or into the under-the-steps store. It could belt into the wash kitchen or the sitting-room or, heaven forbid, into the converted barn. Neither did we want to admit, even to Denis, that we had a mouse and we did not want spectators when we pulled the two Welsh dressers from the wall. We suspected there would be a hole, cobwebs and you-know-what which magically get behind a cupboard in the kitchen.

So we waited for everyone to go out for the day, and for Denis to go to Leeds, and we carried sheets of hardboard from the out-buildings and old doors and a few corrugated tins and we sealed off every diversion so that if, when we moved the cupboard, the mouse ran out, the only way ahead was the open door. We were aware that any early returning guest would have hysterics at our performance! Then we emptied the dressers and the whole place looked as if we had had thieves. The chaos was appalling and I know the reader will be saying, 'What was wrong with buying a cheap mousetrap?' but Margaret does not kill, even a fieldmouse. So we went through this performance and pulled the dressers from the wall and searched in vain for the trespasser's entry. We found no hole the little fellow could get in.

We were reasonably sure he had not made a dash for it and fled through the front door. Thwarted in our scheming we sipped hot coffee and contemplated the mess. We could not let it stay! We swept away the cobwebs and collected the debris and put back the cupboards and returned their contents and we staggered out with the array of our barricade, piling the pieces just outside the door in our frantic efforts to restore order in the kitchen. I hurried for the vacuum to pick up all the stray pieces and we began to relax. I was just winding up the flex when our nearest neighbour paid us a social call. Kay is not a farmer, neither is Fred, her husband. Their converted farmhouse is an immaculate mansion. It used to have the same problems of straw and manured wellingtons, mucky coats, buckets and veterinary medicines as ours, but the last generation of owners sold the surrounding fields to a near-by farmer and made a dream home. When Kay and Fred bought it they improved it still further and it is beautiful. When we have coffee with them we sit on the terrace or in their elegant lounge. They are lively people and they like us as much as we like them.

We would have been mortified if Kay had found us in the predicament we had escaped by the skin of our teeth. Only by seconds had order been restored. We assumed an air of welcoming calm, comparable with a friend who was shouting at her teenage son who expected her to wash his football strip. 'Don't shout,' he had said.

'I can shout louder than this,' she screamed as the door bell rang. She changed from irate mother into welcoming hostess in a split second.

Kay had come to tell us they were emigrating to Florida. We all had another coffee and forgot to have lunch that day. We would miss Kay. One day Margaret had wished to identify a calf to which she had given one jab of antibiotic and which would need to come for a second, and a third. She touched the little fellow with a red spot on its white forehead with her marking stick. It amused us to remember that Kay had phoned to say one of our calves had been shot through the skull. We were really sorry to hear they were leaving.

Next day the little piece of bread had been eaten again. The mouse was still there. Margaret went into town and bought a live trap. In the morning the little creature was safely in it. It was not a mouse, it was a shrew, and she took him well down the Five Acre and released him into the wood. Those who live perpetually with animals treat all the same, tame or wild. At least this has been our experience and we know a lot of farmers.

* * * * *

The cattle did well that summer. The heifers were so fat they appeared to be making bags. Emilie was the one which alarmed us most. Bought for £5 to please Denis, two years ago, she was now enormous.

'She can't possibly be in calf,' Margaret said. For one thing costing only £5 she must surely have been a twin and infertile and for another we only kept bullocks. There was always a possibility she had found our neighbour's bull but she had never been missing and he had never run amok on our property. Nevertheless Margaret kept an eye on the heifer and suddenly all the fat heifers looked to be in calf and making bags. Our reputation was at stake if she unwittingly sold in-calf heifers to unsuspecting buyers so we called in the vet and put every heifer through the crush. They were only fat! The vet agreed Emilie certainly looked to have the makings of a bag and prophesied that we would be millionaires if we had found a way of inducing lactation without the birth of a calf.

Otherwise the summer was peaceful enough for me to toy with my autobiographical manuscript which was attracting cobwebs. I had asked a friend, owning a photocopier, to make me two more copies of the seven hundred pages. Because I had been so possessive about it, it had lain idle for a long time.

Christine Gale, from the National Trust, gave me the push I needed. I had tried co-operative publishing once before and it had worked so I decided to give it another go and wrote to Pentland Press and I tested the interest of my guests.

'If I write a book will you buy it?' I asked. Their promise did not finally make me decide. It was our guests, Joe and Pauline Walters, both wheelchair users, both artists who finally tipped the balance in favour.

'If I write a book, will you illustrate it?' I pleaded. They said they would be delighted and then I could not back out.

* * * * *

We sold heifers in August but the bullocks had to wait until the September sale, just two weeks before our holiday, because of MAFF retention rules. We could stagger sales no more. All would have to go at once on the seventeenth. There were thirty-nine. Samson and his friend we would keep over the winter. Samson could never be put through market trauma which is not the horrific experience some people profess it to be. Sighted animals quite enjoy it because they are innately curious and they like travelling. We have to load animals with no previous experience of transport other than when they were week-old calves. Mostly they climb into the wagon in a group, with the minimum of persuasion. They will need none to re-enter the wagon belonging to the buyer. Should we not sell, because the price is not acceptable, the animals will be as eager to travel home as Lusky is to get into the Range Rover.

*****

There was a recognisable change in the pattern of people's holiday-making. The Saturday-to-Saturday week was splintering. For years we had been used to the regular habits of the British family.

Saturdays were diabolical. When we were finally able to wean satisfied guests from the magnetism of The Currer, when they had, at last, put away their cameras and climbed into their cars and pulled away up the hill, waving all the while, we had to do the weekend super clean, remake all the beds, wash nigh on forty sheets and pillowcases and prepare for the new arrivals.

Saturday was the day we discouraged callers. 'Not on a Saturday,' we would say to friends. We forgot our own meals and nearly died if a load of straw was delivered. But it had its compensations. The rest of the week was as straightforward as it ever is within the unpredictable environment of The Currer. The booking chart was always readable and I could cope with the menu. At the end of the week the bills were easy to calculate.

Imperceptibly all this was changing. The demon called Choice was upsetting things even in the tourist trade. I have always maintained that the seeds of an undisciplined society were sown by the educationalists who persuaded teachers to give children an opportunity to choose, years before they should. Even when I was being trained in 1948, the school timetable for infants included Free Choice. It was comparable to opening a shed door on cattle or the lock gates on the canal. There was no stopping the chain of events which have resulted in educational chaos. When what children wore and ate and watched and learned was the choice of parent and teacher children were secure and happy. When the responsibility of choice became theirs there was an explosion of unrest.

When the demon called Choice started upsetting the accepted holiday week, our booking chart began to look like a jig-saw. People no longer think of travelling on a Saturday. Indeed they purposefully avoid it and the result is a confused floundering proprietress. Not a day passes which is not a mini-Saturday. We are virtually imprisoned for new arrivals come every day and even those coming for a fortnight come Wednesday to Wednesday. Only self-catering cottage bookings remain Saturday to Saturday and long may that state of affairs exist!

One would expect that, with experience, things would get easier. In all the different professions I have known, I find they just get more confusing. A young teacher knows all about the children in the first class he or she teaches. The longer one teaches and the more children crowd the memory box, the more difficult remembering becomes. The first experience of Cattle Identification was the easiest. The more times one has to send in the forms, the more ear-tag numbers one has to read the more one is likely to forget. Similarly with holiday bookings. The first few years I knew who was coming without looking at the chart. Now I constantly look at it and make mistakes. People I expect do not arrive and I see it is tomorrow they are coming and, there again, people walk into the yard I am not expecting until the day after and we have to hand over our bedroom and sleep on the floor of Auntie Mary's.

The change from 'nationally acceptable' to 'do as you please' must have been gradual. It was not really noticeable until 1993 when we were suddenly aware that the length of people's stay varied from one day to a month and I was expected to cope. It was little wonder that we waited in happy anticipation for the one week every year which is not only normal, in that it begins with everyone coming on a Saturday, but it houses the same people year after year. We call them The Gang. They met here many years ago, liked each other's company and re-book as they leave. This incredible gathering of friends gives us almost a holiday week for they cover their own beds, clear and re-set the table and make their own entertainment. Amongst them are able and disabled and good humour and laughter pervade in the dining-room. They do not need me other than to cook their food. If I join them they do not notice I am there and I slip out again and have time to think of other things in my own little world.

We think longingly of this moment of calm which comes every year and never more so than at the end of the 1993 disjointed season. It comforts us and refuels our engines for the final five weeks before our holiday.

If I have given the impression that this particular group of guests ignore us that is not entirely true. Until 1993 the group included our lovable Mr Van. In the nicest possible way our Belgian friend Franz Van Caeneghem was always referred to as Mr Van. He was approaching eighty and never brought his car so we did quite a lot of collecting him from town.

An actor by profession, a weekend yachtsman by choice and a born philosopher, we found time for him to sit at the kitchen table and generously impart his wisdom and one day, every year, we took him into the Dales, to Dent or Malham, or wherever his whim suggested.

I think we looked forward to that early September week more that year than ever. Should anyone ring we were already full, we need do no washing, all our winter's hay and straw had been delivered, the heifers had been sold and it was still a month before we need think about bullock sales. Heaven was round the corner and would stay for a week. We are optimists.

Mr Van arrived looking well. We collected him from the railway station getting kissed three times in the traditional Belgian way. He had first come to The Currer in 1981, our first year and he, like Liz Day, Philip Oakes and the Vinces had been every year since.

He always arrived a few days before The Gang and was so different from them yet so happy with them, pleasure was generated for all. When Mother was alive he brought a little drama into the house which suited her. Arriving late for breakfast every day, he would excuse himself in song. He would waltz into the kitchen singing, 'I could have danced all night.' She loved him.

Because of his long-standing appearance at the end of August he often occupied the spare room or, if we were over-full, even Harry's room and brother had to make do behind the curtain in the family sitting-room. One evening the good man prepared to descend for the evening meal and was unaware that Harry was in the bath. The bathroom light switch is on the passage wall and Mr Van switched it off leaving Harry in darkness.

We do not leave Harry wallowing for long, in the deep water he prefers, and when I called up the staircase, 'Are you alright?' he shouted 'No!' Usually he calls back, 'I'm fine. Just another five minutes,' after he has already been in half-an-hour. On this occasion I galloped upstairs to find him laughing in darkness.

Our Belgian friend was appalled at what he had unintentionally done. 'I would never do that to Harry,' he said. 'To Margaret, perhaps, but not to Harry.'

To believe that our Bank Holiday week with The Gang was to be plain sailing was naive. We should awake each morning expecting disaster to be the norm but we are perpetually optimistic. On the Monday morning of that week Mr Van took ill and was taken to hospital with pneumonia and we spent all week visiting him each afternoon. He was allowed a brief return on Friday. Carol, the trained nurse member of The Gang, went for him by car whilst we got out the 'red carpet' and made a delicious tea. All our guests returned for his brief visit and with great affection he was bid 'Good-bye'. Very untunefully Margaret and I sang for him, 'Will ye no come back again.' Better loved he could not be! It was one of the most moving occasions of our experience. We again visited him in the afternoon of Sunday and the memory will be with us always. He looked like some Roman Emperor in a nightgown the hospital had provided, and in his sandals, he spoke with the wisdom of long experience. We talked, or rather we listened to him talk all the allowed visiting hours aware that we had been extremely lucky to have known this man. Next day he was flown home to Belgium.

* * * * *

Whilst market, in our locality at any rate, is not a distressing experience for animals that can see, it was quite unthinkable, even illegal, to take a blind animal. So Samson and Friend stayed and we prepared to sell the other thirty-nine bullock stores on the seventeenth. The things we said about the government and EEC and MAFF are unprintable. Our thoughts were even worse as we strove to get so many bullocks taken to market in time for the sale.

Margaret decided to do the gathering the previous night and hold them in George's fields which we had continued to rent. Anticipating this she had allowed a growth of grass to accumulate. The cattle would be eating at day-break and go to market content, with full bellies. She had forgotten that an overdose of new growth makes cattle splatter loose manure all over the place. Before the arrival of the wagon, which would have to make three journeys to the local market, we had brought the beautiful, clean animals into the shed. The messy road should have warned us there would be an equally messy wagon.

When the last dozen were loaded we pushed our complete family into the Range Rover, and the picnic basket, and the dogs, and followed. Selling had already begun.

Our cattle are seldom dirty. As calves they are always spotless. Winter can make the cattle sheds mucky but Margaret spreads straw twice a day and our animals sleep dry. All summer their coats are spotless and need no grooming. Imagine, therefore, our horror in arriving at market to find our animals the dirtiest there. Standing close together in the wagon, even on the short distance necessary, they had splattered loose manure over each other. Snowy, who had been persil white for almost two years, was wet and yellow. We had no cloth or water to clean them for the ring, or time to go in search. I was wearing a pullover two sizes too big and I managed to wriggle out of my shirt inside it. The cotton garment was used to scrape off some of the half inch of manure new grass and excitement had loosened. Some headmistress, standing there in wellies and sloppy pullover trying to make animals bigger than herself look presentable for the ring by wiping them with her cotton shirt! Denis would have said, as he has on more than one occasion, 'Jean, you continue to amaze me!'

Margaret repeated over and over again, 'Never again will I put bullocks for sale into new grass. Never!' No doubt they had had a midnight feast. Buyers are professionals and are not put off by a bit of honest to goodness cow muck. If they had smiles on their faces I don't think it would be because of the dirt but because the vendor was so small she was lost amid them whilst showing them in the ring. Ben, the auctioneer, called out once, 'We can't see you, Margaret!' and there was a sudden response of laughter from the ringside.

The bullocks sold well and we were bursting with foolhardy confidence. We had made the right decision when we decided to rear calves from week-old once more. There had been moments when we had longed for the weaned Hereford heifers which would have needed no yellow ear tags, no CID's, no dehorning or castrating but with such a good sale and the MAFF promise of subsidies bordering on £100 per bullock we were more than satisfied. Margaret is capable even though she cannot be seen in the ring!

* * * * *

We turned our attention to thoughts of holiday. We had not re-booked the pleasant town house in Llandudno because that fast dual carriageway had frightened us. The only way of avoiding stress, on the morning we had to be at Beeston Castle at 8.30 a.m., was to stay overnight in a nearby Travelodge.

It seemed stupid to drive sixty miles from Llandudno on the Thursday evening when a seaside resort was not really our choice of holiday. I had, therefore, scoured the map for a coastal village further south and I had found Aberporth. I had booked a cottage and left it at that. We reserved two rooms at a Forte Travelodge at Wrexham and hoped we had chosen wisely.

Pre-holiday chaos never changes. Why we think it will I do not know! The calf sheds were all ready for the first time and the week before departure seemed relatively free of holiday-makers. All we needed to do was to get ready for holiday. 'Things are changing,' we said. 'We are in calmer waters and in for a run of good luck'. We were audacious enough to think we deserved it. 'If we can prepare for holiday over-worked, surely with no building project and fewer guests we will find packing cases "nay bother at all".'

Then Isobel, our TV star goat, became so arthritic she could not get up of her own accord. Three or four times a day Margaret gave her a helping hand and, when I rose at six each morning, my first job was to go to her shed and give her a heave.

We began to be worried about leaving Dorothy with this problem and hoped it might resolve itself, one way or another, but death did not seem imminent. Once up, Isobel tucked into her meal like nobody's business. Occasionally she would go out into the paddock with the other two goats and eat until dark. Mostly she just ate the grass which Margaret pulled for her by the armful.

Our goats were getting to look quite geriatric. We had had Nanny for twelve years and she had not been a young goat when we had bought her. She trotted out each day on legs which would never have supported her without the bandages Margaret put around her knees. Any day, we feared, some interfering person was going to report us to the RSPCA. Charlie was too fat. We could not be blamed for we did not feed him. Nor could we feel any remorse at keeping the two happy old ladies. If you did not look at Nanny's legs, just at her shining eyes, you could just see how alert and content she was and once Isobel was on her feet she was fine. She could talk, Isobel could. As soon as I entered the shed, before my morning cuppa, she started to struggle as if to say, 'I can do it for myself!' All she needed, really, was her head steadying. Always when she stood she said thank you, 'Bleat, Bleat!' and gratefully spent her penny.

All the same we felt embarrassed at leaving Dorothy with an animal needing such attention, such loving care. Having sold the cattle Margaret looked at Isobel and said, 'We've a fortnight to decide what to do'.

Then our decision-making time was reduced to a week and Isobel got no better, no worse. All she needed was a friend who would squat in the field and pull armsful of grass and steady her so she could rise every few hours. The pre-holiday week began to disappear and then, on Thursday, just two days before we were due to leave, Nanny went down and a helping hand was no good. A heave was useless. A vet, working in the local abattoirs was lodging with us at the time and promised to give Isobel an occasional lift but all of us together could not get Nanny back onto her feet. Nanny was down, and she knew, and we knew, that she was not going to get up again. Given time she would have been allowed, like an old soldier, to quietly go to whatever goat heaven there is up there, in her own good time, loved and cherished. But we did not have time. We were leaving on Saturday and on Friday Margaret took the unprecedented step of taking her to be shot. We lifted the aged lady lovingly into the back of the Range Rover and Margaret drove off alone, no Harry, no dogs, just her and Nanny and she talked to her all the way. She told her we loved her and what a good friend she had been and how she would never walk again in this world but, like maimed human beings she would walk again for, in the vast scheme of things, if there is a hereafter for people, how much more so do animals deserve inclusion. I think all farmers must believe this for their relationship with animals is often more rewarding than with people. Animals, like my grandfather, do not fear death for, as he said about himself, they have done nothing right wrong. When it comes to birth and death all in the animal kingdom are equal and all are vulnerable, none can escape, none is superior.

The end was easy for Nanny. There was no question then about what to do with Isobel. It was immeasurably easier for Dorothy to look after her than to cope with a mourning billy-goat. Charlie would cry all the time and Dorothy would prefer to pull grass than put up with his bereavement. We had less than twenty-four hours to find another friend for him so Isobel was given another lease of life which she responded to so positively she could eventually get up again herself and clip her own grass. It cost something in veterinary prescribed pills, of course. It follows that all moments before a holiday are fraught with some trauma or other and that autumn it was the goats.

To a lesser extent it was a sudden influx of guests. Believe it or not we had twelve breakfasts to make on the morning of our departure. Five came from South Africa and three from Australia. We were terribly delayed. I cannot imagine how we had so much dustbin rubbish at the last minute. At midday we were at the local tip getting rid of it before heading for Wales.

* * * * *

Since booking the new cottage in the unheard-of village of Aberporth, I, for one, had scarcely given it a thought. It is unadventurous not to seek out new places but it is impossible to anticipate a different experience whilst working as hard as we do. Had we been returning to Grange and the Lakes or to East Witton and the Dales, or to Fenham and the Northumbrian coast, or Letterfrack in the Connemarra or to any one of the Hebridean Islands, anticipation would have accompanied activity as comfortably as quiet background music. Because we had no idea where this village was or what it was like it was not a recipe for daydreams. I had not even planned a route and I had no idea of the distance.

We were throwing out our rubbish at Sugden End household tip at noon and we did not know how far we had to go. Could we be getting too old for this? I remember Father saying, whilst struggling with the terrible winter of 1947, 'Ah'm wrong side o' fifty for this.' Could it be that I was the wrong side of sixty to cook twelve breakfasts and deal with all the aftermath and then set off for the unknown? Surely not! A cup of coffee on the summit of Cockhill and we would be rejuvenated.

Sipping the life-giving stimulant I looked at the map beyond Beeston Castle. We knew the route so far. 'It looks as far again,' I said with some misgivings for, without stops, it would take us four hours to get to Wrexham.

Of course, other people would have gone to the motorway on the moor top beyond Ripponden. It is an ugly monster snaking its way across inhospitable landscape and we would not touch it with a barge pole. It is lovely to drive through Derbyshire in the autumn. It is full of deciduous trees ablaze with colour and the rolling countryside and the empty B road in front are tranquillisers. I am not sure which holiday we need most, the one after winter before the summer tourist season or the one after summer before the calf rearing. Thank God for holidays!

It was 5 p.m. when we reached the outskirts of Wrexham and headed south. We were little more than half way. 'No more stopping for food or toilets,' I stressed knowing I was unprepared for night navigation. I had forgotten our torch. Map and torch and spectacles were essential. Fortunately if we managed to get onto the right road for Aberystwyth and then headed for Cardigan we would not need to be always consulting the map. Nevertheless I rank that journey to Aberporth the worst-planned excursion we have ever made.

Darkness fell and we were miles away. We always vow and declare we will never drive in darkness. At Aberystwyth I took over the driving. I was tense and tired but there were no deviations from the A487 until we saw the sign for Aberporth soon after 10 p.m.

The village is so small if you blink you have missed it. The owner of the cottage we had booked had sent me instructions of its location but they were somewhere safe with the driving licences and MOT I had a vague memory of being told to turn into a private road beside a house painted red and there it was in my headlights.

'I think we are here,' I said, turning right onto a private track between two houses.

I was reproached from behind. 'Where do you think you are going?' Margaret said as the Range Rover, bumping over a 'sleeping policeman', began to climb steeply.

'Instinct!' I replied.

'Stop and ask,' Auntie Mary scolded.

'Who?' There was not a soul in sight but at the summit was a cedar wood chalet and the key was in the lock. This time, because we were not on an island, there was a note saying our proprietors were out for the evening and would we just let ourselves in. Their house, standing back, was in darkness. Tired out, we were infinitely pleased to be able to do so unwatched.

Had Margaret and I come alone, I am quite sure that is all we would have done, just let ourselves in and curled up, slept till morning and unloaded then. But we do not travel alone and we have dogs and bed is always two hours away. We crawled in about 1 a.m., not knowing what was our environment, so dark was the night.

Harry wakened me at dawn to tuck him back under the sheets and I drew back the curtain of the sitting-room and uncovered a window which filled the entire west wall. I saw a wide vista of open sea. In the foreground the road descending steeply towards it was bordered by neat bungalows, well-kept-gardens and wide lawns. I have said before that, for people who live on a farm, even living in a small town would be claustrophobic but on holiday there is something uniquely comforting about well-kept lawns and middle-class respectability. We experienced it in Grange. In the Hebrides there is no escape from the real world in which we live. We are surrounded by animals and talk with crofters about their husbandry. The seagulls on Tiree compete with the sound of Hughie's tractor. I knew instinctively that there would be no reminders of home in Aberporth. In Grange 'Vacancies' were advertised in the windows of the guest-houses to torture us. Here, I was sure, there was a complete escape from reality, an environment so different, so peaceful, so beautiful, it was bound to be the tonic we needed.

I went back to bed and we all slept until coffee time and the sun was nearly as high as it ever is in October.

We had a lovely holiday in 1993. When do we ever have anything but? Almost always you can organise a successful holiday if your needs are simple and not dependent on the weather. Maybe it is because we are not obsessed with the need for good weather that we invariably get it.

On holiday it is a case of 'little cattle, little care'. The difference between all those animals and all those people and our holiday contingent of four and two dogs is unbelievable. That Auntie Mary likes our sort of holiday is one of life's blessings. Give her a cottage by the sea, a beach to walk on, close to the ebbing tide and she is happy. Only one person loves it more than she and that is Harry, so aren't we lucky people! As Colin from Australia said, 'If they're happy, we're alright.'

According to Auntie Mary mid-October is St Luke's little summer. I have never had time to check her out but am willing to take her word for it, for as sure as night follows day we have good weather in October. That year when the rest of the UK had gales and floods, Aberporth was bathed in sunshine. It is not difficult to understand why we have continued to find this village, on Cardigan Bay, so perfect for our October escape. It would not tempt us in the summer for we like deserted beaches and empty car parks. If the Hebrides are our aperitif then Aberporth is the coffee and the After Eight chocolate. It gives us more personal space than anywhere. Almost everywhere we go we have taken children or family and we have friends and we reminisce. In the Hebrides we remember when my friend Jean's children, Paul and Julie and Vivienne were small, when Donald John at number eight was the heart-throb of the Sea Rangers, when Hazel was in camp and Joan had the only civilised loo.

In Aberporth there is an emptiness very necessary after the months of people. It holds no memories. Resident dogs are household pets and there is not a farmer in sight. Wonderful! For a whole fortnight we can forget who we are, who we were and what is the important errand at the end of our stay and the rude awakening we will get on our return.

We found a different beach everyday. Tenby and Pendine and Saunders Foot, St David's and Solva and Newgale, Newport and St Dogmaels, Tresaith, Penbryn and Llangranog, New Quay and Aberaeron. We sat in blissful contentment. We are not tempted by such luxury to retire but we enjoy a sample now and again.

Our map was useless on the maze of single track lanes which tempted us off the Aberystwyth – Fishguard road and it was fortunate that the sun shone almost all the time and that we knew how to keep it in the right place and ourselves travelling in the right direction, otherwise I fear we may still be lost. To ask one's way, should one unexpectedly spot a human being in an otherwise unpeopled landscape, was no help at all. There were so many turnings, road junctions, and impossible to read signposts that even if the lovely Welsh accent could be understood the directions could not. Moreover those who could walk unerringly through the maze themselves found the sudden task of directing strangers quite beyond their ability.

There was one lovely evening when we were tempted to leave the main road back to Aberporth, to walk the dogs on Newport beach and see the sun go down in Cardigan Bay. The sand there is almost black but cleaned and ironed twice a day by the tide. It is spacious compared to the twin mini-bays of Aberporth and gives us the opportunity to stretch our legs and the dogs to exhaust themselves.

We presumed Harry and Auntie Mary would watch the sea turn red from the comfort of the Range Rover driven right down onto the empty beach but he wanted to go in the wheel-chair and she wanted to walk down to the water line, and the sun had set and dusk was falling before we got all our animal and human menagerie back into the car. We should, then, have taken the faster road home and resisted the temptation to follow the coast along roads which are narrow and have no landmarks and infrequent hamlets.

Margaret was driving and was unusually quiet, hardly daring to tell us that the petrol gauge was registering 'empty'. Darkness fell and we had no more idea than fly where we were or even in what direction we were travelling. We could believe that we were doing a circular, that the road junction ahead was the one we had just negotiated. It was not funny! We ought to carry a petrol can. We ought to stick to the main road. We ought to refill the tank when it is half empty. There are so many things we ought to do which we do not. The darkness became impenetrable. All we could see was the ribbon of the road in front and high hedges on either side. Every now and then we came to a parting of the ways and we did not know which one to take.

We could feel Auntie Mary preparing to chastise and Harry was quivering like electric when we turned in the darkness, in the middle of nowhere and there was a minute village filling pump.

'Petrol', we all called together. There might have been a little shop attached to it. There did not seem to be anything else other than this one pump and when we returned the following year, in the daylight, to look for this godsend, all we found was a flattened space where something had been demolished. We still think that one might have been an apparition. The petrol was real, though.

On the second Thursday we packed our bags and drove to the Travelodge at Wrexham. Next morning we donned farm clothes and wellingtons and went to the cattle market. Our moment of calm was over. Thank goodness we did not know what was in store for us! We bought seventy-five calves and had to order a second wagon to take them home. We paid an average of £126 each. One hundred pounds more than we had done just three years ago. So much for the Government subsidy we had been promised. It had all been spent in buying dearer calves long before any initial payment was due. So much for all this CID, IACS, EEC and ear-tagging lark! 'What's just a little paper work and an occasional struggle to gather cattle for inspection?' the public and the Ministry said. 'You are going to be well paid for your work!'

As far as we were concerned our subsidy had gone to the dairy farmer whose calves we had bought and he is limited to a controlled milk quota. Quite frankly it was all a shambles. The initial cost of calves was £7,000 more. With a bit of luck we would draw even and all the extra work and danger and worry of conforming to bureaucracy be without reward.

But Gillian, a five-year-old I admitted early in my village school headship, had said, 'Ah don't grumble. Ah jus' ge' on wi i'.'

Chapter Twelve

What is good luck but the courage to overcome bad luck?

Anonymous. From Anglia Anthology, GGA

Those seventy-five calves were the most expensive week-old babies we had ever bought and they were disaster-ridden from the word go. Perhaps it could have been averted. Perhaps it was human error. Perhaps some calves had been taken from their mothers without having had their immune system boosted with colostrum. I am not intent on laying blame for what happened in any direction or on anyone for, like all disasters, some good came out of the strife. If a mistake was made it was comparable to the loss of the horseshoe nail or the flicked domino. The battle was lost. The line collapsed. The whole episode was horrendous beyond belief.

The first week, as is usually the case, was trouble-free except for the almost human agony of one Murray Grey calf which got separated from three more of the breed bought from the same farmer. It came in the second wagon and ended up in the old shed pens instead of with its relatives. It was sad that, having a new calf shed, it was still too small and the old shed had still to house about twenty of the year's buy.

We noticed almost immediately that the dun-coloured calf was stressed. It was pacing backwards and forwards continually along the closed, twelve foot gate which imprisoned it, like a caged, demented tiger. Feeding seventy-five babies, three times a day is a mammoth task and Margaret has a twenty-four hour job similar to that of the shepherd at lambing time. We hoped the calf would settle but it continued to march across the gateway and nearly drove us crazy, too.

We had a Brownie Pack on holiday and they made a human fence across the yard so that we could persuade it down to its own kind in the new shed. The difference was immediate. The stress disappeared. The little creature settled at once. I include this small story to demonstrate the sensitivity of animals to family and breed and friendship. All four were survivors. Many humans lack the ability to be so loyal to family, wife or children. It would be more accurate to describe bad animal behaviour by comparing it to human waywardness rather than saying an errant human is behaving like an animal. A farmer is a professional who understands more than the physical needs of the animals in his care. His sixteen hour, seven-day-a-week job pays him less than a minimum hourly wage. I know because I am the accountant in this partnership. Should the farmer be merely interested in financial reward he would quit immediately for a forty hour week and leisure to enjoy it.

Any firm, or school or community of more than two hundred souls will have a sick member. A farm always has one. Sometimes it is a valuable animal. Sometimes it is a geriatric goat, sometimes it is an old dog. Sometimes it is a pedigree calf, sometimes the runt of a litter, an orphan lamb or a day-old chick. In spring the shepherd is midwife to a thousand lambs, a modern dairy farmer calves a hundred cows a year, competently and caringly. Farming, which used to be everyone's background is now the experience of the few and the many profess to know all about it. Strange but true.

Quite frankly I do not think farmers are one bit interested in feeding an overfed general public and there are a million better ways of earning a living. It must be, therefore, that other farmers are just like us enjoying our work with animals.

Making that small Murray Grey calf happy brought a 'feel good' feeling and everything seemed fine as the week drew to a close. The November routine became familiar. Things were going to be okay.

A sudden high temperature in a human or a bovine baby is not unusual. A dose of antibiotic usually does the trick when a calf is ill. By morning it is better if the medication has worked.

Market stress may be responsible for calf temperatures during the first few days. All this is normal but sick calves, however few and however brief the illness, make us nervous. There are so many babies and germs breed far more quickly than rabbits. The babies are so new and they have been separated from Mother and auctioned at market and transported in a wagon. Margaret scarcely goes to bed for the first few weeks and never properly undresses. She sleeps in a sitting position so that she will wake frequently and inspects the wards during the night.

In the first days of November 1993, she worried about two calves whose temperatures resisted the antibiotic and she thought it wise to send for the vet. It was a measure of her concern for, like all farmers these days, she can usually cope with medications herself and sending for the vet once can double the price of the calf.

The two feverish babies had runny noses which is not uncommon but the vet diagnosed infectious bovine rhinotracheitis.

'So what?' we queried having heard of but never experiencing the disease.

'There's a reliable vaccine,' he said. 'Very effective. It not only prevents the disease spreading but it also helps recovery. If you lose these two calves that will be all.' He told us he could ring for it to be sent by train from York and it would be here by three that afternoon. 'It'll only cost a couple of pounds per calf and be well worth 150 quid.'

People love spending our money! Margaret did not hesitate and shortly after 3 p.m. we rang to see if the vaccine had arrived at the surgery. There had been some misunderstanding. The parcel from the veterinary suppliers had not been unloaded at the station. It had gone forward, probably to Carlisle and would have to be dropped off on return.

We were eventually told to collect it ourselves at the station at 7 p.m. We were early but on arrival we found the booking office area in total darkness and not a British Rail employee in sight. The proprietor of the privately-owned station café said there was no evening staff, ever. The whole place was eerily deserted. One solitary lady waited on the platform for a Bradford train. When we joined her she visibly relaxed. She must have felt vulnerable standing there alone in semi-darkness with no escape route should someone undesirable come along.

When I was using the train, forty years ago, to go to evening lectures at Leeds University, those waiting for trains would be sheltering by an open fire in the platform waiting room. The Friendly World 'that is good for you and me' is no more. We waited impatiently for the train but the first one to come was from Skipton. We stationed ourselves on opposite platforms and Margaret queried the guard on north-bound trains and I tackled those on the south-bound ones. When the train came from Carlisle the guard admitted he had left the parcel at Skipton but assured us that, even if he still had the packet, he would not have been allowed to hand it over to two unidentified ladies. What, then, was the use of waiting longer? We tried to phone Skipton on the café phone but there was no-one there either so, at 8.15 p.m. we went home to feed seventy-three healthy calves and to tell the two sick ones we were doing everything possible to get that 'life-saving' vaccine.

'Hang on,' we promised. 'It will come.'

We phoned Leeds and were told they would look and would we ring back later? At 9.30 p.m. I slipped indoors and re-rang Leeds. Yes, they had found the parcel and hadn't we been told to collect it at 7 p.m.?

'Yes,' I said. 'We've been on Keighley station an hour-and-a-half.'

'Keighley?' he said. 'You should have collected it from Leeds.'

It was addressed clearly to Keighley. Having been taken forward to

Skipton we were supposed to pick it up from Leeds, half-way between here and York! I could not believe what I was hearing! However they promised to find a way of getting us the packet that night. Nothing happened. Eventually we phoned again but those dealing with the parcel had finished their shift and gone home. We were told to ring York. We did so three times and at 1.30 a.m. went to bed. A quarter-of-an-hour later the phone rang. The caller said the parcel had been located and was being sent by taxi. We got up and dressed determined to administer the vaccine that night, come what may. We made a cup of coffee and sat by the Aga, talking and waiting.

When it came we thanked the driver and opened the box with its rows of minute bottles and looked for the instruction sheet. The vaccine was Trachering but the instructions were for Borgel, an antibiotic injection. The vaccine had to be administered through the nose. How, we had no idea. Margaret rang the vet, middle of the night or no, but he had no instructions to hand and advised us we would have to wait till morning.

You can do nothing with hungry calves until they are fed unless they are kept in separate pens. Ours are not and to attempt to handle them when they were ravenous was impossible. So first we had to feed them and then we had to wait for the vet to tell us what to do and by the time we had coped with seventy-five doses it was mid-afternoon.

Perhaps we should have flatly refused to give a vaccine which arrived with the wrong instructions. Perhaps we should not have called the vet in the first place. Perhaps we should have asked for a second opinion. We had a resident vet working at the local abattoirs who had worked in research. He, later, said the calves had been too young and possibly too market-stressed for the prescribed vaccine. Perhaps we should have ... should have ... should have. With hindsight we might have done differently. We did what we had been told to do and there was no turning back the clock.

The effects of the vaccine were traumatic. More than that they were devastating. The healthy calves staggered to the back of the shed and fell about like drunken men, flopping on top of each other dazed and docile. Forty-eight hours, the vet had said, and they will be as right as rain. He did not say they would all collapse.

During the next seven days a third of those babies died. The vet queried the vaccine. Had anyone else had problems with that batch? The answer was negative. The trouble was the minute bottles had such microscopic print. We had to get a magnifying glass to read the pharmaceutical mumbo-jumbo. The colour coding was a light shade of orange which was hardly visible in the dim light of the calf shed. We blamed the vaccine, we blamed

Mike, our electrician, for not giving us enough shed light. We blamed our fingers for being numb and cut and sore. We blamed the vet for his prescription and British Rail for being so slow. We blamed ourselves for aiding and abetting the massacre of so many innocent babies. On the plus side it must be admitted that the Practice sent vets daily to blanket-inject frantically, but Margaret must be given sole credit for keeping two-thirds of the animals alive. Those that died did so quickly. They went suddenly cold and lived barely an hour. As Mike says, 'It'll pass,' and that awful week of dragging dead calves to the Range Rover and ferrying them to Jerusalem Farm did pass, but at the end of it, we were considerably older and greyer and lighter on the scales. And during it a miracle happened.

Reading the *Farmers Weekly*, briefly over lunch on the Friday, Margaret's interest was aroused by an article about a North Yorkshire farmer. The calf rearer's perennial problem has always been air-flow. If the air is humid so many warm calves create just the right environment for germs to thrive. Pneumonia, salmonella and scour will quickly pass from calf to calf. That year October and November were misty and warm and there was no movement of air. Though we opened every window and door all that happened was the mist came in.

The miracle which prompted Margaret to open the magazine at that page came whilst the vaccination disaster was in full swing.

The North Yorkshire farmer was using a very efficient and inexpensive air system which had virtually eliminated pneumonia. Many rearers have fans in their sheds but this farmer had fans outside blowing fresh air through a perforated plastic tube. All that was needed was an air straightener to direct the air and prevent the tangling of the tube. This simple apparatus was feeding clean air continually into his shed.

Before we had finished eating lunch Margaret had pushed me into phoning the farmer and then the supplier and before we knew where we were we had ordered it and arranged with the electrician to install it immediately it arrived. With hope at the end of the tunnel Margaret went bravely back to fight the extraordinary after effects of the vaccine.

Halfway through the following week the air system miracle arrived. We unpacked it in the dining-room, helped by a Canadian living and working in Germany. He was here on a Grass Management Course at the nearby Turf Research Institute.

He had been a farmer in Canada and when the packing was removed he said, 'This has been in use for twenty years in Canada. The Government won't give planning permission without it.' He showed us how to install it.

We kept the frightening thing which was happening outside, from the ears and eyes of the guests inside. The resident vet said the stress of a live vaccine would have been sufficient to trigger a reaction of subclinical bacteria brought from the market. He then kept his own counsel and stayed well out of it.

Six days later the shed was filled with fresh, clean air and the remaining calves began to recover. We picked ourselves up and prepared to 'Buy another on Monday'.

Our little herd had been reduced to well below what we termed a viable minimum. The resident vet advised against re-buying. Our own vet likewise said be it on our own heads. Margaret argued that she needed to test, scientifically, in an endeavour to establish what had gone wrong. She dare not wait until next autumn to find out, if she could, just what had happened. The pharmacy insisted they could not have been at fault. Well then, she was prepared to try the IBR vaccine again and asked for a multiple one with additional prevention against two strains of pneumonia. It came in similar minute bottles with their indistinct shade of orange and the unreadable print. She ordered salmonella vaccines for safety and we went to the Monday sale and bought fifteen calves. The vet said she must on no account keep them in the same shed as the Beeston Castle ones although they were now going ahead at six weeks old.

But Margaret had had enough of professional advice. We made pens down the passage of our new shed and we put the calves where they would benefit from the new air system but be vulnerable to infection. I can not pretend we relaxed. On the contrary we were petrified. We had bought the cheapest calves in the market but even if we had been given them we were playing with life and we value it just as highly in animals as we do in humans.

With her heart in her mouth, Margaret administered the vaccines and nothing happened. There was no falling about at the back of the pen, no sudden drop in temperature, no ice cold body and death within hours. The new calves never ailed a thing. The resident vet was wrong, we concluded. The vaccine was safe. Our vet was right. It did protect calves. If those that had died had ailed IBR there would be plenty of germs in the new shed. The vaccine must be making the new calves immune. Our confidence began to return.

'I feel I can buy again, next year. We'll just use these three immunisations: IBR, salmonella and pneumonia and with this wonderful air system our worries will be few.'

She was so confident we discussed building another two bays onto the calf shed. 'It's a nonsense to have some in the old shed,' we argued and we warned John that we wanted an extension in spring when the calves went out.

Having pens down the passageway of the new shed made life, for Margaret, almost impossible. She could not use the ATV. Carrying bales did not pose the problem, it was pushing past the extra pens with a load that was so difficult but she was prepared to suffer anything. The price was small compared with the knowledge that the vaccine was safe. Something else must have been responsible. It was a mystery. Whatever the problem it had come to pass.

The text over the old man's bed read, 'And it came to pass.'

When questioned about his choice the old man said, 'It pleases me more than if it said, 'And it came to stay'. The more you go down the more times you come up, so there are advantages in everything. Auntie Mary talks about the 'ads and dises' of living with us at The Currer but insists the ads outweigh. If anything, I think we come up more times than we go down even though it sounds scientifically impossible.

Years later we were listening to a television programme about a farmer in the Yorkshire Dales who had lost £40,000-worth of sheep by a rogue vaccine. It put our calf disaster into the shade and he must be hurting more than we.

* * * * *

I had written an autobiography for the simplest reason of all. I believed I had a story to tell. Everyone has and it is arrogance on my part to think mine is interesting, but I think our way of living is fast disappearing in this modern world and as a piece of social history it might have value. Besides having a story to relate I also feel we have a philosophy worth handing on to the reader which is again presumptuous. It is pretty obvious I am basically a teacher and a Guider.

There was one more reason I felt the need to put on record the way we have lived. More than anything I would love to know how my ancestors lived, what was their struggle and their philosophy, their beliefs and their faith. We know the names of all the tenants who lived here since 1571 but, apart from what his will tells us about the character of William Currer in 1605, we know nothing further than a name about our predecessors. Perhaps those who reside here in the next millennium may be interested to know how we brought back the house from near ruin. We have no children or grandchildren to speak for us. I hope that those who read it might say, 'That was a good way to live, a certain way of being happy.'

Writing it was like taking a photograph and, on receiving the developed print, finding there were details there I had not noticed, a cloud formation, a pattern of a field and wall or flying bird. When you put your life under a microscope you find there are things you were unaware existed. Re-reading your own life story is an education. Out of the pages a story steps, a pattern evolves, a philosophy grows. It is an exercise worth attempting whether or not it is read by others. Personally I wanted the children I taught and those I took to the Hebrides, to know how much I loved doing so and I wanted those who have misconceived ideas about farming to know what it is really like.

For a month before Christmas my day was lengthened by two hours following the return of the proofs of *We'll see the Cuckoo*. I rose every morning at 4 a.m. to read and correct almost six hundred pages. Each morning I stirred the embers, threw on more logs and curled up in an armchair with a cup of coffee and my manuscript. I had two copies of proofs and I gave one to Hazel to read and correct and Margaret followed me, reading into the small hours. The job was an imposition on all of us coming just before Christmas and at the same time as the buying of the fifteen replacement calves.

Nothing was further from our thoughts than Christmas. We were totally preoccupied with calves and proofs. Men in the Mistal were working on the cable bringing a new choice of telephone and television to those who had time to talk and to watch. We had neither but the let cottage paid for the second buy of calves and the men were no trouble. That is until they cut through the electricity main in the village and severed power completely. We were without electricity for thirty-six hours.

Having a solid fuel Aga which also heats the water, cooking and heating water for mixing calf milk caused no problem. The plastic tube in the calf shed collapsed and we hastened to open all the doors and windows. By night-fall everything was in total darkness. Auntie Mary had no television and could not read by candlelight or play her tapes and Harry could not see to do his jigsaws. Harry's competence and obsession with jigsaws gives him pleasure unequalled during the long hours Margaret and I are both outside. They were bored but that was tolerable compared with the impossibility of feeding calves in total darkness. There was no full moon and a mist obscured any borrowed light from the highway in the valley below or the houses on the opposite hillside. We had only one source of adequate light. We drove the Range Rover to the open, huge north door and put on the headlights knowing we would probably have to recharge the battery next day. Who wanted cable television anyway? Not us! That was for sure!

We were grateful though a few days later for the presence of one of the workman. Because of bad weather the yearlings had been penned for several days and were bawling to go out one fairer morning. Margaret wanted to persuade them to go up onto the moor and thought she could drive them up the road, between the wall and the fence. She went up and opened the gate, at the summit, onto what used to be moorland and is still so called. Then she returned, let out the herd and shooed them up the hill. They did so happily enough but something must have startled them before they reached the open gate and eighty head turned and stampeded back towards the steadings. Margaret heard the thunder of hooves, saw a cable man tinkering with his van and yelled for help. Together they managed to pull a gate across the cattle grid. Had they failed the leaders of the stampede would have hurtled onto it and legs in plenty would have been broken.

I was making beds and heard the thundering hooves and dashed out to see the stream of animals veer away from the gated grid and pour into the shed. We waited for the herd to quieten and for us to get back our breath. Then Margaret went to open the gate halfway up the road and stood guard whilst I turned the cattle back into the lane. Margaret intended to guide all of them through the halfway gate but they were out to cause bother and half chased past her up the road so that there were cattle running uphill on both sides of the fence. The ones in the field were winning and we could see what was going to happen. They would reach the top gate first and come to meet those climbing the lane and as sure as anything the whole eighty would hurtle down again and we would be in their path. So we started to shout and wave sticks in an effort to hasten those we were following.

Almost at the summit all eighty converged. A man, coming along the top road in his lorry, towing a compressor, must have seen the mass of animals on our side of the entrance cattle grid, but he kept driving into the herd and amazingly tried to reverse amongst them. He could not hear our shouting for him to stop but the little man out walking his Jack Russell could. We ran to hammer on the window of the lorry and the driver braked. Cattle were milling all around more curious of the intruding vehicle than intent on stampeding again. They were out for trouble, mind, and his side mirrors were knocked away and we were in imminent danger.

'Don't move,' I think we called him an idiot. I think we might have sworn, a thing we seldom do except when being trampled on. He, to give him his due, did not move, but the little man with the Jack Russell continued to walk into the fray. We arrested him on the cattle grid.

'And don't you move either!' we ordered in a most unladylike manner. The little man, unlike the lorry driver, knew his rights. The driver was trespassing on the road but the little man was on the footpath, a right of way and, by golly, he was claiming that right. No-one was stopping him taking his dog along the footpath.

'Oh, but we are,' we shouted. 'Until we sort out this madding crowd you do not cross the cattle grid!'

'This belongs to the National Trust and is a right of way!' he shouted.

'It's covenanted to the National Trust but it belongs to us and there is no way you are crossing until we say.' I know we said much more than that but it would be unprintable. We knew the confrontation was ridiculous. He was probably a reasonable man if we had approached him gently and said 'please' but we had just had the sort of fiasco with cattle that few people ever experience. We were out of breath and afraid. It was a public footpath and for all we knew there would be others, out walking their dogs, climbing up the lane. Should anything cause the cattle to stampede down the hill a second time someone could be killed. The man in the lorry, I am sure, was thoroughly enjoying seeing these angry women holding at bay a little man with a Jack Russell.

We won! We got the herd safely onto the moor and let the lorry be driven out and the little man enter. It was not fair that we should feel so guilty of losing our temper. We feel the same when a jogger, causing havoc, will not stop and a reasonable request from us turns into a volley of shouted condemnation.

* * * * *

That one silly hen was sitting perpetually in one place went unnoticed. A few days before Christmas, when we had far too much to do, Margaret came to me in the calf shed to say that the hen had hatched one chicken and deserted it in disgust. The chick was wet and dead, she said. Indeed when she'd lifted it out of the nest its lifeless head had hung from her fingers.

We had arranged for the vet to come and castrate and de-horn the fifty-one survivors of the Beeston Castle calves. His car arrived at the same time as she dropped the lifeless chicken into an ice cream carton and covered it with a handful of straw. There were some things she needed from indoors and she absent-mindedly took the carton with her and left it on the Aga top. Don't ask me why! It was dead! She had not only said so but she had remarked, quite out of character, 'Thank Goodness.' The last thing we wanted was to cope with a noisy hen and one chick in mid-winter. The silly biddy had been wise to abandon it and re-join the flock.

The de-horning and castration went on until well into the afternoon. When we eventually struggled out of wellingtons and dirty clothes, all we wanted to do was to sit by the fire. Unfortunately women farmers have no wife preparing dinner. When our man's work is done the cooking and the cleaning begin. Harry and Auntie Mary were impatient. They had been waiting a long time for lunch. She had been biscuit making and the finished delicacies were cooling on trays on the table. I filled the kettle and approached the Aga and I heard a distinct, 'Cheep, cheep!'

Having just rid ourselves of the vaccine problem and the backwash the tragedy caused, the significance of that voice from the dead filled me with foreboding. I stood as if stricken. Margaret was coming through the front door relieved of her morning's activity, tired, bloodied and ready to flop, like her dead chicken, into an armchair.

I said in a monotone, 'Your chicken isn't dead. Listen!'

The whole family paused and waited. Sure enough from the white plastic box on the Aga came a voice, 'Cheep, cheep!'

Margaret removed the straw and there, still bedraggled but standing and shouting, was the 'dead' chicken. It was like a Phoenix, risen from the ashes. We definitely did not want one ailing chicken, weighing less than an envelope, abandoned in mid-winter but we gathered round the minute, re-born baby with all the wonder of the approaching Nativity. It would not stop screaming its, as yet one-syllabled language so Margaret poured some milk onto a saucer and dipped its beak in and tipped it on end so that it swallowed. Forgetting our own hunger and weariness, we all gathered round and watched.

We are too experienced not to know how time-consuming that baby, if it lived, would be. We knew the attention it would need, the frequent feeding, the constant warmth. The job would be Margaret's as usual for she is Saint Francis and always has been. I knew, and she knew, that the little ball of improperly dried fluff would have to be fed many times a day, on the table out of the way of the cats. I knew that Margaret would attend to it before she came for her own food and how the wearisome job would be her last before midnight and her first before dawn. Nevertheless no-one could take eyes from the chick. We watched its beady eyes accurately locate the oatmeal Margaret scattered. Every time she nail-tapped on the table, the little miracle imitated with its beak and picked up the grains. It knew instinctively how to drink and raise its head to swallow.

At night it was kept in the plastic box in the cistern cupboard in our bedroom and the angelic baby never made a sound all night but was wide awake every morning. I can honestly say that Christmas joy emanated from that minute baby. Watching it was more pleasure than television. We monitored its growth, its developing plumage, its quick learning of skills and routine and the most fascinated of all was Auntie Mary. Margaret called the little thing Angel as soon as the tiny wings became feathered but Auntie Mary called it Charlie-Chick as soon as its masculinity was obvious.

It is no exaggeration to say that the chicken provided the Christmas entertainment for all our visitors. Very soon the large kitchen table was too small for the adventurous chick. It found shoulders desirable places from which to view the domestic environment. It was expected that Margaret would soon tire of spending so much time at the table deprived of sitting room comfort and that she would bring the chick to the comfort of the hearth-rug and guard it from her armchair. We could not complain for the winter life of the farmer is cold and hard and in any case we were all captivated by the bird.

Of all of us, Charlie preferred Auntie Mary. I was pleased he did not favour me. I am not at home with feathered things and always send for Margaret if a bird comes down the chimney. We heard a cooing, once, coming from the flue box above the Aga and were apprehensive about lifting it down for there was, without a doubt, a pigeon in the chimney. I will never forget our surprise on revealing a white dove sitting on the floor of the box. How it had remained so white and why the bird of peace did not fly in fear but allowed Margaret to lift it down and release it outside, we never could fathom.

Auntie Mary was completely at home with the chicken and said he preferred her because she was the only one who did not wear trousers and had, therefore, a lap for him to rest upon. He used to shuffle between her knees until comfy and would stay there preening himself until she began to eat her evening orange. He was fanatically addicted to fruit and could not wait for the peeling to be completed before he was pecking at the orange. The laughter round the fire circle bordered on hysteria as the old lady attempted to share her nightly luxury with a very determined bird.

I think it must be the variety of our very full lives which intoxicates us. Life is never dull and affords more comedy than tragedy, as many miracles as disasters. We have equal amounts of hope and despair, calm and storm and surely, the little chick was sent to make us laugh again. We were all wondering what would happen if the little fellow made a mess on Auntie Mary's skirt but he never did and that was a miracle, too.

* * * * *

Hilda came for Christmas and on the holy day itself we went to Downham and Pendle Hill. It was a white landscape through which we drove, slowly, and silently. What a year lay behind of us. We could not get it out of the way quickly enough but we did not want Christmas Day to end it was so beautiful. The white scattering was glistening everywhere in the watery sunshine and there was a considerable depth on Pendle. The witches had had a bonzer time terrorising us at The Currer all year but December was ending on a quieter note. We had, as always, survived.

We parked the car on the plot beside the river and Margaret fed the multitude of ducks whilst I unpacked the picnic and unscrewed the flasks. No cloud lasts so long that the sun cannot break through. The Wilson family had converged upon us for the second year and there was a lively Christmas atmosphere in the house when we returned. I had pre-cooked a shepherd's pie and put it into the bottom oven of the Aga and we ate it by the fire watching the antics of the chicken. The white beauty outside and the peace within were as comforting to the soul as the warm food was to the stomach. Having eaten we went out to feed the calves who are not weaned until the new year.

* * * * *

January did not bring the relaxing atmosphere we optimistically hoped for. Charlie MacLean had been losing ground for some time. His brother wrote to say he was very ill. I wrote to the hospital immediately but he died before the letter arrived. The news was not without its joy for tributes to his courage and his character came from everywhere. Charlie had been one of Tiree's finest sons. We remembered his visits to The Currer, never too immaculate to mucky his creased trousers and Persil-white shirt to help lift a cow which was down after calving or feed a pen of calves.

The frightening thing was that the monster we had believed surgery had removed permanently from Joan, reappeared in both lungs. As was to be expected she faced the ultimate challenge with cheerfulness and courage. We talked endlessly about the children we had taught. There was always hope. Roy Castle himself suffering from cancer, was singing 'High hopes.' We would not accept that they might be 'apple pie in the sky hopes'. Lillian, one of our best loved guests had been coping with cancer returning for twenty years!

Winning or losing a great inner joy seemed to radiate from all of them.

They seemed to recognise there were two battles to fight, one of health and one of character and if they were losing the one they were all winning the other. We were in danger of getting a hang-up about cancer. There was far too much of it around. Our January work-load was a blessing and a consolation as we mourned for Charlie and feared for Joan.

We had decided that an extension to the calf shed was top priority. Carrying milk and feeding equipment between the two sheds was just a nonsense. Since we have spent close on forty years at The Currer, making things easier, how come it is still so ridiculously hard? It ought, one might think, to be a bed of roses by now.

The spring of 1994 was not, mainly because milk feeding went on until February. The calves bought to test the vaccine caused us infinite work, carrying bales and struggling between pens down the passage. We nearly went round the bend, but the confidence those healthy calves gave us was worth all the hardship. We were game to buy calves again come October. Margaret had not lost her nerve. We just needed that shed extension and we needed a hot water boiler in it and a corn bin of some kind. Carrying boiling water through and from the house for seventy calves, in the nineties, and sacks of corn from the store we used for the bullock sheds was ridiculous. Enough is enough! We told John to go ahead with the building, to get it done in the spring so that Johnny could put in the water heater and Mike could put up the lighting and everything would be ready months before it was needed.

Besides the extra calves out of line and two months younger, there was another time-consuming labour of love which Margaret performed all winter. That was to feed and water Samson and his friend who were housed in the Dutch barn, a long way from corn and water. Otherwise the accommodation was palatial. Bedded in comfort, among the diminishing walls of hay and straw, the two animals lived in luxury. Samson really was the most beautiful animal we had ever reared. Big, black, benign creature, we felt inferior in his presence. Completely sightless but incredibly serene he was let out, most mornings, with his inseparable friend and some heifers we had kept to fatten. One of these was Precious. She was a lively Red Hereford. When she was a baby the vet had said she had a dicky heart and would, one day, drop dead in the field. Watching television one evening, Margaret had seen the Duchess of Kent lift a terminally ill child into her arms murmuring, 'Oh, you precious darling!' and thereafter the baby calf was called Precious Darling. Precious did not die. Vets say these things sometimes. Nevertheless Margaret would not sell her to anyone else just in case. And there was Emilie whose udder had grown suggesting she had had a calf and a buyer would not class her as geld.

All, except Samson and Friend, returned at night to others in the bottom shed. The blind one could have been mistaken for a Smithfield champion. He, and Friend, went to the Dutch barn and one Saturday Margaret saw a small boy struggle bravely through a ventilation gap. It was pretty obvious he had been dared, by his peers, to enter the domain of the big, black bull.

Margaret caught him in the act. He stood his ground frozen by her sudden appearance and said in an awesome whisper, 'Is it a bull?'

'It's not him you need be afraid of. It's me,' she answered whereupon the trespassing child fled back the way he had come and we put a bar on the ventilation gap.

We believed Samson to be harmless but he was enormously heavy and blundering. He always turned away rather than charged but we began to grow uneasy. Animals are unpredictable and as we never provoke we do not know how they will respond to provocation. We knew that benign Samson would not attack but his weight was a danger to himself and to anyone in his path. Sadly he had to go before we left for our spring holiday. We could not leave the responsibility of him to Dorothy, with George out at work all day. Samson was already bigger and heavier than any other animal we had ever had. It was unfair to deprive Friend of a summer on the hill so Margaret introduced a heifer into the luxury of the barn so that, when Samson went, Friend would not fret unduly.

With great reluctance and sadness Margaret sent for the man who deals in casualties and he came and shot Samson, at home, in the barn where he had lived. Dear, trusting Samson. He had given Margaret more work and more pleasure than she could calculate.

Where are these people who, in their ignorance, believe that farmers do not love and respect their animals?

The choking feeling we had in our throats lasted all day and returns as I write this. We worried about Friend whose devotion had never faltered, but he seemed happy with the heifer and was allowed another summer as reward.

As soon as the calves were let out muck was taken from the shed and the foundations of the extension were laid. This was going to be the first time we would take a holiday in October and not worry about an incomplete shed. 'The supplier will come and measure for the two bays,' John promised.

'We'll need another fan,' Margaret said and in order that there would be no mistake made, she sent for the expert salesman to whom she had frequently talked, on the telephone, when the air system had been installed. There would be a fee for his visit but she asked him to come, thinking his advice would be worth paying for.

The expert came and he and Margaret spent time measuring and calculating and he came to the conclusion that one fan would be enough. Personally I do not trust computers.

'All you need is a longer tube. A ninety-foot one. I've got one in the car. You might as well have it now. It will save sending it by carrier.' The man advised a solid gable end, no Yorkshire boarding and for that, and that only, we thank him. He persuaded us to buy four expensive wall heaters which have never really been worth the cost.

John said we should get a proper silo for the corn. No wonder farmers are always penniless. No wonder they have been urged to diversify. The tourist trade has been recommended as a diversification because of the lovely places in which most farmers live, but catering and farming go together like chalk and cheese. The two jobs are not compatible. One is dirty and one is clean. We have to greet new guests, shedding muddy wellingtons at the door and taking wisps of straw onto the parquet tiles. I leave the shed I am mucking out and walk across the yard pulling stray hairs from my eyes and wishing that I looked elegant. Guests come when we are loading cattle, worming, injecting or taking off a lorry-load of hay and I am ashamed of the way I look and apologise for the way I smell.

Our immaculate guests climb out of tidy cars and insist on shaking hands. I glance at my rough palm and the broken nails and wonder what the guest is thinking. That same hand has to make their evening meal. We privately find it amusing when the phone goes and one of us hastens from some manual chore, dripping with rain, white with snow, dirtied with manure, or streaming perspiration. On picking up the phone the voice at the other end says, 'Is that Reception?' or, 'Can I speak to the proprietor?' We like the 'Reception' bit best.

For some reason nobody seems to mind. No-one recoils from a handshake, no-one cancels the evening meal. When he who is asking for Reception is told we are farmers and this is a farmhouse, he says, 'That sounds lovely.' I still maintain the two are not compatible and to suggest that diversification, for survival on the hill farms, should be to tap the tourist trade, is to be ignorant of the implications.

However we survive! We mop up the pool of liquid manure left by our wellingtons and return to the shed. We respond to all phone calls courteously whatever the predicament we are in, unless we are being asked to buy a closed circuit television to which we react, 'Whatever would we want one of those for?' or if the phone goes at precisely 7 p.m. and a friend is calling or, strangely enough, an annual guest who ought to know better, then we answer with an indignant, 'Don't you know what time it is? I'm just serving the evening meal!' The paying, annual guest is full of apologies. A friend seems to get the message that it is our personal mealtime, chats with unstoppable energy and finally says, 'I'll let you get your meal.'

We change from clean to dirty clothes and vice-versa several times a day and one might think we had an obsession with washing hands. We seldom sit. When we do something forces us up within seconds.

Some guests are more demanding than others and intrude upon the sanctuary of our sitting-room. We had just flopped into armchairs after serving an evening meal when a guest came to the door and into our momentary privacy.

'Can I have a word?' he asked. You have no idea of the sinking feeling experienced when such a request is made. I know, when a cottage holidaymaker walks determinedly towards me, that something is unsatisfactory, coins are stuck in the meter, the electric kettle has blown, water is dripping or the television will not work. When a bed and breakfast guest asks to speak to me, personally, something is wrong and my heart sinks to my boots.

On this occasion, fortunately, we had not begun to eat. I think we were too exhausted. We are always embarrassed when a guest invades our privacy when we are eating from a plate on our knee and luxuriating in an armchair.

'Go ahead.' I answered.

'Can I speak in private?' he asked.

I did not know whether I was capable of getting out of the chair. 'Can't you tell me here?' I begged.

'I had rather tell you in private. It's about the dog.'

Nothing can be private about a dog, I thought. Margaret was immediately roused. 'Have you run over Lusky?' she accused.

'No, no. It's not your dog. It's mine!'

'Has someone knocked yours down?' Margaret asked.

'No it's not that.' The man really was embarrassed. 'I left it in the bedroom and it's eaten the window frame.'

'Has the glass fallen out?' I said.

'No.'

'Will it mend?'

'Yes.'

'Can you repair it?'

'Yes. Don't you want to come and look?'

'Not particularly,' I said. The dog really had gnawed away the wood round the frame but the man was able to fill in the damage with wood filler and re-paint.

We are always afraid someone will maim our dogs for collies chase cars. We cannot always be there to hang on and we refuse to chain them. Most drivers will not wait until we have grabbed a collar and those who do err on the slow side and leave us hanging on in a crouched, uncomfortable position whilst they laboriously fasten seat belts, wave a prolonged good-bye and slowly engage their gear. Bed and breakfast is not compatible with collies. We blame hanging on to them for whatever backache we have.

Perhaps the reason we forget so many of the colourful incidents our diversification affords is because of the sheer quantity of them. Not a day passes but what we collapse in laughter at the strange behaviour of our numerous guests. The incident is quick, the humour is explosive and the memory of it is short.

'I saw three donkeys,' the elderly lady said.

'Two,' I corrected.

'There were three,' said the observant lady. 'Two brown ones and a white one.'

'If there are three, one is a stray. We only have two.'

'I can count,' the lady assured me insisting I went with her. On the green behind the house the three goats were fending.

'One, two, three,' she said. 'I told you!'

Chapter Thirteen

These things I ask of thee, Spirit Serene,
Strength for the daily task. Courage to face the road.
Good cheer to help me bear the traveller's load
And for the hours of rest that come between,
An inward joy in all things heard and seen.

Rev. Henry Van Dyke

If there was the remotest chance that pre-holiday chaos could be avoided by not going on holiday I think we would opt to stay at home. Certainly the coincidence is too regular. We could almost think that there is something sinister at work, that we are not meant to go and are being warned.

After each traumatic departure Mother would say, 'That's the last time we go on holiday!' and we could very well do the same. However our holidays are such joy we re-book as we leave. We are beggars for punishment!

The wonder of it is we always think we have things organised and that we are not going to have that awful last minute rush. 'Dare I say it?' I voiced, 'I think we're further ahead than usual. We've finished the decorating, we've ear-tagged and wormed and injected copper. The Range Rover has had its service, we've paid our last visit to the Cash and Carry. We don't need new clothes and, touch wood, all our paying guests go out on Friday morning. If we don't breathe, things are in hand this year.'

I have come to think it is fatal to have things under control, to have even just the anticipation of plenty of time. An elderly friend has critically told us, several times, that she and her sister always went to the pictures on the night before her holiday because they were all packed and ready and had nothing else to do. That is organisation! But I know from long experience that the more time we have the greater the chance of disorder.

Take the spring of 1994, for instance. John's men had already laid the foundations of a shed we did not need until October. 'While you are away the man will come from the suppliers of kit sheds to measure up.'

This was one for the books. Never had a shed been ready before it was needed.

Everything went smoothly until the last week. We even got a strong fence erected round the foundations to keep cattle away and, to make things easier for Dorothy, we had the paddock fence brought nearer. Everything was moving fast.

Then, suddenly we got a spate of spring foul. One after the other bullock went lame. Margaret could not count the cattle without bringing one home for the necessary Streptopen injection. One jab is enough but leaving paying guests for Dorothy to care for is one thing and leaving eighteen-month-old bullocks to force into the crush is another. It gives us a nervous breakdown even to think about it.

Five days before departure date, when things were hotting up in the house and we were realising we had been disillusioned and were nowhere near ready, one very big bullock got wooden tongue.

It was the very bullock we had found impossible to persuade into the crush since its ear-tagging and copper. Once bitten it was twice shy. Because of its refusal to go into the crush we had wormed it with 'Pour on'. We don't really want to be killed. Now somehow or other we had to give it an injection each of the five days we had left. We took three animals into the collecting shed, in which stands the crush, but there was no way that bullock was going to oblige so we staggered into the confined space with an iron five-barred gate and the unsuspecting animal did not seem to know we had pinned it against the wall. It never felt the jab Margaret gave it. We managed all five injections in the same way.

* * * *

We had a phone call from Joan Blamires, a more important friend than most. She had been a pupil, Guide, Sea Ranger and had, with her parents been caretaker of The Currer, whilst we took holidays, before we diversified. She had been staff nurse on the ward when Margaret had had an operation and she had been working at a hospice for terminally ill patients when we had needed advice for Mother.

She was ringing to say she wanted to come and see us, for coffee, during our pre-holiday week. We do not drink coffee that week, or eat even. There is not time. We couldn't tell Joan that, so we said of course she would be welcome, nobody more so. 'I'll bring my mother,' she said.

Three ladies arrived and I did not recognise the third. Joan and Phyllis had smiles from ear to ear. 'Guess who we've brought to see you?' they said.

Incredibly it was Monica, a refugee we had taken to camp at Budle Bay, Bamburgh, in Northumberland thirty-four years ago. I had not seen her since she was ten years old, a frightened little girl from Germany. I had always had reservations about the hospitality we had provided for eight children during Refugee Year. They had stayed a week at The Currer, then gone to camp for three weeks, then into the homes of individual Guides. Monica had gone to the Blamires family.

Here was my opportunity to question her. 'Did we do harm or was it a good thing to give holidays in this country to such young refugees?' I asked. It had been a very traumatic summer holiday after which none of us could decide whether the experiment was right or wrong.

'It taught me to speak English,' Monica replied. 'When I went back to Germany I had another language. That was a good thing and Joan and I have been friends for thirty-four years. It was, for me, a good thing.'

We were interrupted by a carrier. Publication day was imminent and the three hundred copies I'd already ordered of *We'll see the Cuckoo* were delivered to me. We were all excitedly opening the huge boxes and handling the heavy volumes. One is widely expected to thrill at the sight of one's name in print. I can honestly say the name on the front is meaningless. The book is the important thing. Even the title has more meaning than the author's name. A book has an identity of its own. It will stand or fall on its own merit and who wrote it is immaterial in the long run.

Phyllis took the copy she had been promised and the next few days were full of close friends collecting their copies and all thought of the holiday disappeared.

Before we go away things happen, in multiples of three. The amazing thing is the variety. We never know what to expect because trouble always comes in a different guise. Just one day before we were to leave Margaret announced she had brought home one of the October calves with a very swollen navel. It was enlarging even as we looked at it. As we were at that very moment packing cases to leave, she decided to send for the vet. The probationer came and we put the afflicted animal in the crush to be well and truly prodded. 'An injection, every three days, ' the young man said. 'Probably all fortnight!' Dorothy could not be left with that job so Margaret delegated it to John who said he would come every three days and do it for her.

The phone rang about 4 p.m. The only man we had in was not having an evening meal for which we were more than grateful. We had told him he was not getting any breakfast either unless he made it for himself for we were heading for Oban before eight and were abdicating. The lady on the phone wanted bed and breakfast for two people that night.

'Sorry,' I said. 'We are going on holiday at first light and the only guest we have is making his own cuppa and toasting his own slice.'

There was obvious disappointment at the other end of the line. The caller refused to accept my excuse. 'Please let me come. I'm in a wheelchair and a friend has brought me to my daughter's to make plans for a wedding and I can't get into her flat.'

I again said I was sorry but we had a long way to go and had closed doors on everyone because we could not make breakfast.

'I don't need breakfast. Neither does my friend. If we could just have a bed. There seems nowhere else to go. My friend is a nurse. I don't need anything except a bed and a ground floor toilet.'

So I said, 'Yes.'

The man from the Ministry of Agriculture, fourteen years before had said, 'My dear, when you are going into this, you are going into it!' How true his prophecy had turned out to be. There was never any way to back out. No escape at all.

'You'll have to make your own breakfast,' I said. Thank goodness for the new kitchen.

'That's no problem. No problem at all.' She was such a nice lady.

I do not really believe that my acceptance of the disabled lady had anything whatsoever to do with the miracle that happened overnight. When Margaret went out to check the calf with the swollen navel she found it had reduced from the size of a football to that of a tennis ball. It only needed John once. After that it was cured. It could not have been because I had opened the inn door. Our guests I am sure were not responsible for its unexpected recovery but they were responsible for alerting us to a problem which would maybe have deterred Dorothy from ever coming again to release us for holiday.

The nurse came to us, late at night, and said the ground floor toilet appeared to be blocked.

'It can't be!' Margaret said, but it was. The fall from the ground floor to the septic tank is not steep enough and, should someone flush away something alien or bulky, we have a blocked loo.

Because it has happened before, we have a manhole up which we can push a drain rod so Margaret said she would attend to it first thing in the morning, as soon as it was light. Exposing the manhole, however, she found that the blockage was not between it and the house but between it and the septic tank fifty or sixty yards away.

This was a major tragedy. A filthy, horrible job! She left me to cope in the house. Without her help we would be two hours late setting off – even if all went well and she did unblock the drain. Eventually she found the trouble to be not a foreign body but nettle roots in the pipe near to where it entered the tank. Suddenly there was a gurgle and fifty yards of sewage rushed into the septic tank causing the most atrocious smell. It was far worse than manure spreading.

'Never mind,' Margaret said victoriously. 'At least Dorothy didn't have to deal with that!'

* * * * *

It was one song of a morning. Utterly, utterly beautiful. Why do we feel the need to leave such a paradise? The reason is obvious. Perhaps, you think we have too many holidays but the beauty is deceiving and we would go crazy if we did not have a break.

We were stowing the last things onto the roof rack at around 10 a.m. Our guests had risen and insisted on making their own breakfast even though we were still around. Suddenly Margaret grabbed my arm and pulled me to the seat outside the front door, the one we had placed for Mother's ninetieth birthday.

'Listen!' she said. 'Listen!' It was the 30th April and the cuckoo was calling his name. Repeating it over and over again and tears streamed down my cheeks. We know great happiness at The Currer.

Similarly we always know great happiness on Tiree. We have recently lost a weekend by travelling Sunday and sailing Monday. The *Lord of the Isles*, like all her predecessors over the forty-five years we have been sailing to Tiree, berths in Oban on Sunday night and leaves early next morning. A night on board is not only necessary but convenient for then an early rise to catch the boat is not needed. We had decided that the loss of a weekend was too high a price to pay for this convenience and had booked our sailing for the Saturday and our night's accommodation at a guest house very near to the pier. The *Lord of the Isles* makes a Friday-night crossing from Lochboisdale and Barra arriving at Oban at 5 a.m.

We have proved time and time again that it is possible to get to Oban in one day even at the leisurely speed we travel. After the 'carry-on' we had before leaving late on that otherwise extremely beautiful morning, we dawdled more than usual, nodding after our coffee and napping after our lunch.

Those who do not see the beauty in spring and only crave for heat wave, scorched earth and silent streams and waterfalls will never relate to the Brown family. We are wildly ecstatic, seeing, smelling, touching, hearing and dancing with the sheer joy of green and white and blue and an open road ahead. We do not want to lie, eyes closed, burning on a crowded beach. We want to hear and feel the wind 'sweet with scent of clover, salt with breath of sea' and to hear leaves drinking rain, 'rich leaves on top giving the poor beneath drop after drop'. We want to travel on a by-road which enables a speed gentle enough not only to spot the red deer on the hill but also the cheeky chaffinches by the roadside. To visit mountains and not raise ones eyes to them for help is a lost experience, a missed opportunity. We are never just getting from A to B. Our journey north is through Wonderland and we were never more conscious of the fact than on that lovely day after loosening a drain.

At Inverary darkness was beginning to fall and we, who hate driving in such, began to wish we had not tarried so. We phoned the guest-house saying we were still miles away and drove nervously from the lighted township. We need not have worried. No-one else seemed to be going to Oban that night. The road was empty. An enormous stag decided we were driving slowly enough for him to risk crossing the road in front of us. There had been several sightings of roe deer on the wooded hillside adjacent to The Currer. Dorothy, walking her dogs in the early morning had seen them frequently. The presence of the wild on our doorstep excites us no end.

Our almost solitary journey ended in the lights of Oban and a warm welcome at the guest-house. We would rise too early for a cooked breakfast. The wherewithall for a light one was set for us in the dining-room. We paid the proprietor and after saying hello we said good-bye immediately for we would let ourselves out before he rose.

Sunshine awaited us as we drove onto the quay and joined the 5.30 a.m. queue for the boat. We were a little saddened because there are noticeably fewer seagulls than there were in those halcyon days when we were accompanied by children. Environmental pollution is doing untold damage. We are losing the larks and the peewits from the moor at The Currer and we use no fertiliser. And where, tell me, are the swallows and the sparrows which used to nest in the laithe porch and make such a mess at the dining-room entrance?

We left the dogs in the Range Rover and took the lift to the television lounge and struggled up the companionway to the observation deck. We are always critical of the one-armed bandits present on the *Lord of the Isles*. 'It's as well we took children all those years ago in the non-plastic age,' we frequently comment. I think we were a little subdued by the early hour and the memory of Charlie's recent death and the letters we had had from everyone, telling us how wonderfully he had run his last furlong to the winning post. Our occupation of his wee cottage would be fraught with memories, happy ones, of course, but angry ones at the unfairness of his terminal illness. We would be frightened too, for Joan who had occupied

Charlie's cottage during camp. I feared to tell her of his death. She had not asked.

Charlie had loved his homeland very much and we too loved the flat, fertile acres, the white, empty bays and the island scattered, sparkling sea. Auntie Mary sat outside the cottage many hours that first day. Harry fell asleep in his wheel-chair and we pushed him, still sleeping, indoors and rolled him onto the settee. Tiree is very kind to us.

So are its people. There are so many of them to visit. Perhaps it is by returning to the Hebrides that we mark the passing of time. At home we are perpetually Peter Pan, not growing old with the years though many are behind us. We function as we always have and the fact that Harry and I collect our pensions is not meaningful at all.

On the islands the loss of our youth is apparent in that we have no children and that Margaret MacInnes, whose wedding we went to thirty years ago, is a grandmother. So is Mairi Campbell at Cornaig Beg. We visited her on Margaret's fifty-fifth birthday. We personally do not think our sister looks any where near her age though her boyish hairstyle is grey. She is small and agile and smiling. She insists she is only greyer than I am because she has had more work and worry!

It was amusing to return to Mairi's a second time and hear that, when we had previously left, her six-year-old granddaughter had said of Margaret, 'She's awful wee, Granny, to be so old!'

'Skipper rules OK' used to be written on the sand. Sadly no more!

Tiree was looking very beautiful. The EEC may be a dirty word among the mainland farmers and crofters have the same struggle with the paper work but grants have made it possible for them to improve their homes which has resulted in a beautifully tidy island. It is so neat and cared for we can partially excuse bureaucracy.

Tiree islanders have achieved modernisation very well indeed. Where possible they, like the Harris people, have used the grants for the improvement of their original homes so that the outward appearance of the scattered, whitewashed townships is unchanged. Where new houses have replaced the old, ruins have not been left but have been dismantled and scattered on a rocky shore.

As I must have said before and make no apology for repeating, there is no point in going to Tiree if you are not besotted with empty beaches and the sea. The land is intensively grazed with out-wintering cattle and sheep and, when we go at the beginning of May, calves and lambs are being born everywhere and our spring visit is to an animal kindergarten. We revel in it. There are times when we are tempted to take a holiday later in the year, when the sea pinks are in bloom, the primroses are at their most prolific and the winter bareness of Tiree has become the luscious, green carpet familiar to us in our camping days. But we would not like to miss the nursery atmosphere of the island when the young sea birds are newly hatched and the eider duck sails with her flotilla of babies close to the water-line.

Every day Hughie's herd is increased and his flock bigger. A very busy, up-to-date farmer, is Hughie who was not even born when we first knew his parents Effie and John Lachie. His tractor has all the modern equipment for doing his work. His Continental bull sires beautiful calves which fetch high prices at the autumn sale.

He had just bought a new crush gate. Our crush had been engineered to suit the requirements of two ladies but we were increasingly aware it would be eyed with criticism if a MAFF inspector came to make sure we were doing nothing illegal.

'That is just what we want,' Margaret said. 'A crush unit's no good to us but we could use a proper mechanical gate. I could tell John exactly how we wanted it swinging.' We went into the crofter's house and Hughie found a leaflet advertising the separate gate and we lost no time in ordering one. I regret to say it stood idle in the farmyard a year, waiting for John to fix it just the way Margaret wanted. Men think you should buy a whole crush. They do not appreciate that women have to do things a different way simply because they are not as strong, or as big. Margaret needs to be above a huge bullock when worming or injecting and could not possibly reach from the ground.

If mankind is not careful it will lose its sensitivity to nature and its adaptation to the seasons. If so the lives of its most peculiar species will be dulled. Only the countryman now, it would seem, experiences the wonder of the spinning planet on which we live. Because we have no longer our own suckling herd, giving birth in spring, we feel it increasingly necessary to spend a fortnight where there is a constant reminder of rebirth. Many of us would welcome a return to sex being gender, and adolescence an outdoor adventure without the problems imposed by the media. On Tiree the joy, the laughter, health and energy of my tough and enthusiastic children haunts me. They were part of an age worth recapturing. My children wrote letters home to two natural parents. They did not travel with radios or play with computers and human reproduction was taught in schools. It was a safe and wonderful era when drugs were unknown and barn dancing was all in vogue. The incredible truth is that it was less than two decades ago. Discipline in school when I retired sixteen years ago, was normal. The ten laws and the promise of Scouting and Guiding were perfectly acceptable.

'Things have gone wrong in the last four years, Skipper,' said Jenny Jones, whose brother Mike canoed down Everest and was later tragically drowned saving a fellow canoeist hurtling down another Himalayan mountain river.

Jenny, a Queen's Guide, is now a teacher. She sat on my hearth-rug saying, 'They now look at teachers and say, 'You can't do nuffink!"

From the things she told me afterwards perhaps she is taking my place and showing children there is a right way and that it brings happiness. I just know her job will be harder than mine ever was.

If we, who live at The Currer, fear that work may isolate us from the full glory of the spring which ladens the hawthorn trees with heavy white May blossom and scatters a hundred different species of wild flower across our hilltop pastures, how much more difficult it must be for the town dweller. The seasons cannot be a part of his diet and his life must be incomplete.

* * * * *

Our return from Tiree preceded, by days, the eightieth birthday of Mrs Wass, our loyal school dinner lady for so many years. Winnie Annan alerted me and I invited several who had been colleagues, one way or another, during the twenty-one years I had been teaching up the dale. She collected Mrs Wass and we had a birthday party. Joan was looking ill, I thought.

Shortly afterwards we received an invitation to a reunion, at the school in which we had all worked. It was such a happy day because we had taught all the returning adults. The school building is no longer meaningful but the grown children are all my family. It was a brilliantly sunny day spent on the grass in the churchyard, surrounded by those we had taught, enveloped by their warm affection. The classrooms, the cloakrooms, staff room, the chairs and desks whatever, meant nothing any more but the obvious pleasure, of those who had known us when they were children, delighted us.

Almost immediately after our return from holiday we began to pack and post three hundred copies of my book and a phone call came from the local BBC radio station asking me to take part in a 'Real Lives' afternoon hour at the beginning of June. I mention this solely because of my unprofessional performance before the interview. I am ashamed of my inability to cope with the city. I think I could cope with the North Pole better.

The studio, I was told, was approached via the City Hall which meant I had to go through the centre and I planned to go by bus. I would be happy enough that way. I am ill-educated to drive and park in a city centre, but the thought of wasting a whole day walking to the village, taking a bus to town and then one to the city and repeating the exercise in reverse was ludicrous mid-season. Our accommodation was full. We had a big evening meal to make.

'Coward,' I reproached myself. 'Anyone else would be petrified of the live interview and you, silly old woman, are just terrified of the city.'

The Cash and Carry, where we buy catering size stores, is fortunately right on the rural outskirts, our side of the city. We had to make a necessary visit there a few days before the radio programme and, Margaret being with me, we decided to go down the road from Thornton, which leads directly into the centre, to see how far we could safely drive. The experiment was successful. The straight road could be left at the Sunwin House car park. We turned into it, paid two pounds and parked to drink coffee and allow Harry an opportunity to view city life for a change. Had we gone into the Co-operative store and made a purchase our two pounds would have been refunded, but we were not there to shop. Margaret and I got out and walked the relatively short distance to the City Hall. Afterwards it was easy. We left the car park turning left and left again and drove back to the countryside without having to negotiate any road complex.

'I'm going by car!' I announced. 'It's no problem at all. I'm not wasting a day timidly travelling by bus!'

So I did. I set off early, for The Currer was full of guests all planning to switch on the radio at the appropriate time. I needed some space, 'A little peace and quiet', we used to say. I took a flask and some sandwiches and as I pulled into the Sunwin House car park I had time galore to spare. That in itself is unusual. I smiled to myself. Without family and dogs I could get to places without being late, without running the last hundred yards! Without family and dogs I could be as dignified as anyone!

I sat for, perhaps, half-an-hour, drinking coffee and eating sandwiches. I reckoned I needed another half-an-hour to walk to the City Hall, find the studio and compose myself. Three hours and it would all be over. I played unconsciously with the parking ticket for which I had paid £2, tapping it rhythmically on the dashboard. Having nothing else to do I read the conditions on the back of the ticket and nearly leapt out of my seat. My £2 only allowed me two hours. That was not enough even if I drove out and came back in again.

I ran to the man at the entrance booth and told him I was due on radio in half-an-hour and needed about three hours parking. I took out my purse but he shook his head. He looked at me as if I were an idiot.

'You can park nearer than this!' he said.

'This is just fine,' I said. 'I'll just have to pay more.'

'You can't do that,' the man said. 'You are on camera. When your time is up you'll be clamped and I can't stop them.'

'Can't you tell them?' I was amazed. Everything, in this modern era, is done by computer. Human help is no longer available. It became apparent that I was going to have to move.

'There's another car park right by the City Hall,' the man said. 'Go right at the traffic lights and right by the ... and left again and right at the next traffic lights and left over the pedestrian crossing ... '

'I can't do that,' I gasped. 'I live on an isolated farm. I'm not a city dweller.' He shrugged his shoulders as if I were something from another planet.

I ran back to the car, drove out of the car park and turned left and left again and began to retrace my earlier route into the centre. There were some side roads. I pulled into one, an unadopted one, cobbled and ill-cared for. I pulled up in front of a dirty workshop of some small welder and I asked if I could park the Range Rover.

'Aye, lass,' he said. 'It'll be all reight.'

And then, by golly, I had to run. Not just Scout's pace, I had to very nearly gallop. 'You silly woman,' I thought. 'You are about to be questioned on air because you can teach and build and farm; because you can camp and sew, decorate and build roads, because you are a plumber and an electrician and a cook and bottle washer; you can fence and build walls, lift and carry and drive to the north of Scotland and you become a nincompoop in the city.' It was pathetic, as eleven-year-old Roddy used to say when he could not do his sums.

I looked at my watch. I must not stumble. I must not be late. I was amazed no-one stopped me and arrested me but I got there with a few minutes to spare. It didn't matter that I looked dishevelled. I was not appearing on television but it was important that I find a toilet. It was important to find it in the City Hall and not to have to ask at the studio. I would have been embarrassed to do that. I was not having an attack of nerves!

It occurred to me that the man about to interview me was taking an enormous risk. How did he know I would arrive? How did he know I would not be tongue-tied? He had never met me and he had to talk to me for nearly an hour! The programme was live! The man was trusting to say the least!

I passed the Gents and a toilet for the Disabled but could find no Ladies. A passing man said ladies had to use those facilities provided for the disabled and that a lady in the office opposite would give me the key.

I have been acquainted with loos for the handicapped ever since they became a feature nation-wide. What possessed me to think that the dangling string above the toilet was the flush chain, I cannot imagine. In pulling it I not only activated the alarm but I also caused a HELP sign over the door to flash and bring someone running. Acutely embarrassed I ran out of the cubicle yelling to those running towards me, 'It's a mistake! I'm alright. It's a mistake!'

* * * * *

So I entered the studio laughing and I was not tongue-tied at all.

Our old goat Isobel died at last, naturally and at home. Poor Charlie was bereft. 'We can't have this.' Margaret decided and went down to the Saturday cattle market in the neighbouring town. Sometimes goats are for sale, usually to Pakistani buyers. She arrived late and the goats had been sold. One man was willing to make a bit of a profit and resold a little nanny to Margaret for far too high a price. The creature was wild and very agile. We had to reinforce the fence round the garden but she managed to eat most of the plants before we caught her. If it had not been for Charlie and the liberal amount of food Margaret was prepared to offer, she would never have stayed. She did disappear once and we thought we had lost her but our accommodation must have lured her back. We called her Rosie.

One of our Guides was married and living on a smallholding on the moor above Heptonstall. We had visited her in the spring taking with us Auntie Mary's chicken. It had grown into a beautiful cockerel and was arguing with the cocks we had. Alison had hens only and wanted to breed so we took her the fine bird we had reared in the house and learned that her nanny was in kid. Knowing Isobel to be past the 'sell by date' we had promised to take any unwanted billy kid. It would be company for Charlie should he be left. However Isobel had died too soon and there may not be a billy kid to be adopted so we bought Rosie and when, indeed, a billy kid arrived and Alison deposited him on us, Rosie found we not only provided food but babies as well.

The Persil white spring baby brought great joy to the guests particularly to Denis whose job it was to accompany any holiday child who wanted to bottle feed Jimmy, as he was named. Fortunately his little hooves were soft for he bounced all over the place, leaping onto the wall of the Loft Cottage steps and cavorting in front of the door.

We had many re-bookings for the following year because children had become wildly in love with Jimmy and thought that he would still be a baby twelve months hence. Alas kids, calves, kittens, piglets and pups mature at an alarming rate. The Christmas baby had become a fully combed and plumed cockerel in the space of five months. Five-month-old pigs are ready for market. Year-old kids are fully grown and eight-month-old calves are monsters. Lambs cavort and gambol in the spring and are sheep in the autumn. It is all very sad but undeniably true.

Harry enjoys Denis so and regular guests arrived hoping to find him here. Children mature much more slowly than their animal relations but even so the boy Denis was becoming a man. Fearfully we looked for any change. We found none. Growing older, going for a year to Russia and yet no-one could say Denis was different. Guests who had met him at fourteen heaved a sigh of surprised relief and said, 'Denis is just the same.'

Denis was quite happy to help us handle any animal, no matter however large. He was very useful when ear-tag numbers had to be checked. Neither Margaret nor I can see the almost invisible collection without glasses. A three-man team is desirable when worming with Dectomax for Margaret must be in the crush and it often takes two to persuade animals, wishing otherwise, to join her in the limited space. Most men would run a mile from Friesian bullocks but they are the most benign of bovine breed.

The accountant had promised to return my books, after audit, in person. 'I'll be there about 11 a.m.,' he had said.

We had decided to worm calves that day and had put the seventy eight-month-old animals into the road ready to pass them through the crush. They were fine animals which had survived the vaccine plague and it pleased us that our accountant would see what good calves Margaret could rear. We were delaying the worming until after his visit but at 11 a.m. he did not come and we sat around waiting. At noon he phoned to say he had been to the top of the road and, seeing so many animals in the lane he had decided not to come and I might as well pick up the books when I was next in town! I did and found a mistake he had made. The accountant had said we reared heifers. It was an error which could have cost us dearly. Had he visited us as planned, he might have spotted they were bullocks, providing of course that he knew the difference between male and female. One man said to Margaret, 'I know which is a cow and which is a bull but what are the ones with tufts under their belly?'

The error caused a delay in presenting the profits to the Inland Revenue. I had been sending balance sheets to the tax office for nearly fifty years, at least forty of them without the aid of an accountant and had never been late.

The mistake, had it gone uncorrected, would have been far more serious than the delay. For us to be claiming subsidy on bullocks the accountant told the Government were heifers would have caused a rumpus indeed. The error had been human failure to tell the computer that we had stopped buying weaned Hereford heifers and begun buying Friesian bull calves at a week old.

The necessity to audit the accountant's work and the returned forms and documents from MAFF has come with the computer. The mechanical robot says the most extraordinary things. Farmer friends received returned documents for young calves allegedly born in 1930.

* * * * *

Guests continued to come and go all summer, filling all bed spaces and necessitating that Denis should sleep behind the curtain. I am sure that owners of normal guest-houses look forward to the end of the season and a break from the seven-day routine of bed-making, laundering, meal-making and entertaining however much they love their work and however nice their guests are. We dare not look forward to the autumn for it means cattle sales and we close our minds to those. If it were not that we must say good-bye to those beautiful cattle pasturing on the hill, we too would yearn for the October holiday and the silence in the kitchen because the washing machine is idle. As it is, we wish to prolong the summer indefinitely, prepared to tolerate the joy and problems of summer tourists for ever if it held back the sales.

We can thank the EEC for the fact that there is no way can we sell before the end of the retention period. There is no selling of the best first, in August. All must wait until September and, after the heifers are sold, we only sit once on the hard seats of the ring.

The response to my book was interesting. There was an overwhelming acclaim from afar and a quietness from close-by, illuminating the fact that those who thought they knew us found they did not. My memoirs sparked dormant ones in others, some of whom I had not seen for ages, many of whom I had never heard of before. Because of our experience of happiness, my recounted stories excited similarly pleasant ones in other people. They remembered good times in the countryside and The Hebrides where, amazingly, so many people have wandered. Farmers, teachers, Guiders, sons and daughters all had something in common with us. There were only a few to whom our way of life was incomprehensible. They shook their heads and said they would have to read it again. It pleased me they were willing to do so.

It was a warm and satisfying experience but we had no time to enjoy it for we were entering another period of struggle caused initially by the long, hot dry summer which lasted right through September and into October.

* * * * *

In September we re-awakened the Council to the fact that Altar Lane needed attention. We had now been periodically alerting them for over thirty years. It was eighteen months since the sub-committee had agreed to do something and to advertise closure. Once again they were roused only because we had to complain.

Walking the boundary one lovely September day, Margaret found that someone had illegally dumped several lorry-loads of evergreens and completely blocked the road. We had no complaint about that for we had wished to do likewise for as long as we could remember. It was the fact that the heap was of evergreens which concerned Margaret. If the cattle remained on our side of the boundary no harm could come unless children threw over the branches. Should someone on a motor-bike breach our boundary then the cattle would have a feast. We have no garden evergreens and we did not know from what tree the branches came. Yew is deadly poisonous. Margaret rang the Council and asked for the lorry-load to be removed. There was to be another meeting of the sub-committee later that month, we were told, so we engaged the help of a local Councillor and he promised to walk the road with us. Three weeks later he arranged to meet us and we walked up the moor and onto the road and noticed at once that the pile of evergreens had not been removed. Anxiety about it had been dulled by guests assuring Margaret that the branches were not yew.

A village lady, exercising her dog, was coming towards us and though I greeted her I kept walking beside the Councillor and it was Margaret who paused.

'Just look at that!' the little lady said pointing to the almost total blocking of the green lane.

'I know,' said Margaret looking, not at the lady, but at the huge pile of cut branches. 'I've told the Council. I rang weeks ago telling them to move it. I thought it was yew. The holiday-makers said it definitely wasn't yew but I'm still afraid it might be.'

She turned at last to look at the ashen face of the timid, little woman who, in a shocked and frightened voice said, 'Why ever did you think it was me. I wouldn't do a thing like that.' I don't think she really understood why Margaret bent double with laughter. She tried to explain but the Council man and I were a long way ahead and she ran to catch up with us and the dear lady might still be perplexed as to why we thought she had dumped a lorry-load which had closed the lane at last.

The pile was never removed but it held back the joy-riders adequately and it persuaded our local representative that legal closure must be. We attended the sub-committee meeting and a small number of the committee opposed the motion and the matter had to be taken to a second meeting and then the go-ahead was given for work to be started. Winter, we knew, would be the next deterrent to hold up the process.

* * * * *

I used to assume that it was all our fault that we were always at the last minute. In 1994 the fault was anything but. It was the fault of this strange society of which we refuse to become a part. There is a lack of competence among those who have succumbed to computerisation. The age of professionalism has gone and mental wards will soon be full.

We nearly increased their population prior to going on holiday. Nothing was our fault. Is it feminist to say all our problems were the fault of men and computers and a man-dominated Government?

On the 16th September we had to sell forty-nine bullocks. The retention date expired on the 15th and, had we missed the sale the following day, we would have been on holiday for the next monthly auction. Two large cattle wagons had to come twice. For the first time we had kept the cattle in overnight so we were not late. Indeed neither were our hauliers. The first two loads went off with ample time to return for the rest. At the mart, however, there was abnormal congestion. The paper work is phenomenal. It is difficult enough to pen animals in order, without having to match CIDs with ear-tags. The cattle wagons bringing in animals for sale were queuing up to unload. Ours, therefore, returned late and we were pacing the gate like the Murray Grey. The second half of our forty-nine bullocks, and we who were following, arrived too late to take our place in the Draw and we had to wait until the last. It did not seem to make any difference to the price but we are not experts at doing nothing, hour after hour, perched on uncomfortable benches, tired and hungry.

We were fed up for a host of reasons. Believe it or not, after laying the hardcore and the foundation of our calf shed extension in April, the project had come to a standstill. It was wholly the fault of the suppliers, John said, waiting anxiously for the delivery of the relatively few iron uprights, girders and roof tins. The original gable end had been dismantled and was to be moved a few yards south. Concrete blocks did not arrive. 'There's a shortage,' we were told. The recession was cutting deep but whoever heard of a shortage of concrete blocks? With builders standing idle one would have thought there would have been a backlog.

Apparently not! When we built our own sheds we could decide to do so in the morning, buy second-hand materials in the afternoon and start building in the evening.

No amount of telephoning hurries the modern man. All one gets is an answer-phone. No wonder a cousin listened to his, the other day, and the only words on the tape were 'bugger it'. In a country where so many are out of work, offices are so understaffed that callers are placed in a queue, told to wait patiently and not put the phone down. Whilst doing so they are serenaded with soft, tranquilising music which adds to rising aggression.

Time was running out. The extension had to be finished before the electrical additions could be made, the air system extended, the corn hopper ordered, the plumbing done and the hot water heater installed.

'It will be ready for your return,' John said and Mike said and Johnny said and the hopper suppliers said. Of the few scripts of the soap opera we call our lives which repeat themselves, I place at the top this assurance that things will be ready for our return with calves. Men seem so unconcerned. Returning home is traumatic enough with seventy babies without having to test the air system, familiarise oneself with the boiler, the shute of the corn hopper and check that the wall heaters are properly positioned.

As the Yorkshire boarding came down I used the wood, got out my alligator saw and made new hay racks which would not be needed until

January. Try as we might we could not keep village children out of the Dutch barn. Hay and straw are always fire hazards so we decided to kill two birds with one stone and fin the air gap of three feet between the bottom bullock shed and the barn. That kept the children out and would, heaven forbid, keep any fire in, but it did mean the cattle shed was deprived of air, so we ordered another fan and air straightener and tube. It was extravagant but we had not had to have an extra one for the extension in the calf shed. Roundabouts and swings is what it is all about.

* * * * *

I was weary of coping with the upstairs shower. Consistent use, many times a day over many years had caused the grouting between the tiles to leak water down into the dining-room. The problem of leaking water was not a new one. Because some guests were careless and splashed water all over the floor, we had fitted solid, sliding doors but found them impossible to keep clean. Denis resealed the well to the wall with silicone but the new problem was identified as leaking grouting. It is impossible to keep it absolutely free from fungus. I was heartily sick of showers.

'I can't cope with this any longer!' I said so we went straight into town to look for a shower unit which was a solid cubicle. Joan had had one for the last twenty years and if I had one like hers I would be sure of success. The plumber's merchant had a kit one.

'I don't want one to assemble,' I insisted. 'Sticking the thing together is the problem. I want one which is without any joinings.'

The second place we went to had just the thing we needed, tucked away out of sight and damaged. Sometimes a second is a bargain. This was impossible so we went to the assistant who said he could order us one and that it would be about a week in delivery.

Weeks went by. We phoned frequently and at the end of a month we were exasperated. No wonder British Telecom is making such a profit! But if there was stalemate re the shower at last things were moving fast on the extension. Once John got the kit pieces his men were attempting to beat the clock and be finished by the time we returned from holiday. With only days before leaving for our second year in Aberporth, Mike was able to come and fix the lights and to put the longer tube on the air straightener and his brother Johnny was working on the water heater.

Early one morning, Mike had already arrived when the shower cubicle was delivered. We were excited. Johnny would be here to do the heater in the calf shed. That could wait a day whilst he put in the shower. We felt like raising the Jack on the flagstaff Mike had fixed for the Royal Wedding in 1981.

'Where do you want it?' the carrier asked. We were courteous. It was not his fault there had been a five week delay and that we were all set to go on holiday in two days time.

From that moment we had problems. We should have been asked, before we bought, whether there was any hope of getting a solid 'telephone kiosk' into one's home even if one was prepared to wait for delivery. 'It will come eventually,' the salesman should have said, 'But you might not be able to get it into the bathroom!'

We have several ways to approach the first floor. We can, if necessary take furniture up the barn steps leading to the Loft Cottage and pass through its sitting-room and fire door. We had brought wardrobes and rolls of carpet that way. A shower cubicle would not go any way. Back door, front door, inside steps, outside steps, this way, that way, no way at all. If we managed to twist it through the door we were balked by the 12 inch by 8 inch principals which supported the roof. The carrier was extremely patient. I must commend him for that. In desperation I ran for Mike to down tools in the calf shed.

'I'll get it in for you, Skipper,' he said. I had taken him to camp. He is good at solving problems. Impulsively he whipped out his ruler and measured the window in the bedroom on the other side of the corridor from the bathroom and then he measured the solid cube and declared that, if all the paper was removed, he could get the shower through the window. So he could. The operation was fraught with danger to life and limb and brand new shower cubicle, now naked of all wrapping. The stubborn box had to be carried up the ladder horizontally. I think the carrier, holding up the stern, was as petrified as I was catching the bow as Mike, under the keel, eased it through the window. It was typical of Mikeonian confidence and the lad beamed with pride as we all stood victorious in the bedroom.

So now what? The bedroom door is not opposite the bathroom one and, though the corridor is wide, there is a principal beam right above the door making it impossible to use any ceiling space. The low beam also means a less than 6 foot 6 inch door. Again it was stalemate. We had a half en-suite bedroom and it looked as though our evening guests would have to share accommodation.

'We'll have to send it back,' I decided.

The carrier said, 'That'll cost yer an arm an' a leg. It's all unpacked. It'd only sell as a second now.'

Margaret was not for giving up. She turned it this way and that. Mike said, 'How much do you want that cubicle, Skipper?' It was a silly question. However much you want something you cannot always have it.

Margaret said, 'A lot, Mike.'

'Okay.' he said. 'We'll take it through the wall.'

Johnny had arrived. He and his brother took out Stanley knives and the carrier thought it was wise to disappear. The two men cut a square in the plaster on either side of the bedroom wall directly opposite the bathroom door. They tapped out the plaster and then sawed through the stoothing and lifted out the criss-cross of 2 inch by 3 inch joists. Then they passed the cubicle horizontally through the hole, across the passage and into the bathroom. Johnny went to the builder's merchants and came home with angle irons to re-insert the stoothing safely, plasterboards to cover the hole both sides and plaster to skim. When Margaret whitened the wall, before we left for Aberporth, the operation was invisible, no stitch marks whatsoever! The cubicle fitted snugly into its rightful corner and we breathed again.

Johnny fixed the water heater but there were as yet no water bowls in the two new bays. Mike fixed up the ninety-foot plastic tube and repositioned the fan. It had been in the south wall. When this was moved two bays we thought it better to have it put on the north gable. It really looked as if it would be ready to try before we left. It was. Mike switched on and the longer tube did not fill.

'I told you we needed two fans!' Margaret was almost violent. 'If one isn't enough to fill the tube, how can it push out any air at all?' she screamed over the phone to the advisor.

'One fan is enough,' he repeated. 'I must have left you a tube with the perforations too near together. I'll send you another immediately.'

She was justifiably angry. 'Immediately' was too late. The tube would arrive whilst we were away and, although Mike would see it was slung, time and money had been wasted. There was another thing. The wall of the extension through which the shute from a new corn hopper must pass would be built after our departure. What's new? NFU. Normal for us. If we are going on holiday the only difference is that our pre-departure problems are new. The same thing never happens twice. We wonder how long our script-writer can think up new hurdles for us to jump.

There was one thing Margaret believed she could control. She had given up hope where buildings were concerned. That was no longer her domain. Rearing calves was. In December she had taken the risk of retesting the vaccine. None had died. She knew just what she wanted. What had happened last year would not happen again. To be doubly sure she had booked a consultation with the vet at which she had submitted her order to prevent three possible infections, salmonella, pneumonia and IBR.

'Collect it the day before you go Wales,' the vet said. To go to the surgery on such a busy 'day-before', is always a straw likely to break the camel's back, but she agreed. She would at least know the life-saving injections were ready in the fridge, awaiting our return. Some sudden emergency call must have attracted the vet's attention on her departure for when she went to collect the vaccines she learned they had not been ordered at all. The vet had forgotten. Of course Margaret is human and was angry but it was not the end of the world. There was plenty of time. Dorothy would collect.

I remember a time, not too distant, when reliance on other people was unnecessary. Farmers could get on with their job unhampered by red tape and computer errors. I can remember when we could build our own sheds because our tractor was small and loads not too high. I remember when we could take the muck out ourselves and spread it with a hand fork and still make a living. I remember, not so long ago, when we did not have to buy all the calves on one day and sell at one sale. They were gentler days and I remember them with nostalgia. Any incompetence was our own and we could laugh and cry and clear up our own mess. In the autumn of 1994 incompetence was laid at the door of the professionals and suppliers and one's attitude toward that is different.

Matthew said to his mother, 'Why are you so cross when I spill something and not when you do?'

'Because,' said his mother, 'When I spill something I have to clear up my own mess. When you do I have to clear up that too!'

That is absolutely how we felt.

* * * * *

But first we were going on holiday. We decided it was foolish to leave at the crack of dawn and travel all the way in one day. It, we believed, was tempting providence. So we booked a night at Wrexham. All desire to hurry had left us. It should have taken us three hours to get there and it took twice that time. We had planned to cook a meal somewhere en route using our camping technique and we ended up drinking soup in the Little Chef at the Travelodge. We cooked the meal in an empty car park next day at Barmouth. Arriving at Dolgellau we had plenty of time so we drove out there to satisfy a whim of Auntie Mary. Harry sat in his wheel-chair overlooking the sea whilst Margaret gave the dogs a wee bit stroll on the beach and I cooked lunch. We are ever so relaxed on holiday.

We had simply glorious weather. We have learned to expect it in spring. The Hebrides are islands in the sun during May when each sunrise heralds one song of a morning. That sort of weather cannot be expected in October so we take Harry's long johns, padded anoraks, thick socks and wellingtons. Even though it be St Luke's little summer we do not expect to be treated in the same benevolent way. Yet at Aberporth in 1994 we wore summer clothes, sat sleeveless on warm beaches, their emptiness, only, telling us that it was not the height of summer. We found a different beach for coffee and another for lunch and if we stayed at The Cedars we had lunch on the picnic table outside.

We phoned home frequently. Yes, the tube had arrived. Mike had fixed the wall heaters. The postman brought a photograph of the newly erected corn hopper. Yes, Dorothy said, the vaccines had arrived and had been collected but the salmonella vaccine was unobtainable. There had been an epidemic nation-wide and the pharmacy had run short. John had put in the new water bowls, all animals were well. The weather was dry and warm and we were to enjoy our holiday and not to worry.

There was, currently, a great deal of public concern about calves being sent abroad to veal crates. It distresses us, also, that veal is produced by keeping calves deprived of space. We know how much calves love to dance around. We wondered how all the unrest would affect the calf sales. We went to Carmarthen and to Cardigan auctions of week-old calves and realised we would never be able to buy in Wales as it was impossible to tell what the auctioneer was saying. We discovered there was really no change. By one means or another dealers were managing to ferry calves to the Continent. The poor little jokers are a bi-product of milk. Reared for veal they have a short cossetted life. Kept for beef they have a longer one, preferably grazed out of doors as they are at The Currer. Deprived of one or other of these outlets the only alternative is to slaughter at birth for there is no lactation without a calf being born.

We spent only long enough at each market to discover that things were as near normal as would make no difference to availability and price. We were on holiday and not yet buying. Margaret bought some capsules from the Osmonds man because those she had ordered through our agricultural merchant had not arrived and she wanted to give each calf one in the market before the long journey home. The earlier ordered ones arrived three weeks later. What is wrong with today's producers and sellers that they are so lethargic?

We found the same lack of enthusiasm to sell in the empty shops in Wales. The supermarkets were busy enough but the high street shops had Depression Fever. There was little to buy and less enthusiasm to sell.

We wanted to buy a chair for Auntie Mary to sit on the sunshine-flooded beaches. We had several at home but had not expected that to bring one would be necessary. We do not normally expect it to be so warm in October. However the beautiful weather seemed all set to continue so we tried several stores. The season was over, the shopkeepers seemed disinterested and their premises eerily empty. We found a hardware store in Cardigan selling camping equipment. Having lived through an earlier age I asked for a deck-chair. The sleepy proprietor said, no, he had no deck-chairs. I persisted and walked amongst the gaudy, modern camping equipment but he was right. He had no deck-chairs.

Next morning, walking out early in Aberporth, I called at the miniature hardware store which appeared to sell everything, as Calum did on Tiree and MacAskill on Harris. Again I asked for a deck-chair and was answered negatively.

'I just want something for Auntie Mary to sit on the beach,' I said.

'I have a folding, canvas chair,' the young man said. 'Would that do?'

Is it possible that it was my own fault all along, that 'deck-chair' is an old fashioned noun? I carried the purchased article jauntily along through Aberporth. My vocabulary might be out-dated but there is still a spring in my step.

We were able to relax. The outstanding jobs at home were being finished. True to his promise John would have everything done for our return. The post which brought the photograph of the new bin brought a circular from BOCM Pauls, our corn suppliers, offering a free, assemble-oneself corn bin. The offer came too late. We had just bought a professional-looking one costing almost £2,000 to hold the calf corn. However we could not ignore this unique offer and decided to accept the free one and put it somewhere handy for the bullock corn. We would have to find somewhere high enough. We were determined to do something. Things have got to get progressively easier as we get progressively older.

Our calf buying went well. Standing by the pen marked, 'Mrs Brown', I thought the calves were a bit smaller and rougher than previously but there were plenty of them. The dealers were buying vigorously and were prepared to outbid Margaret for even the smaller calves. Her face was grim when the sale was over and she came to the pen to look at the calves she had bought. Two of them should never have been put through the market. She would not have bought either had the pace been slower but there is little time to hesitate. One had pneumonia and one was scouring. The man helping us to check the calves against the bill of sale said she should not take them out of the market but Margaret will leave nothing to die alone without trying. Buying seventy calves is a gamble and if we are not prepared to take it then we must quit.

Arriving home, two hours after the first load and close on the heels of the second delivery, our spirits rose. Well-bedded in clean, deep straw the 1994 batch looked better than they had done in the market and every one of them was drinking before we went to bed not too many hours after midnight.

We had one unease. The instant salmonella vaccine was unavailable due to the epidemic which had used up stocks. We could get a substitute from the agricultural merchant but with it calves would not be immune for three weeks. There were little annoyances. The heaters were positioned too high. Had they been installed before we had left Margaret would have had them put at a different angle. Because it had not been properly tested the water heater, our long awaited convenience, stopped filling and we resorted to carrying heavy buckets from the kitchen.

'It's an air lock,' Margaret said. The plumber said it couldn't be. Everyone always jumps to the conclusion that Margaret, being a small woman, is wrong. The plumber poked around and then announced with confidence, 'It's an air lock,' as if the idea was his.

I was reminded of my behaviour on the football pitch with my eleven-year-old boys. I never knew what errors constituted a 'corner' or a 'penalty'. I waited until I heard a boy shout 'Corner!' and then I quickly blew my whistle and repeated, as if I knew all along, 'Corner.'

The water heater saved us hours of heavy carrying. We should have had one installed four years ago. Just to turn on a tap and get instant hot water was heaven.

* * * * *

On the morning of our return, before the roof rack was emptied of cases, we prepared to give the IBR vaccine. Margaret took the twenty-eight small bottles into the calf shed. We had intended to do exactly what we had done in December when we had asked the vet to repeat the disastrous formula and nothing at all had happened. No calf had been unwell. Not one of the test cases had died. All summer we had ridden on a raft of confidence. Suddenly, with £150-worth of vaccine in our hands we trembled.

'Let's just give it to five,' I said. Each little bottle had to be mixed with another and vaccinated five. 'Let's be sure.'

By some miracle Margaret had retained one bottle of each of the previous year's supply. With her spectacles on she noticed, for the first time, that the colour coding was a slightly different shade of orange. The October one which had preceded disaster had only been recovered from the floor of the calf passage when we had cleaned it in spring. The December bottle Margaret had kept purposely. She had put both into her drawer but had not, months later, compared the colour coding. Now she did and saw a very slight difference. Even wearing spectacles she could not see the small print but comparing the colour coding on the 1994 bottles there was again a slight difference. We thought it might be due only to batch but we hunted out the magnifying glass to bring up the small print and discovered that, although the December bottle looked the same and was administered the same way it did not in fact contain the IBR vaccine, only the pneumonia one. We were shocked and horrified. In December we had believed we were vaccinating against IBR and we had been doing no such thing. We had left our test calves vulnerable to whatever infectious bovine rhinotracheitis germs had been prevalent in the sheds. The December test had not been a test at all. It had all been a waste of time and money. We had struggled through winter with calves in the passage. We had had a staggered feeding of milk which had lasted nearly into spring. Our last year's work had been doubled and we had proved nothing. We were furious!

How dare the pharmacies use such big words, such small print, such almost indistinguishable colour codes! Perhaps the resident vet was right. IBR vaccines may, after all, be dangerous to week-old calves wrenched from mother, stressed after the ordeal of separation and market and travelling. How dare we humanely administer the live vaccine again? Margaret went down to the vet's surgery with all twenty-eight full bottles and the two retained empties and came home clutching the two empty ones with which she will never part. She went to bed, that night, so angry she could not sleep. We went into the season's rearing with calves unprotected and relying totally on our air system to prohibit the multiplication of germs. The two ailing calves died. We wondered what diseases they were already guilty of spreading.

Three days after our return Denis's father, Bernard, and no-longer-little Emilie came to stay. It was so pleasant as to be a balm. We had not seen Emilie for four years and now she was a delightful ten year old and it was like having one of my Guides to stay. She loved the calves and helped morning and evening, ate well and laughed a lot. Dear friends from Lyon. Their presence was a bonus we had not expected. We are only grateful that their ten days with us were over before the 1994 ordeal really began.

* * * * *

If in the next few pages I have nothing other than criticism to offer the reader I apologise and emphasise that I do not aim it in one direction but scatter it widely. There was no one person to blame for the tragedy. If anything wholly was to blame it was the weather. All summer, rain had deserted the UK and suddenly it came with vengeance. Cattle had hovered near the water hole drinking the sluggish flow of water and leaving cow claps galore. The stream had become little more than a bog.

Farmers, everywhere, were worried knowing that a milk-producing cow might drink twenty-six gallons a day. Ours were not milkers but they were emptying the bath Denis had sunk to collect water before it had the remotest chance of refilling. When the rains came the hard ground would not accept the water and it stood on the surface as thin, manured mud.

Work began on Altar Lane. Huge, huge piles of hardcore were tipped and JCBs moved in to shift and spread it on the mile-long stretch of road. The motor-bike men and their wheelbarrows never appeared so we were deprived of entertainment. The pile of evergreens had never been removed. The tractor swept it to the wall-side and branches of it hung over the boundary wall unbeknown to us who were busy feeding calves. Hippies came and parked their conglomeration of vans on the road but could not stand the tractors passing through their midst. They left their litter and moved on.

The day after our return from Wales Margaret had checked the yearling herd and the ten remaining two-year-olds. The latter just would not get fat and kept growing up and up. Margaret had intended they should all go straight to the butcher but they eventually had to be sold as stores and only three, including Precious, avoided auction.

Margaret's inspection had found one panting bullock. The *Farmers Weekly* was reporting outbreaks of pneumonia everywhere. She brought the animal home for medication and kept him in. The weather was just right for the multiplication of germs causing influenza in humans, pneumonia among cattle and making the unusual commonplace.

We were not unduly worried, for problems are few when cattle are still outside and we had an excellent air system in the calf shed. The trial year of it had been a success. We had no calf pneumonia at all. During the initial weeks of rearing, when babies are so small, we have scarcely time to take a breath so we only vaguely worried about the efficiency of the 90 foot tube. Occasionally Margaret would say, taking deep breaths, 'Do you think there is enough air?'

Then, shortly on the heels of Bernard and Emilie's departure for France, there came a very warm spell of weather. The high temperatures were most unusual for November and Margaret questioned the freshness of the calf shed air more frequently and more and more cattle locally went down with pneumonia.

Checking the cattle on the hill Margaret found one scouring. Thinking it must have worm she brought it home and gave it a second dose. We had wormed all the yearlings before going on holiday. Perhaps I had not filled the 'gun' properly when I had fired the liquid wormer into its mouth and it had not received the required dose. She brought it, and one bothered with New Forest eye, home to be company for the one recovering from pneumonia in the shed and where she could check all three frequently.

The unseasonal weather continued. The new calves were growing apace and the huge shed warmed and we nervously tried to remember how noisy the fan should be, how great the draught had been last year. On the Sunday we began to have grave doubts and we rang the Electricity Board to see if there had been a reduction in power but were told there was no change.

'There's not nearly enough air here!' Margaret said. 'The fan isn't working properly, I'm sure!' What's more a few calves began to run high temperatures and little lungs began to wheeze.

We sent for the electrician saying it was an emergency. He was overworked and had been too busy to install the fan in the bottom bullock shed. There were too many out of work yet those who had a job were working long hours overtime. Margaret had specified that the new fan had to be working before she let in her cattle because our fire precautions had excluded air.

Our electrician came, as he always does if there is an emergency, but he did not appreciate the extent of it. We rang the advisor who had been adamant no extra fan was needed. He would not change his theory. 'Take your sick animals out of the shed and isolate them,' he advised.

'That's what you should do,' Mike agreed. Everyone else knows best.

It is never the answer! Ever. Ventilation is the key to the problem. Margaret stood her ground. We opened all windows and doors and made Mike disconnect the regulator to see if we could induce more power. We needed desperately to make the fan work harder. Mike stood on a dividing gate and touched the tube. It was soft. No wonder little air was coming through the holes. He squeezed it in the middle and reduced the length to half and the tube became rigid and clean air flowed out. We could feel it cool upon our faces. 'You need two fans!' Mike said but we could not get the advisor to agree. We phoned the workshop direct and the man who answered the phone said, 'You need two fans for a 90 foot shed.'

We already had one waiting for Mike to install in the bullock shed so John was sent for. He came on Monday and Mike connected it immediately but the damage was done. Several calves had pneumonia. Two had lung damage which stayed with them always in the form of wheezing. We told the suppliers to send another fan and straightener but Mike disappeared and it lay in its packing until after Christmas.

Re-worming the skittering yearling had no effect. Margaret doubted the efficiency of her wormer. A check on the grazing animals found another scouring in the same way. Margaret brought it home and used a different wormer but that was useless too. What was more the two animals were passing pools of blood.

'This isn't worm,' she said becoming really worried. We went together up to the boundary wall and waited for a word with the man operating the JCB. The branches his shovel had pushed over the wall had been eaten bare, presumably by our cattle. We were horrified. The branches in the road were still green so we took some and we phoned the Council complaining that they had never removed the evergreens and that their bulldozer had thrown some over the wall and cattle had eaten it. We phoned the Council tree man and he said most evergreens were poisonous! In a panic we took the green foliage down to the veterinary surgery.

The vet was certain the foliage we showed him would not have caused the scour. 'Yes,' he said, 'Yew would have caused death. This is cypress. Eaten in quantity it might cause an upset stomach. If it has caused the scour it will be a passing thing.'

What about the hippies, we wondered later in the week as the two animals became weaker and a third bullock showed the same symptoms. Had they thrown something over the wall, oil or paraffin? In desperation we sent for the vet to come and see for himself.

'Look at the blood these animals are passing,' Margaret said.

'The weather is unusual,' the vet said. 'I'm sure it's only worm.'

'But I wormed little more than a month ago,' Margaret stressed.

To re-worm the whole eighty strong herd did not excite us or our purse.

'Whether it be antibiotic or wormer, sooner or later animals become immune and a different one must be tried,' the vet continued. 'I think you should try another.'

But Margaret did not believe it was worm which was causing the trouble and, at great expense, she asked for a laboratory test and the young vet took a sample. We gathered the herd into a field close to the house and prepared to wait for verification and then, if necessary to re-worm, a quite dangerous thing to do for the animals were wild with summer freedom and would prefer not to be driven into the crush. Ours was getting increasingly rickety. The new, super-efficient crush gate was still leaning against the yard wall waiting for John to execute Margaret's multi-purpose installation instructions! Men!

The new giveaway corn bin had arrived in multiple pieces and with no bolts or instructions. The carrier had neglected to leave them so it, too, was leaning against the wall, a useless pile of hardboard. We reported the missing pieces to the maker and he promised to chase up the carrier. We waited three days for the laboratory result. We had decided that, if it did not come that day, we would go to the surgery and get an alternative wormer. We were beginning to panic because every time Margaret walked amongst the herd she found another scouring animal and the ones she brought in were quite ill and deteriorating by the hour.

'They are going to die on me before the result comes,' Margaret feared, with reason. The animals had stopped eating and were rapidly losing weight. On the third day, however, the result came. As regards worm it was negative, as Margaret had predicted but, said the vet at the other end of the phone, the sample showed more than usual coccidiosis, red diarrhoea, caused by a spore. Perhaps we had an unhealthy water supply? He added that it was relatively unknown in fourteen-month-old cattle. He would have thought they were too old to be thus affected. This laboratory test cost £200!

Everything about the seasons had been unusual. An English summer is seldom so hot and dry, an autumn never as warm and wet as we were experiencing. A sea of manured mud lay deep around the water hole.

'So what?' asked Margaret. 'How do I treat?'

'There is no known cure. Hopefully it will pass,' was the only comfort the vet could offer.

'One at least is going to die,' Margaret told him.

The older vet suggested she inject 100 ml of sulphonamide under the skin, but it made no difference. The first two animals she had brought in went very thin but did not die. The third became rapidly worse in spite of the huge injection. He was a beautiful, fat red Friesian and from his bowel he passed disintegrated intestine. He went down and could not get up. He would neither eat nor drink and we poured water into his stomach through a tube and sweetened it with honey and glucose. We turned him over frequently but he died. Every day there was another new starter until the number of ailing animals in the shed grew alarmingly. It was quite the most horrific disease we have ever experienced in a lifetime of farming.

If the vet could offer no help Margaret decided she had nothing to lose so she asked advice from Mr McMillan, a homeopathic practitioner from near Chester, and recommended to us by guests who have a large dairy herd in Cornwall. Yes, this man said, he did have something for coccidiosis. Regardless of cost Margaret sent for one-hundred-and-forty doses.

This was all happening against the background of the air system fiasco in the calf shed. Post went unopened. Eight identical envelopes arrived as usual from the Inland Revenue. I did not open them. Payment was not until the New Year. The instructions and bolts arrived for the bin but we waited for the arrival of Tommy McGinty, an autumn guest who would erect it for us. We struggled with a falling-to-bits crush. We condemned the air system advisor, the electrician, John, the vet and the carrier. They were standing in a circle, as it were, saying, as five-year-old David had said, 'It worn't me!' We were heard to say such things as 'I'll swing for him, yet!' and very nearly mean it.

We gave the hundred-and-forty doses knowing nothing was going to cure those already destined to die but hoping, against hope, that something would halt this awful disease.

The previous year we had made an amateur cowl to protect birds from entanglement in the fan and it had been a necessary defence against snow. We had decided to buy proper cowls from the maker rather than mess around making our own. Each cost £60 against the £20 our home-made one had cost but we were too harassed to care. Both cowls lay waiting for John. We decided to have just the south gable one fixed to see how it worked. John sent his man and the cowl went up and on Sunday there were gusts of 74 mph. They bypassed the cowl completely sucking out the air from the tube and arresting the fan. A fuse blew in the box. We hastened to open all windows and doors and sent for our electrician. He came at once but every time the wind blew the whole thing fused again and again and we had to take down the cowl.

We showed Mike the fan awaiting fixing in the bullock shed and we showed him our very sick cattle. 'Please don't be long,' we begged. We organised John to make the hole in which to fit the fan and clean air from the gale blew in that-away.

We rang the suppliers of the cowl who could not understand why a gale could fuse the fan but who, under pressure, agreed to a refund if we returned the cowls. Were we totally fed up?

The instructions re the corn bin could have been written by someone from outer space. They were incredibly difficult to follow. Tommy and Veronica, his wife, and I bent over them perplexed. It took three days of Tommy's holiday to sort it out. We were assembling it at the entrance to the old shed we had used for calves in the experimental year of 1990 and, because it was barely high enough for the bin it followed that it was too low for the tractor to scoop out the muck. I had, therefore, taken a three foot depth of it out with an old fashioned muck fork and the ATV and trailer. I was pretty pleased with myself for still having the stamina required for that job but without Tommy's muscle power we would have had no success with the bin.

Whilst he and Veronica were here the second bullock died.

In late November the Inland Revenue sent me a second batch of envelopes. The first had remained unopened in the chaos of baby calves and ailing cattle. I opened them and was reminded that the half-yearly payment was due at the end of December. I opened the first batch. All were identical. Both editions were incorrect, forgetting that Harry was over sixty-five and no longer paying National Health contributions and taxing me other than as the sleeping partner I had been for nearly fifty years. Worse even than these computer errors was the estimated payment, a result presumably of the delay over the heifer/bullock mistake.

It was as if every professional we employed was relying on incompetent computers. I am afraid I protested loudly. The errors were corrected and the estimate changed for an actual, before payment was due. A computer error at the corn merchants resulted in the delivery of the wrong brand of calf milk powder. A change can upset human baby and calf alike. Only a few years ago our corn was delivered from a local warehouse. Amalgamation being today's disease, corn now comes from Newcastle-under-Lyme. A few years ago the driver of the wagon was a man who had been delivering to us for twenty years or more. Now the driver is Welsh or from Chester and he does not know how to find our address. He rings en route from a crackling mobile phone which goes dead before we have completed directions. One-hundred-and-fifty miles is a long way for corn to travel, because of a mistake. No wonder motorways are full and traffic jams frequent. A huge wagon broke down in the farm-yard and the driver phoned to Preston for a mechanic to come to repair it. We heard him giving directions on the line for the man to go ten miles out of his way to get to us. We tried to take our phone from the driver's hands to direct the repair man straight to the farm but the driver would not release the phone believing that he, who lived in Cheshire, knew best how to direct a man from Preston to the farm in which we had lived since before he was born!

Everyone seemed to be floundering in a sea of bar and colour codes, unreadable small print, long words in biochemistry, unpronounceable antibiotics and verging on computer madness when we were struggling with the simplicities of life and death, lack of appetite and bovine diarrhoea. What we most value is having each other. In a maze of paper work and numbers, together we can remain sane. We weep for those many farmers now struggling on alone because the invention of the tractor has taken away the need for a farmer's man.

The heavy rain had filled the ditch we call McDonalds and it had burst its banks and water again poured down the eight-acre. We sent for John to clean out the ditch and loads of wet mud were spread on the moor, too close to where the solid track ends.

Tommy and Veronica had had to stay an extra day to finish the bin and they left one dreadfully cold November Sunday. The cattle came down from the hill. We checked them in their cluster nearby and spotted a Piemontese making a pool of blood in the gateway. Margaret opened the shed door and let it in. In all we had almost a score of infected animals. She counted the herd twice and found one missing. We decided to let the herd through the empty top bullock shed and count more accurately, first as they went in and again as they went out into fields more restrictive than the open pasture.

One was definitely missing and darkness was coming pretty quickly. By the time we drove out the Range Rover, and heaved Harry and the dogs into it, daylight had gone. We left Auntie Mary by the fire and drove up on to the moor. We dared not venture beyond the firm track because of the scooped out mud from the ditch. We left the vehicle and wandered everywhere, amazed at how many walkers there were crossing from Bingley or just taking out their dog. We flashed torches hoping the animal was mostly white but we could not find it. Since the building of the trunk road in the valley the lights are dazzling, so much so, I think, had the animal been there, we would not have found it. Eventually we gave up trying and returned to the Range Rover only to find that we either had to reverse the whole quarter of a mile beside the ditch or turn at the end of the track which is normally easy. The mud we were trying to avoid was too close and, in turning we encroached upon it and began to skid all over the place getting rapidly into a worse and worse predicament.

The landscape, green and pastoral in daylight dazzles in darkness and we dare not focus our eyes on the valley, for it blinded us to the road and the mud. We were angry with ourselves that we had brought the Range Rover. We had gained nothing. We had not found the missing animal and we were in danger of stranding the vehicle and having to drag Harry home, in the darkness, through the mire. We were admitting defeat when the wheels found firmer ground and we slid onto the track. It was a silent journey home.

Behind The Currer is our small plantation. We had the Scots pines sent from Spean Bridge long before Father died two decades ago. They form a shelter belt and cattle gather beneath. The grass below them is no more and the earth is the same colour as the wall which encloses them. Margaret found the missing animal, a jet black Galloway, standing hunched and miserable and almost completely camouflaged against the dry stone wall. She had counted and re-counted cattle around it in the dusk and never seen it. As expected its tail was wet. God help us, we thought. When will this plague leave us?

We had a score of infected animals. Three had stopped eating and drinking. Margaret went down to the shed hourly day and night. One of the really sick animals began to fit, its rigid limbs jerking constantly. The vet, on the other end of the phone, said it was not unusual for coccidiosis to affect the brain. Margaret tried everything then sent for the man from Jerusalem Farm and his humane gun, a thing she seldom does. The man, a kind one, said, 'You should let me shoot the other two as well, whilst I'm here.'

Margaret, of course said, 'No. Some of the others are getting better. I don't know why unless it is the homeopathic stuff I've given them. I'll go on trying a bit longer with these.' Margaret's trying means turning the animal over often, pouring electrolytes down its throat and giving it the same tender loving care she would give human members of our family. On the table in the calf-shed can be found glucose, eggs, treacle, honey, ginger, Ionaid, cooking and cod-liver oils, cough medicine and paracetamol as well as syringes and antibiotics. Mostly she wins. In November 1994 twenty animals survived and four died. Four animals can represent a farmer's profit for the year.

Margaret was dissatisfied with the diagnosis. The animals were really too old for the disease which usually, the vet said, attacks four-month-old calves not fourteen-month-old bullocks. We decided to accompany the vet to Jerusalem Farm and see the last two animals opened. It proved nothing. Again the vet took samples from the intestine. The lab found no worm, just the coccidiosis spore and with that we had to be satisfied. It remained a mystery. Much, much later Margaret began to suspect it might have been acorn poisoning. Now when green acorns are falling she closes the gate on land well scattered with oak trees just in case.

Something, be it time or the homeopathic remedy, had arrested the disease or the poisoning and had aided recovery. We soon had a relatively healthy, though depleted herd. Christmas was approaching and we had little interest for it whatsoever.

* * * * *

Having cleaned out the depth of four-year-old calf bedding, in order to erect the giveaway bin, we needed a new frontage of corrugated tin and we went to our old friend who deals in 'ex-army anything'. We were facing this shed on the weekend when the house was being used by a family whose daughter was marrying and there was great activity and a lot of coming and going of cars.

We had divided the shed into two so that the bin was separate and donkeys could be housed in the other half. On the morning of the wedding I called Margaret to help me hang a fairly heavy door I had made with tin. Having taken the muck out, the floor of the shed was still wet and the entrance to the shed was sloping so overnight rain had entered and it is no exaggeration to say the wetted manure was all over me as well as the floor. Touching a stray hair I had painted my face with it and my hands were black.

In our busyness we had forgotten all about the wedding. Had we been reminded of it we would have said that all the cars had left ages ago. We had been told to make sure we saw the bride in her white gown and veil but we had been too busy and too dirty and had left Auntie Mary and Harry standing at the window to watch. That seemed ages ago so when we heard a car approaching we believed it to be some caller and both of us stepped into the path of a white Rolls Royce, wedding ribbons fluttering. The chauffeur on his way to pick up the bride must have caught a glimpse of two of the dirtiest ladies he had ever seen. We have never disappeared more quickly proving, once more, that farming ill befits one to provide respectable accommodation.

The wedding, nevertheless, gave us pleasure. We approve of married status. The family is the most precious of all institutions. A lady phoned in December asking if, by any chance, we had the house free for an almost immediate weekend. We had and she was overjoyed.

'Can I book it?' she said unbelieving of her good luck. There were to be many people. Their visit filled us with hope for mankind and thinking of them kept us sane throughout December. It transpired that a surprise family party was to be held for parents returning to their homeland in South America after many years of living, and raising their children and grandchildren, in England. Almost all of them were black, all were beautiful, the elderly, departing couple were regal, and the family affection and interdependence was so moving it gave us the warm feeling missing in casual relationships. We approve of marriage and family.

Our first guests arrived at the beginning of the week prior to Christmas. The older couple were closely followed by their pregnant daughter and her husband, home from America for the expected, mid-January birth of their first baby. The Cottages became alight with festive cheer and there was a meaning to the Nativity. A baby was expected! We tried to shake off the pall of sadness and enter into the seasonal cheer.

* * * * *

The vet was due to de-horn and castrate on the Tuesday. I had been rising at 5 a.m. to send off the many cards demanded of a tourist trade. I put all aside and we hurried the morning chores to be ready for his arrival. It is a long and arduous day on our annual calendar.

He did not arrive at the appointed hour and eventually rang to say his car had broken down but that someone would bring him and collect him. He could not stay long, he said, for he must pick up his repaired car late afternoon. 'I'll do as many as I can and come again in the New Year,' he promised. Margaret wanted to talk to him about the 'plague', and the present health of the herd. Was it unfair of us to think he was more worried about his car? Even caring men, whose profession it is to look after four-legged friends, are miserable if four wheels let them down.

We were desperate for the electrician to fix the wiring to the fan in the bottom bullock shed. Mike, we knew, would come for an emergency but, try as we might we cannot ever convince him we have an emergency unless lights fail or a television breaks down.

'Soon, soon,' we kept saying, 'We will be independent. Once the electric is connected we'll need no-one! We've finished! No more building or extending. Nothing!' It is a pipe-dream. Even as we said it we knew we would have to bring a town supply to water holes in the spring. We could not let this tragedy happen again.

With the job half completed the vet left and a semblance of Christmas peace was suddenly shattered by the blaring of the fire alarm whilst we were feeding the calves next morning. The man in the top cottage came running to tell us.

'Is your wife frying bacon?' I shouted above the noise.

The smoke detector in the Loft passage was lit and the good lady was causing a breakfast aroma fit for the Palace. Within minutes four fire engines could be heard on the top road. Since we had put in the cottage smoke detectors this had happened several times but this lady was making such a delicious breakfast I could not stop the noise or put out the warning light. The firemen are always so tolerant and I am always so embarrassed. Wishing us a Happy Christmas and a fire-free New Year the firemen jumped into their appliance and disappeared out of the gate. I went to the switchboard and tried again to get it back into neutral and somehow triggered the alarm again. The automatic phone alerted the Centre and the firemen got the call and came hurtling back down the road much to my horror and confusion. Our life would make a 'Carry on' film any day.

We were inundated, as usual, with pre-Christmas callers. Cousin Michael phoned from Harrogate to say he and his wife were coming. He was recovering, splendidly, from a heart transplant. We appreciate friends and relatives who phone to announce their proposed visit for it gives us time to do essential jobs, whilst they drive over, and then we can relax a moment and enjoy their visit.

We thought it was his car pulling into the yard but it proved to be an ambulance followed immediately by a police car. I went to the door with a puzzled look on my face saying, 'I think you must have come to the wrong place.'

'Haven't you phoned?' the policeman said.

I was told there was a smashed car, on our land, at the top of the road. Someone had presumably been driving up our road and had accelerated into the wall and written off a car. Someone had dialled 999 but there was no-one to be found. No-one had been here and our guests' cars were still in the yard.

'Were you expecting anyone?' the policeman wanted to know.

'Cousins from Harrogate but they would have been travelling down not up.'

Indeed they arrived announcing that there was an empty, concertina'd car in a heap of wall stones at the top of the road. The paramedics and policeman left and next day a young man came to apologise for knocking down our wall. He had been taking his girlfriend out for a drive and thought ours was a through road, though a notice on the gate says it is not. When he realised he was heading for a farm he turned in the gateway halfway down and accelerated up. A fox at the summit was transfixed in the glare of his headlights. The young man had just passed his driving test and put his foot on the accelerator instead of the brake. He had been unhurt and had carried his girlfriend away from the car and his parents had taken her to hospital. We never discovered who had phoned for the ambulance but we did find the young man to be a competent dry-stone waller.

Grants for rebuilding dry-stone walls were still available and we were employing two men to build a beauty across the foot of the intakes. It is ambitious to think we might one day leave a well-walled property covenanted to the National Trust but we just keep going with Brown determination.

The Wilson Family were expected for their third Christmas at The Currer and were intending celebrating a seventieth birthday on the day following their arrival. Sadly a week before Christmas the grandfather died. The united family decided to keep their arrangement to spend Christmas together and to spend it here so there was mourning in one half of the house and an expectancy of birth in the other.

'Let's get this crook year behind us,' we said. 'It's spooked.'

We collected Auntie Mary's ninety-two-year-old friend, and a second Christmas tree for our own sitting-room, on Christmas Eve. It had been reduced. We had so many jobs to do we were late collecting the old lady.

She rang asking if we had forgotten her. Poor soul! We had had to take a mass of uncollected rubbish to the tip. A coating of ice only and we see neither bin-man or postman for weeks.

I have a very dear and long-standing friendship dating back to student days. Since I left college, nearly forty-five years ago, I have phoned my Northumbrian friend on Christmas morning. I picked up the phone to do so and it was dead. I could wish no-one a Happy Christmas and no-one could contact us. Knowing that, should the fire alarm go off for real, the system would not operate, we went as soon as we could to a call box and recorded our line dead. We had had enough of 1994!

* * * * *

We desperately needed the kind of tranquillity provided by the Yorkshire Dales. Upper Wharfedale, given to the National Trust in 1987, has healing properties for those who have been struggling.

We headed up the dale beyond Grassington, through Kilnsey and Kettlewell, Starbotton, Buckden, Hubberholme and beyond. It was a soft gentle day. We could easily feel sad. To be otherwise seemed almost hypocritical. I had been unable to phone my friend and her husband was ill. The coming year proved to be a fight with cancer and a battle lost. Joan was still fighting, wonderfully brave and cheerful.

We ate our ham sandwiches, squashed together, parked beside the baby River Wharfe whilst tears from above washed the windscreen and there was no-one in the world but us. No people, no animals (the dogs were silent in the back of the Range Rover), no cars, no phone. No-one in the world but us and sometimes that is a very nice feeling indeed.

We did not even get out to take the dogs for a walk. It seemed silly to get them and us wet and we admitted unnatural tiredness. It wasn't the tiredness of six decades of work and winter, of caring for disabled and elderly, of talking and organising, of feeding cattle and fighting disease. All these are acceptable and tiredness is like an aspirin. It heals. Our unnatural tiredness we acknowledged was different. It was the tiredness of being afraid and very, very angry with people and computers and the weather which surely man is changing.

We made a premature New Year's resolution. 'We are not going to shout, or grumble, or criticise or be angry in 1995,' we vowed. 'It's too debilitating. Vowing we'll swing for someone is killing us!' Sadly we did not keep the resolution long!

On the way home we had an extraordinary experience which was not extraordinary at all for the sun rises and sets every day and has done so since the world began. It had been such a grey day, soft with celestial tears and different from the normal crispness of most Christmas Days. It had matched our mood and we had not found it unpleasant.

We climbed out of Burnsall and headed for Bardon Tower and Bolton Abbey and suddenly the clouds parted to give us the most brilliant of sunsets. Driving into it was almost blinding. God was in His heaven all right. I think He was just reminding us.

* * * * *

We had no phone for a whole week. The technician came the day after Boxing Day and checked the indoor terminal. 'It's outside,' he said. 'It'll have to be done tomorrow. It's too dark.' Tomorrow came and the rest of the week passed and we had an emergency and we had no phone!

Beulah's baby decided to be born! It was not expected until mid-January. On occasions such as this one definitely needs a phone not only to get one's daughter to the hospital but also to keep in touch with the maternity ward throughout. Beulah was rushed into hospital and twenty hours after the birth she was back at The Currer with baby Holly Isla in her arms. It was the first baby to be born at The Currer for a very long time and she was soon to accompany her parents to Papua New Guinea.

* * * * *

We had torrential rain. Perched on the hillside we are never going to experience two feet of water in our farmhouse as people do if they live on the flood plains. We watched news bulletins reporting unfortunate home owners everywhere. Mud is now kept at bay at The Currer but there was no mistaking the fact that the earth had consumed all it could.

The vet came, on the Thursday following Christmas, to finish the castrating and de-horning. During the work there was a mighty thunderstorm and a torrential downpour. We abandoned activity until it stopped for we were under tin and against metal gates. The baby horns were too big and we declared that never again would we wait two months.

The calves were bigger and stronger than ever before. They had looked so small and poor when we bought them. The live export trade was becoming an issue and rightly so. Three-quarters of the meat display at the Cash and Carry is filled with meat from Zimbabwe. Beef in the supermarket comes from Australia, lamb from New Zealand and bacon from Holland and Denmark. The system is wrong and farmers are blamed for a practice which is political.

We finished the de-horning late and went wearily indoors for a cup of tea. Harry said, 'Someone wants you in the Snug.' I felt less than gracious. Perhaps they wanted to pay for they were leaving in the morning. They had had a lovely Christmas convalescing after their bereavement. Together as a family they were coming to terms with their sadness.

Perhaps they just want to pay, I thought hopefully. Reluctantly I went to see.

'Half the carpet is saturated,' everyone said at once. It was indeed. My stockinged feet sank into half-an-inch of water. It was so long since it had happened before, that we really thought we had solved the problem. Obviously we had not. The heavy rain must have so saturated the earth that water had risen above safety level. The problem we had had in the eighties was here again! I went for a pile of towels and padded the place where the water was seeping in and when the guests went early to bed we moved the furniture and lifted back the soaking half of the carpet. We scooped up the saturated underlay and were near to tears.

All next day we caught water in towels. We sent for Brian to bring his humidifier and plugged in every heater we had. Miraculously the inflow abated and finally stopped. Our next large party for the New Year was to arrive on Saturday. We had just thirty-six hours to dry out and it was a miracle that we were able to lay back the carpet. The underlay, being sponge, was still wet a week later and remained hanging inconveniently in the kitchen. A sudden frost petrified the river running down the road and we had to take ashes up to the bend and walk to the Post Office for our post. I do not mind the steep climb to the cattle grid and the saunter down the top road to the village Post Office. Kevin the postmaster is a cheerful fellow. I am seldom on my feet in the village. I am not often walking alone these days, an experience I value, so I never hesitate to jump at the opportunity. It is my brand of meditation. My thought slot. Many thoughts pervade the mind when an old year ends and a new one begins.

The most persistent one was a nagging fear for Joan who had, somewhat tardily, been offered chemotherapy. The decision was to be hers, she did not reach it easily. Her life still had quality and if quantity was to be achieved with feeling rotten for six months it was a frightening prospect. I knew, of course, that she had the courage and the will and that she wanted more time and that she would eventually say yes and do exactly as she was told in a brave last bid to defeat illness. I think, perhaps we were more frightened for her than she was for herself.

Faced with such major anxiety at the end of an exhausting year lesser problems lost their importance.

Chapter Fourteen

Take things as they come
Eat crust as well as crumb.

Grandmother Smith

I suppose it is too much to hope for to experience a New Year which is serene so we do not expect it and are therefore prepared for whatever might befall. 1995 started well. Men are competent when motivated. It took Mike only half-a-day to wire up the fan and clean air flowed into the bottom bullock shed. It took the British Telecom man only half-an-hour to reconnect our phone in time for a call to come from Vi MacAllister, our friend from Ardrishaig. Her last visit to The Currer had been in 1958 shortly after our removal from the village. She was staying with her son in Darlington and they would come down to see us next day, January 2nd.

Cousin Stuart and his wife were out walking the same afternoon and Sandra, the social worker currently working in the locality and having bed and breakfast at The Currer, returned after her New Year break and our sitting-room was full for afternoon tea. The following day two elderly ladies from the Lothersdale women's meeting rang to say they were coming to collect a book. We had ashed the road well but the lady driver caffled, as we say hereabouts, and decided to reverse. In doing so she burned out the clutch.

I had seen the car descending and Margaret and I climbed towards the stationary vehicle already filling the clean air with a pollution which was definitely smoke. Here we go again, we groaned. Two days peace and trouble rears its familiar head.

We got out the Range Rover and brought the two old ladies to the house, grateful to have a phone to send for the stranded car to be towed away to the garage. The lady driver insisted on accompanying it and I took her passenger home after we had all had tea. The two ladies were great friends of Hannah Hauxwell, our extraordinary contemporary, whose lifestyle by the Hury reservoir excited the world. Boredom is never present at The Currer.

Our next visitors came and went without a hitch. We had long been inviting Mary Davis, from Crossapol, Tiree, to come and visit us. Her job was to care-sit for the elderly, literally all over the UK. She was assigned to a very sick old lady in Leeds and came out from Tiree. As the job was within twenty miles of here, she arranged to come and see us. Unfortunately the old lady died and Mary went to stay with her sister in the north-east and the whole family came to see us and it was lovely. There was no unexpected snowfall, nothing whatsoever to mark the occasion with anything but pleasure. We took them to see our hefty calves. Almost everyone on Tiree is interested in cattle.

Then, on Saturday morning, when we were feeding the calves, Margaret saw one of them bump its newly dehorned forehead on the manger. 'What have you done?' I heard her say. It had knocked off the scab over its right horn stump and blood was pouring out. It was a lively calf and was frantically dashing round the pen, shaking its head and showering blood all over the place.

In the next two hours Margaret tried every ruse we knew to stop the bleeding. The artery was pumping too quickly for coagulation. This had happened to us before. A pad and bandage may do the trick. A flour and water paste will sometimes stem the flow. Even a cobweb from the shed roof, used as a net to clot the blood, brings occasional success. Once, now long ago, nothing had worked and we had recalled the vet who had come prepared and cauterised the artery instantly.

At 11 a.m. Margaret decided she would have to do this again. She phoned the vet whose experience was, after all, greater than ours. He was reassuring. 'It will stop of its own accord, Margaret,' he said. 'I am sure it will.' We are grateful that our vets do not make expensive visits unnecessarily. 'If it doesn't stop, and I'm sure it will, ring again.'

He was about to go off duty but he promised to tell the young probationer that we had rung. 'Let's hope he's right,' Margaret said remembering that the probationer was the one who thought she was 'more of a hobby farmer'!

It was impossible to think that a calf could bleed to death simply by taking off the scab after de-horning. The vet would be right and it would stop. We returned to the shed. The poor little fellow was staggering under the weight of flour paste and pad and bandage but blood was still finding a way out and was dripping onto the bedding straw.

Shortly after 2 p.m. Margaret again rang the vet. It was the young one this time and she was curtly told the bleeding would stop as soon as the blood pressure went down! 'Apply pressure for twenty minutes and it will stop,' he advised. Thus rebuked Margaret returned to the calf shed and put what pressure she could on the now docile calf.

'It's going to die,' she prophesied, 'and no-one cares!'

At the end of the application of manual power, Margaret's hands were stiff and bloodied but we did think that perhaps the bleeding had stopped. It was difficult to tell with all the red bandaging. We enclosed the weak baby with sheep pen fences to protect it from inquisitive companions and waited. Eventually, from another angle we saw the drip of blood become a steady flow. Margaret laid the little fellow down. He did not protest. 'Time me,' she said. 'I'll try another twenty minutes more. If that doesn't work I'm sending for the vet again.'

At 4 p.m. she was ringing the surgery.

'It's no good me coming,' the young man arrogantly replied. 'I can't do anything more than you have. It won't bleed to death.'

'Then,' shouted my desperate sister, 'I insist you come and do nothing, but come you must!'

We waited for an hour for the arrival of his car. We were really worried. Youth is often indolent these days and I blame the car. Doctors and vets, in my childhood, did not have its convenience. They had to walk to their destination. Whilst delivering milk with horse and float I have seen my father spot our Dr Tom Spencer, an old man in the thirties, hurrying on foot to some emergency. Father used to hail him and Dr Tom would jump aboard and the horse, Dick, would be urged at a trot to wherever was necessary.

Now even the young have cars, mobile phones, amazing equipment, proven drugs and a laziness which borders on discourtesy as they shrug their shoulders and jab with a needle. It is difficult not be infected with rudeness for all relationships are an echo. Silence receives silence. We tried to put warmth into our welcome. We needed help. It generated no radiated heat from the young man who thought we were calling him out unnecessarily for something we could handle ourselves.

He walked into the shed in silence. Amidst seventy calves one bloodied head stood out from the rest, grotesque because of the enormity of the bright red pad. 'Which calf is it?' asked the sulky boy.

He sounded just like the doctor in the casualty department of the local hospital who had looked at my bare foot from which dangled a dislocated big toe and asked, 'Which toe is it?'

The boy went to the calf and with sudden efficiency ordered warm water and a towel. I ran to his bidding. He wiped the blood from the calf's face, stood back and said, 'Now what do you want me to do?'

'You're the vet,' Margaret said. We were getting angrier.

'Well,' said His Royal Highness, 'There are two things I could do. You must choose. I can leave all this mess of padding and paste and let nature clot the blood naturally, or I can disturb everything and let the blood flow. Which have I to do?'

I remember asking him if he were married with children and when he shook his head I said, 'Young man, when you are and your child is dying and the doctor asks you what he must do, you will be very angry!'

I must have shouted. Or perhaps Margaret did. She is the one with the loudest voice.

'I've never been shouted at before,' the young man complained.

'We can shout louder than this!' we assured him.

He was still adamant that we should make the decision. The calf was dying at our feet. 'You haven't done anything,' Margaret said. 'Check its mucous membranes. Its mouth and nose are dry. It is completely dehydrated. Look how white it is!'

'How much blood does it have to lose, anyway?' I asked.

The young man was getting flummoxed. He did a professional calculation, something to do with a percentage of body weight and came up with the answer. 'Forty pints.' Deeper and deeper the boy was floundering.

He tried to make the calf blink. 'It's been unconscious a while,' Margaret said. 'I'm going to lose a calf because it accidentally knocked a scab off a wound. I sent for help six hours ago and if you don't do something I am going to the phone to send for someone who will.'

The young man said, 'You'll be glad to know I'm leaving in a month.'

'Well, you're here now and you'll do something. A human would be given a blood transfusion. Have you a saline drip?' Margaret took command.

'Not with me.'

'Have you brought your Calor Gas and cauterising poker?' I asked.

'No.'

'Then hurry down to the surgery and fetch both,' Margaret ordered.

From that moment, we must admit, the boy became a professional. When he returned from the surgery the calf was dying but he set up the drip expertly and found the jugular vein accurately. He stripped off the padding and in seconds cauterised the artery. He stayed with us until all five litres of saline drip had entered the vein and moisture had returned to the mouth and nostrils. We were able to phone the surgery next morning and report a much improved animal. We had poured several pints of Ionaid down a tube into its stomach and it was conscious.

We were weary of fighting against those who counselled, 'Do nothing. You do not need another fan. Being without a phone when a baby is on its way is not an emergency. Sorry about the calf milk. Tomorrow will do.'

The strange, commendable thing is that once motivated all can do a good, quick job. Human spark plugs seem to need a good squirt of WD40 these days.

The awful winter of 1994/95 was not over. A few days later a fifteen-month-old bullock did not get up for its corn. Margaret forced it to its feet and drove it into the crush. 'It's its stomach!' she said. She took its temperature and found it so low it should have been dead.

'Try another thermometer,' I suggested. She did but the result was the same. We drove it into the bottom shed and we were nervous wrecks. The animal refused to eat or drink or stand and Margaret got a handful of blood and mucous from the rectum. Within two days it was dead. We were very badly shaken. I sorted out the CIDs for the dead and sent all five to MAFF.

The fire alarm went off again. This time it was caused by a young woman having a shower without properly closing the bathroom door. Again appliances screeched down the lane. The poor naked woman was in a panic. 'The shower is getting too hot,' she said. 'I can't turn down the heat.'

I thought of all my colleagues in retirement and wondered how long it would be before we had a nervous breakdown but our escapades fell into insignificance compared with news of a terrible earthquake in Japan.

* * * * *

In a brief period of calm weather we planted ten fruit trees in front of and to screen the septic tank. We rushed to beat the frost and bought new fencing posts and wire and worked until dark fencing an orchard area safe from cattle and goats. Our neighbours had sold both farms and we had lost our annual supply of apples and plums. We should have planted our own trees years ago when, to build a new shed in 1981, we had desecrated the unproductive apple trees we had planted in 1960. We are not confident we can grow fruit but, even so, we are trying again. We planted these in the semi-darkness for January days are short and next day I carted barrow-loads of mulch from the wood pile to surround the trees.

Some problems have a habit of recurring. A developer was thought to have bought or to be buying the land across the valley and suddenly women here, there and everywhere started to protest. Things do go in threes. First the return of the river of water in the Snug. Then the revisitation of the coccidiosis plague and now the resurrection of the Council's determination to build on the valley slopes and spoil a square of woodland which has been a landmark for a long time. Like flotsam tossed up onto the beach with a high tide, a higher tide retrieves them and loses them in the ocean only to bring them back another time and in another place.

Europe was experiencing the worst rainfall and floods of recent years. The Low Countries, France, Germany and Great Britain became seriously flooded. It was a disaster never even envisaged. Thousands left their homes as rising water entered houses and buildings in towns and villages everywhere and farmers fought to save the lives of their animals.

The heavy rain, in comparison, caused us a problem so small I am ashamed to record it when people in their thousands were seeing flood water wreck their homes and destroy their belongings.

In Yorkshire the first of the rainfall fell as snow. Here on the hilltop it was about a foot deep. The valley receives less, always, but two or three inches these days causes havoc until quite recently unknown. Rush hour traffic is horrendous everywhere and we, in our isolation, have no real experience of it and only see pictures on television.

The chaos rush hour traffic caused on Wednesday 25th January, will be remembered in the locality for a long time. It took George Winup ten hours to do five miles from Bradford to Bingley and the Cathedral housed a congregation entirely unexpected. Police advised motorists to leave their cars and opt for its warmth. All this was because there were four inches of snow! We who have lived and worked and walked in two feet of snow and seen drifts of ten feet deep not long ago, before the earth started getting warmer, are astonished at the havoc caused, now, by so little.

No longer do people dress for the snow unless they are on the mountain slopes with their skis. They dash from their heated homes to their heated cars to their heated workshops and offices without winter clothing being necessary. Then, sadly, when stranded they freeze in their cars or venture out and risk dying of exposure.

We pronounced our road as impassable. In an emergency no doubt, with a day's digging, we would have got out with the Range Rover. Four-wheel-drive vehicles have a better chance than most but we did not try. No postman and no dustbin collection is a small price to pay for a short period of no casual visitors or bed and breakfast winter guests. Our top cottage occupants left their car on the top road and walked back and forth. Sandra, the pleasant social worker, did the same. No-one grumbled.

The sudden snowfall was no more exceptional than the sudden thaw. Our twelve inches melted into eight inches of slush in record time. The already saturated fields could not drink it and McDonald's Ditch became a raging river falling down Currer Lane in a noisy rapid and hurtling in several waterfalls into the diverted stream below the Dutch barn.

Water again saturated the carpet in the Snug!

Margaret and I moved the furniture to the dry half of our beautiful sixteenth century sitting-room and draped the carpet over the lot to dry.

We sent for Brian our ever reliable builder and he spent all day making a hole below the two-feet-thick west wall where the stream of water continued to appear even as the river flows from under Malham Cove. He made a well, below ground, thinking it would contain the water till morning and promised to come next day.

'We'll put some drains under the floor and lead it away,' he said. 'Empty it when you go to bed and it should last until morning.' As Margaret goes to bed late and I rise early, the plan seemed feasible. We did not bargain for rain to fall so heavily and so continually.

'The reservoirs will be full,' we said baling out our mini one which began to fill more and more rapidly. We were soon emptying it every twenty minutes. Sandra was left in charge whilst we went out to feed the cattle. We could hear the heavy rain beating a continuous battery on the tin roofs of the sheds. Water ran in a river down the sloping yard and Margaret got a soaking forking manure to the calf shed door to prevent its entry. Water splashed from over-full guttering and, singing noisily, dropped down the drain pipes so quickly the ground drains could not take the deluge.

'I'm emptying it every seven minutes,' Sandra reported when we came in. We could not both go to bed. Margaret sat up until 3.30 a.m. when I crawled from my bed to relieve her. I sat before a log fire, scribbling. People are always asking when, in this busy life, do I have time to write? Opportunities do arrive, I assure them.

When I put on the early morning news the BBC told of people suffering a thousand times more than we were.

Brian laid land drains as it seemed the only sure way of dealing with water we might expect to appear whenever the heavens opened drastically and the water levels, in the banking behind the farmhouse, rose above the level of the Snug floor. The flow stopped before the operation was completed. Brian led the drain into the cellar where a pump is permanently working to empty the well there. Since then there has not been any appreciable rainfall and the drain has been, so far, unnecessary.

Water had also seeped inside the only double-glazed windows in the house. When Father was ill in 1976 we had put an extension on our family sitting-room. It had enabled him to look out onto the farmyard and onto the hillside which had greened under his farming competence. We had not had enough time, or money, or incentive to match the mullions with those of the east wall and had bought Magnet Southerns windows supposedly double-glazed against the weather. We had made a feeble effort using stone sills and lintels but we were secretly ashamed. When we eventually converted the barn we really made an effort to do the job properly.

When we covenanted to the National Trust we made a vow that, as soon as we could, we would replace those awful standard windows.

With an elderly lady in the house we had refrained, thinking it a job we could do when we would be the only ones to be inconvenienced. The cloudy double-glazing, however, began to make seeing through it impossible so we took the plunge.

'Whilst we've got Brian let's have mullions put in the sitting-room,' we decided. And we did. How beautiful they looked. Amazingly no-one noticed for Brian did the work so well, with old stone, that the five new mullions looked authentic. Of course the work was not done overnight but no-one could grumble.

* * * * *

The first Saturday in March is always Guide Reunion Day. We do not alert our ex-campers but rely on them remembering. We do not know who will come or how many. It does not matter. They are all our children and they all enjoy seeing each other again.

I had not done any necessary baking for the evening buffet so I rose early on the Saturday morning and prepared for a busy day.

Cousins were coming to take Auntie Mary to see a show in Bradford so when the phone rang at noon there were only the three of us, drinking elevenses whilst still working. I had scales and cake ingredients all over the table.

With floured hands I picked up the phone. It was Mr Booth, who farmed in St Ives, reporting that all our cattle were rampaging round the estate. Work had been abandoned in November on Altar Lane, and men had left it impossible for wheeled traffic and dangerous for ramblers. Consequently they were using our land rather than the lane and were climbing the wall at the far extremity where a triangular field reaches the Altar at a very acute angle. We had rebuilt the wall no more than ten days before but spring-like weather had brought out the walkers and by accident or intent they had reduced the wall, again, sufficiently for the herd to go a-wandering and a-vandalising in the estate.

There was nothing we could do but down tools, get out the Range Rover and take Harry. Margaret walked overland with the dogs and I went and parked in the estate. I tried to get out of the driving seat door and it would not open. Something inside the mechanism had broken. With Harry in the passenger seat all I could do was an acrobatic contortion over the back of the driving seat. In wellingtons and warm clothing that is an almost impossible feat for me but Margaret had not arrived and I had no alternative.

There was not an animal in sight but the wall was down almost to ground level. The Council does such funny things. Solid stone steps give walkers an easy route over the wall. The pillared gateway used to be closed against cattle straying from The Currer on the the outside, and from the estate home farm on the inside. Now an iron barrier is swung across the gateway. It prevents cars from passing from the estate onto the recreational common ground of the Druid's Altar but allows plenty of room for horses and walkers and cattle. There is now nothing whatsoever to stop straying cattle from having a field day in the open-to-the-public grounds, Nature Trail, Children's Adventure Playground, Golf Course and all.

Saturday afternoon walkers were eager to tell us that big Friesian bullocks were dancing all over the place. Our nearest easterly neighbour, Mr Davidson, came to help and the Booths assisted with the round-up. We drove the cattle to the entrance knowing they had left hoof marks all over the estate. We hastened to pull the wall down wider for we know that animals eagerly escape through the eye of a needle but will never return home unless the gap is enormous. We recruited as many walkers as we could for, at that particular grid reference, there is a meeting of the ways and cattle always choose wrongly. In the midst of the fiasco we had to almost physically restrain men on mountain bikes who must surely believe they have a divine right to keep cycling even if it means their own danger or dispersing the herd or knocking us down in the process. We find them frequently a menace to the countryside, coming up behind us, at speed, silently and passing unexpectedly without even a greeting. The younger generation of mountain bikers are responsible for several of our dismantled walls. Following the footpath they cannot get through the ramblers' gate without dismounting and lifting their bikes over the gate. Children cannot do this so they pull off the top stones instead. Caught red-handed they innocently say, like all children, 'It warn't me!'

By the time we had all the animals safely home and the wall rebuilt it was dusk. The house was a mess, the cakes were unmade and the naughty cattle had galloped home ahead of us. They were waiting at the shed doors eager for food and absolutely unrepentant. It was half-past-five and we had a Guide Reunion at 7 p.m.

I beat up the cakes and shoved them into the Aga. I had cooked a piece of ham. I took bread rolls from the freezer and I whisked the untidiness out of sight and at six o'clock the first flakes of snow began to fall. The fields whitened quickly and Guides from Holmfirth and from Lothersdale rang to say they had intended coming but dare not.

'Please don't,' I begged. The prospect of people coming in cars and being unable to get back up the hill frightened me. 'If it continues,' I said,

'I'm going to ring round and cancel the evening. We can postpone it until next week.'

I did. I phoned as many as I knew might be coming but even so twelve who-cares-about-snow stalwarts came, cars and all and it was a miracle that the snow abated and all re-climbed the hill. The small gathering was a more intimate one. Large numbers exhaust hostesses. Those we had discouraged came the following Saturday and a good time was equally enjoyed.

We could not allow people to pull down our wall continually. Once the Council had completed the expensive, tardy surfacing and closing of the Altar Lane we suspected our troubles would be largely over. Until then it was easier for people to walk on our side of the wall and scramble over it at that acute corner. In driving snow on the following Monday, we drove over to the field at risk, with fencing posts and barbed wire and secured it against vandals. We doubly-fenced with sheep netting and we anchored the end posts to trees. We had had just enough of chasing cattle. We did the whole job in a driving blizzard knowing that cattle had preferred to stay in. There was no such luxury for us. Wet, cold and filthy, we drove in the posts with a heavy mallet and fumbled with staples and live barbed wire.

Returning to Harry, in the warmth of the Range Rover, we finally opened the coffee flask and poured its steaming nectar into plastic cups. Margaret had had to climb over the driving seat back because of the broken mechanism in the door lock. We started up the engine and warmth blew into the vehicle but we remained shivering. We folded hands over the coffee cups and all windows steamed but we continued to freeze.

'Come on,' we said, 'Let's get this silly door fixed!' and Margaret drove over to Simmonites. Because she could not get out, I went into the office looking like a lady tramp.

The man was amused. I think I might have had mud splashes on my face. I was soaking wet and had filthy hands.

'The Range Rover door won't open,' I said. 'Can someone do something quick before we die of hypothermia!'

The dirty mechanic looked positively elegant compared with us. He fixed the door and we drove home and stoked up the fire but only the hard work of feeding the cattle defrosted us.

* * * * *

By the 1st of April we had finished the spring-cleaning. Margaret had whitened all the walls in all the rooms of the converted barn and I had papered the kitchens in cottages and in our own accommodation. Brian had finished work on the mullions and we had bought material and made new curtains and chair and cushion covers.

Taking down the old curtains I could not help remembering first hanging them in the then new extension to the sitting-room. They had not been new. The New Do, as Mother called the viewpoint we had made so that Father could watch activity in the yard, had cost so much money we dare not spend more. Curtaining is expensive and Mother had insisted we did not need any when we never drew them. I was adamant that the three new windows, which did not match the east wall mullions, looked bare but Mother would not admit it. We could not afford any and, 'What we can't have,' she'd said, 'we won't miss.' I had waited until she and Harry had gone to the aunts for Saturday tea and then I had searched for some material which had been left when I had made the living-room curtains some years ago. The material was very strong and the yellow, floral design had never faded. When I had compared the curtains with the remnant there was no change. There had not been enough material to make proper curtains for the three new windows but sufficient to make what we called 'cod-on' ones which were barely long enough and, if drawn, did not meet.

I had taken down the yellow curtains from the living-room mullions and exchanged them for the green ones hanging in the sitting-room and I had pulled out my machine and made the imitation curtains. We weren't ever going to pull them on so no- one would ever know they would not meet. The effect had been commendable and I had swelled with pride in my achievement. Praise is not something we give generously within the family. Margaret had nodded her head and said, 'Yeh, that's alright,' and Father had been more interested in what was going on in the yard.

When Mother and Harry returned I could not wait to exhibit my afternoon's work. I had shepherded her into the front room and halted her before the newly 'curtained' windows and said, 'Well?'

There had been a moment of silence and then the old Yorkshire lady had said with predictable reserve, 'They're not foul'.

I cannot type this without laughter. How we loved our Mother. When I think of the present-day parents treading softly with their children, giving too much praise, being too liberal with their applause, I think of our wonderful mother and of David Gibson's grandfather who lived to be nearly one hundred, up at Denholme. He never commented on the meal he had just eaten. Annoyed by this one day, his daughter, David's mother, had asked, 'Did you like your dinner today, Father?'

'Why?' the old man had replied. 'Wot wor wrong wi' it?'

Having replaced the sitting-room curtains twenty years later I found that the 'cod-on' ones fitted the Loft Cottage kitchen, that the sitting-room mullion curtains were good enough to hang in the Loft lounge and that there was enough to re-curtain Harry's single bedroom. It's good material, that!

* * * * *

So we were all ready to entertain my college year who descend on me for a reunion every two years. We served a midday meal to twenty mostly retired teachers who had trained with me forty-five years ago. One of them I had not seen in all that time. It was a little too near to Easter for comfort but we had a great day and followed it up with the annual visit of residents from White Windows Cheshire Homes the very next Thursday.

The few days interval between College Reunion and their visit promised to be plain sailing. Then Margaret noticed that our brown donkey had a swelling and a soreness in his groin. We had a Greek vet staying with us at the time working a spell at our local practice. He did not have a car and was collected every morning by one of the vets, the one who is the horse expert. Margaret called him over to look at the donkey which was walking very badly. We had had both donkeys for fourteen years and the holiday-makers were very attached to them. In fact we always said people came back year after year just to feed the donkeys.

Chocolate was really very wobbly and our vet suspected there was a back problem but urged Margaret not to keep him in or his muscles and joints would stiffen. So Margaret let out both donkeys on the Wednesday morning and we all went into town to buy food for the Cheshire Home visit and for the influx of guests arriving on Saturday for the pre-Easter week. Brian was still clearing up the aftermath of building and the man employed dry-stone walling was working in the Footpath Field.

We really felt quite relaxed. Spring was coming. Get over Easter and within a few days we would be travelling north on holiday. We had time, we decided, to go to the swimming baths. Too frequently in the past six months there had been no time to take Harry on this, one of his favourite outings.

So we all went out, something we rarely do and we did a little holiday shopping. Margaret bought some new trousers and she and Harry bought new shoes for the beach. Then we took Auntie Mary to the hairdresser's and we went swimming. We were in holiday mood and I believe we went up to Redcar Tarn to eat our sandwiches by the water and Margaret walked the dogs round the perimeter.

Then we went home to bedlam! The waller was just re-climbing Currer Lane in his car. He got out and came to tell us that the brown donkey had fallen into the bog which forms on the upside of the boundary wall behind our duck pond. The land belongs to our neighbour but we had been renting it for some time and the gate between the properties was open. When we first began to graze the land we had fenced the bog for Margaret foresaw that animals might sink. Our own moor pond is fenced for just such a reason. A fence round a bog needs frequent attention for the posts are in land where the water level is high. Eventually they rot or shake and Margaret had put this job on the list to do when Denis arrived at the weekend and before she let the calves in, prior to going on holiday.

Too late! Chocolate had stumbled owing to whatever back trouble he had. He had caught his leg in the barbed wire and the fence had collapsed under his weight. His entangled leg prevented him from getting up again. His weight caused a hollow in the ground which quickly began to fill with water. A lady solicitor, not dressed to pull out an increasingly muddy donkey, actually saw it happen. She called the dog she was walking and ran into the farmyard and found Brian. He had whistled for the waller and the three of them had managed to pull the donkey onto dry ground but had failed to get him to his feet. They had wiped off the mud and covered him but he was going to die. The waller offered to drive back down the hill with us and a passing jogger (most won't stop), offered his help and together, immediately for time was running out, we got the donkey onto a sack, heaved him onto his feet and half beat and half carried him onto straw in the shed. I did not notice how much I had filthied my anorak. When next I was in town I stepped out into the supermarket car park and only then did I see how awful it was. I hastily took it off and went shopping without it!

The bog wasn't going to kill Chocolate. From that he would have recovered with normal Currer Care but something which had caused a donkey to stumble, ailed our brown friend. He ate happily and the front half of him remained cheerful for a time. The back half of him stopped functioning. His hind legs seemed paralysed and the swelling would not go.

The next day, Thursday, a double bus contingent came from White Windows. For over thirty years residents have been spending a day at The Currer and, during that time, we have grown to know the long-livers very well. We look for them eagerly among the smiling arrivals. Some are new to the home, some have been coming for several years and Eric and Derek and Val were already there when I first took the Rangers in 1962.

Eric is mobile and wanders all over the house and buildings. He is capable of summoning his own transport and visiting us independently and he arranges his own summer holidays, often going abroad. His appetite remains healthily enormous and his memory as acute as Harry's. Derek has been blind and helpless in a wheel-chair as long as I have known him.

He will not come on a Sunday and only accepts the invitation every two years, and there is Val who must have been very young in 1962 and does not seem any different from when we first met. The staff have changed many times but the quality of their service remains the same.

Such activity there is before they come, such a pushing-out of chairs and arranging tables to accommodate so many wheel-chairs. There are many hands to help serve and wash-up and re-set the tables for tea. There are many tongues talking and singing and laughter is unlimited.

The Thursday after Chocolate fell into the bog was similar to all such annual occasions except that the muscular members of the party were enlisted to help with the frequent lifting of the donkey onto his feet, the steadying of him in that position and the lowering of him onto his alternate side. Well-fed and happy the guests eventually left us, waving if they had hand mobility and smiling widely if they had not.

Their departure left us just one day to prepare for the arrival of the pre-Easter guests. I battled on almost alone for the donkey commanded much of Margaret's time. He was like a member of our family. We wondered what would happen to Jasper when the end came.

Father used to say, when an animal died, we may cry just as far as the house door. We might mourn but as long as the tragedy was outside the door we must be thankful. We had had too many losses in the last eighteen months, it was difficult to shut the door on our grief over the donkey. We were miserably aware that no matter how Margaret struggled it was going to be a no-win situation but instinctively we went on trying.

Jasper was tearing at our heart strings. He was mooching round, outside the shed, barely eating and wary of all the coming and going and the immobility of his paralysed friend. His playmate was ill and there were too many strangers. For fourteen years the two had been inseparable, untethered, wild on the hill. The door of their shed had never been bolted save when the blacksmith came. They had playfully reared and fought and brayed and enticed our guests to return year after year. It was not easy to obey Father and shut the door on tragedy thankful that it was not within. For Jasper's sake, as well as Chocolate's, Margaret fought to the end.

On Saturday Denis came and he and the Greek vet helped to hoist the animal to its feet. Collecting the young man from Athens, our own horse vet said the donkey undoubtedly had a tumour on the spine and would never recover. The house filled with holiday families and there was a sadness in their arrival for they had all been before and loved the brown donkey. The Easter marathon, which would culminate in our departure for Harris, had begun. The weather was glorious.

On Sunday evening kidney failure was causing so much pain Margaret sent for the vet. She and Denis cradled the donkey's head lovingly whilst he was put to sleep and allowed to go to his Maker for as Denis said he had done no wrong. The conceit of man is deplorable that he thinks himself above and apart from the animal kingdom. I would lay my odds on man displeasing God far more.

There was silence at The Currer.

Few, other than the medical profession, nurse sickness more than farmers. I will warrant that there is not one in the whole country not caring for a New Forest eye, a foul of foot, a wooden tongue, a mastitis, a scouring calf. Not one who is not worming cattle, administering an antibiotic, giving an old dog his pills or dusting a wound with aureomycin. Most are curable and more patients live than die but only because the farmer is caring and competent and even then, sometimes, more die than usual.

* * * * *

Denis was a comfort. English is his second language but in it that boy from Lyon can say just the right thing when you have almost had it up to here. On Monday Jasper stood with his back to the wall in abject misery. 'How long can we stand it?' Margaret asked.

We had one spare room and Sandra rang to say she was doing a spell of night duty and needed day accommodation. Denis had a room until Wednesday and then he would have to go behind the curtain. According to my chart four people would go out on Wednesday and six come in. They were people who had been coming every year for a long time. A retired headmaster and his wife made the annual booking for themselves and for Jill and Moss, who was confined to a wheel-chair following a road accident and Marlene and Fredrich from Germany.

Jimmy, the goat, had begun eating the door, the very expensive front door leading to the dining-room. His naughtiness coincided with John's men actually building the crush, to Margaret's specifications, and hanging the excellent new gate which had lain in the yard for nearly a year! We have proved that crush to be the most versatile and useful construction Margaret has ever invented. It is just right for us. It answers all our needs, safely and without distressing the animal. I give her full marks for her idea and John's men equal commendation for interpreting her orders accurately, even if tardily.

Whilst they were here with their equipment I asked the boy to make a gate to keep Jimmy from the front door and he was doing it on the Tuesday before the Wednesday change-over in the house.

We had a moment of calm. Denis was out on the ATV. Sandra was in bed. We had nobody going out, no-one was to come in. There were no beds to make. No washing to do. So we sat after lunch and talked about getting a replacement donkey. Margaret planned to ring various contacts and see what could be done.

The evening meal was to be plaice. I had recently bought a new box. Had it been meat for the evening meal and the joint already in the oven I would have been in a pickle.

I heard the boy soldering in the laithe porch and I went to make sure that he was fixing the gates so that wheel-chair users could manipulate them. There was a lovely pastoral peace pervading. A man was walking towards the house to explain why we had a herd of fallow deer on the hillside. Margaret welcomed him in. 'Did we mind?' he wanted to know. They were causing quite an attraction, these beautiful animals on the the hillside, frequently in our fields. We had wondered how they had got there and suspected they had been placed. Of course we did not mind!

I failed to notice there were three cars in the yard as I approached the young man fixing my gate. It was a beautiful day. The boy and I were mirrored in the glass of the dining-room and behind our reflection I saw people. With a horrible sinking feeling I opened the door and saw Tom and Joyce, Jill and Moss, Marlene and Fredrich. They were sitting there chatting as calmly as you please.

My mouth dropped open. 'I didn't expect you until tomorrow,' I said.

'Well,' said Jill, 'We are not going home!' I hugged them all wondering what on earth I was going to do. 'Help me, help me,' I prayed. 'What am I going to do?'

'Don't panic,' I said aloud. 'I didn't expect you until tomorrow but we'll do something.' I had not a room spare in the house. We could hardly vacate ours for Harry was already using it and a stranger was in his. Denis could go behind the curtain as he had expected to do next day. Sandra was sleeping peacefully but she was on nights, so when she woke up she would have to move out and she could use our accommodation during the day as long as was required. Miracles frequent The Currer daily. In the ground floor twin, allocated to Jill and Moss from Wednesday, were two girls whose parents occupied the double room which had twin bunks. On previous years their daughters had shared the family room but had now been deemed old enough to have a room of their own. The family was leaving the next day and would not mind, I felt sure, if their girls used the bunks. The mother must have the power of telepathy. When their car drew into the yard and she saw all the extra vehicles she said 'How much do you bet you'll be in our room tonight?'

I extended the table by six places and we had twenty to serve with plaice and chips for the evening meal.

After that nothing worse could happen at Easter. Geoff and Sally came, Richard and his family. Always when entering our sitting-room he walks to the photograph of Mother and says, 'I liked her'.

John Skeggs came as usual to do his ten days voluntary service on the Worth Valley Steam Railway. He came laden with jigsaws for Harry to occupy himself next winter. Our guests take the place of family since none of us married and of the older generation only Auntie Mary is left.

Margaret was unable to bear Jasper's loneliness and phoned all over the place asking for help in finding a new donkey. She was eventually directed to ask a man at Otley Auction Market. She phoned him and he said he would make enquiries and he found Joe.

'Come and look at him on Monday,' he said. We would find him at the sheep pens at midday. He had a patch over his eye and was called Jack Pack.

We decided first to call on Olga and Arthur, some friends we knew only because they spent three months of the year at Horgabost, on Harris. John and Mamie MacKay stayed with them when they came off the island and ventured south to Yorkshire. They were not going to Harris for they had sold their caravan believing the journey had become too far to travel. We thought it terribly sad. As we were leaving for Harris in a week's time we had promised to take up the Christmas cake Olga always made for Mamie's freezer. 'Come over on our return,' we begged, 'hear all the news.'

Richard and Denis came with us and after coffee we went to look for a man with a patch over one eye who would be at the sheep pens at midday. He was easy to spot. He told us how to find his allotment on the outskirts of the town and we drove there and parked on the banking. The man was waiting for us outside a field shed and Margaret, Denis and I climbed out and joined him. He opened the shed door and led out the animal. The donkey was quite a fellow. He was so shaggy he looked like a miniature shire. Haltered and led out he reared like a circus pony. He had so much hair on him he looked bigger than he really was. He had been shut in the shed and was festooned with straw and there was enough dust on him to fill a vacuum bag. My heart sank to my feet and Denis stood well back.

I know that having seen an animal, any animal, Margaret does not turn away. Whatever we go and look at be it goat or pig or poultry or cattle or donkey, she never rejects it. If it looks at her she is hooked and it is essential that she is kept away from situations from which she can find no escape. Had Joe only had three legs she would have bought him. 'He needs me,' she would have said.

In this instance we needed a donkey and Jasper needed a friend. Donkeys, it would seem, were hard to come by but it was essential that the animal was safe. We had a lot of adoring children to entertain. This shaggy, dark brown monster, rearing like a circus pony looked like one big problem.

'Is it safe?' I asked as Denis took a step further back.

'Friendly as a kitten. Lovely with children,' the man said attempting to control the performing beast!

Margaret asked, 'Will he befriend Jasper?'

Donkeys should never be kept alone. They mourn, like humans, the loss of a friend. Jasper was not eating or leaving the farmyard. Others might have sent him to a sanctuary. We have enough problems and wild donkeys add to them. If cattle stray we are reasonably competent to get them back. We have no control over donkeys used to infinite freedom and allowed to go wild on a hillside.

Our ancestors were skilled horsemen. Grandfather was a professional cabby. His father owned a livery stables with twenty-two horses. I was reared in an era when milk was delivered by horse-drawn float. As a child I could harness and drive a trained horse but that was a long time ago and any skill I had possessed had been forgotten. All that was left was a latent fear. We are cattle farmers and have no equine knowledge.

I knew Margaret would buy it. I had known since the moment Chocolate died that, like Father, she would 'buy another on Monday'. Father used to say more frequently than anything. 'Something will turn up.' He used to believe it and sure enough it would.

Margaret of course bought Joe. There was no way I could have talked her out of it. The man expected us to find transport. His own vehicle he said, was too old to climb the necessary hill to The Currer. I personally suspected it was too fragile to contain this wild thing for which Margaret had just paid an enormous price. Donkeys, it seemed, were scarce and expensive. Sanctuaries would not part with single animals and would not sell where there was already a stallion donkey. We had bought Chocolate when Jasper was only a foal and they had grown up together. We had thought we might have to buy a Jenny but Jack Pack assured us that Joe, who was not a stallion for all his monster appearance, would settle fine with bereft Jasper.

I, at least, was reserving my judgement. There was nothing I could do to object. Margaret was buying the donkey! I think the vendor was sensing my reluctance so he quickly said he would transport the donkey himself. He could not risk losing his buyer.

We paid, returned to the Range Rover and left the man to cope with the loading of Joe himself, believing that the less we saw the better.

When the donkey was unloaded at The Currer we penned him in the yard shed, behind a five-barred gate so that he could see out and we, and Jasper, could see in. It was not long before our white donkey was alerted by smell and instinct to the presence of Joe. He galloped to the gate. We could not decide whether or not the meeting was friendly. Jasper ran frenziedly round the yard approaching the gate on each circuit whilst Joe brayed and reared. Animals can certainly talk to each other. Whether the communicating donkeys were amicable or not we did not know. This excited banter looked aggressive to me. The floor show in the yard continued all day and we went to bed somewhat anxiously.

The next morning, very early, before any guests were stirring, Margaret opened the gate and ran to the house for safety. The two animals met, reared, brayed and galloped into the Five Acre. We hardly dared to watch as they careered down the field, dancing, performing, talking in noisy adjectives. They did not kill each other and have been inseparable ever since. Joe lost his long dirty coat and has never again looked like a miniature shire. Of the two I fear him the least. He is wholly benign and gentle. Is it possible that we did not get a monster after all, that the vendor brought us a different donkey? I think our first impressions were all wrong and my fears all unfounded. The man was right. Joe is no problem at all.

* * * * *

Our Easter guests left and we were suddenly inundated with family and friends who wanted to come. Not living on the roadside, friends and neighbours do not just call. If they come, they come to stay a while and have at least afternoon tea here. We like them to phone first so that we can organise our work to leave time to sit and converse in the welcoming manner for which we are noted.

Unexpected callers catch us in the middle of a job which has to be finished before we can break off, for our home is our workshop. The phone rang continually during the three days before our departure for the Western Islands. We were continually keeping tidy and everything we baked was eaten instead of packed for holiday. We never got chance to 'bottom' as Mother called a good clean. If I appear to over-record pre-holiday activity it is because its variety fascinates me. It is impossible to predict. That we would be receiving guests and making afternoon tea, daily, for two groups at a time of family friends who phoned to say they were up from Cornwall or Timbuctoo, could not have been expected in the four days before we left. Everything had to be done on the last day and there was no way I would leave without visiting Joan, coping so admirably with fortnightly chemotherapy. It was not as awful as she had expected and my visits were always enjoyable. We really believed the treatment was working. She was eating well and putting back some weight and regaining confidence. All her friends and family had a growing respect for her. On my return I found Margaret with a bullock problem which had to be solved immediately.

With so many bullocks we must expect the odd one, here and there, to be ineffectively castrated. This was not a problem with Herefords and Angus but a Friesian bull is bad. He is a dangerous animal liable to do someone severe injury. Besides the one which had thrown Margaret over the fence, we had had two more where surgery was needed because one testicle had not dropped. One had made history by being the first to be operated on in the newly-built veterinary surgery. It had had to wait outside, in the trailer, whilst last minute touches had been made in the operating theatre and had had its photograph taken to be put in the Practice archives. This had been so successful we had sent two four-month-old calves down in February believing, too, that they had been improperly nipped.

When Margaret checked her cattle mid-afternoon, another eighteen-month-old bullock charged her, head down and making aggressive bull-like sounds. Investigation proved he needed attention. If we had done the job ourselves we would have believed it was our incompetence. We had had it done professionally and it was just one of those things which cannot be left whilst one goes on a two-week holiday! We sent for the vet as an emergency. He came at once and he assured us that the animal would very soon lose all its aggression.

So we were at the last minute as usual and I was still getting my accounts up to date before 6 a.m. on the morning we were to go. I have already commented on the conglomeration on the kitchen table 'ere we leave. Every last minute thing which must go is there. I cleared a place, close to my first cup of coffee, and proceeded to add up the accounts to date. Then I took my briefcase, filled it with receipt and expense books, bank statements and invoices and took the lot into Auntie Mary's bedroom which is not used in our absence.

Everything else went into the final hold-all and we left for the north.

Chapter Fifteen

O God, my master, should I gain grace
To see Thee face to face when life is ended,
Grant that a little dog, who once pretended
That I was God, may see me face to face.

From the French of Francis Jammes.

Twice a year, mercifully, there is this moment when we pull out of the yard and leave The Currer which causes us so much joy and sorrow. One of our friends said, 'We'll look after things for you for a weekend any time. Just give us the nod and we'll come.'

Thankfully we have friends who will come for a fortnight, less than which would never, in a month of Sundays, be worth it. To do all that preparation for one weekend would be out of the question! To do it for one week, even, would be likewise.

For a long time we had failed to convince our Land Rover specialists that there was something wrong with our Range Rover. When it had needed a new clutch they had politely blamed us for using it incorrectly. We said the fault was elsewhere, that it was inconceivable that both of us should feel like novices when letting in the clutch and doing a learner's kangaroo hop. You cannot win if you argue with the experts so we presumed we were both tired out. Once out of the low gear the car drove well so we relaxed. Restarting on a warm engine alleviated the problem which niggled us and it is sufficient to say that the Range Rover did not let us down or give us cause for concern all holiday and brought us safely home when it was over.

That we arrived in Inverary very late at night was due to our missing the hourly Clyde ferry by seconds. We saw it pulling away from the jetty at Gourock and knew nothing would reverse it and that we just had to wait patiently. We rang the lady at the guest-house and warned her we would be an hour late. We were looking forward to our stay there. She had a mews of converted stables and barns made into self-catering cottages and she was renting us one for the night and she promised breakfast in the main house next morning. The use of one of the cottages overnight was sheer private luxury as undemanding as a Travelodge but much more comfortable and not nearly so silent. It was warm and we could isolate ourselves and curl up for a good night's sleep.

By morning we were willing to be sociable again and walked through a sea of daffodils to eat breakfast in the farmhouse. The sun magnified the brilliance of the golden carpet. I must plant some daffodils in the new orchard at The Currer, I told myself.

The day worsened and by the time we got to Ullapool it was perishingly cold. We found our truly delightful overnight bed and breakfast and, in spite of the cold, we left Auntie Mary with the old lady hostess and pushed Harry down to the quay.

Margaret and I have been going to the Hebrides now for more than half a century. The excitement has never waned. There is magic for us in Oban or Mallaig or Uig or wherever we find a pier which leads to a steamer sailing to an island. The last evening chore before berth or bed has frequently been a stroll along the waterfront. It has often been chill, sometimes wild but always with that peculiarly salt-scented wind which announces the very near presence of the sea. We breathe it ravenously for it is familiar and we know that, with sun-up, so will anchor be and every sailing will be westward. There is a smell of fish boxes, wet ropes and nets and oilskins and no matter how freezing the night air might be it intoxicates us and warms us and tempts us more than the deep armchairs of a comfortable sitting-room.

On our return our hostess joined us and told us she used to run the hotel and that four bed and breakfast guests were 'Nay bother at all.' We did not stay up late for the RMS *Suilivan* sailed early.

When we were finishing breakfast someone said, 'Have you got the tickets?'

'Of course.' The bright red Caledonian MacBraynes wallet had been amongst the table clutter at The Currer and I had folded it and put it into the black wallet which contained the driving licences, MOT and insurance documents, plastic AA card and Barclays Connect.

The table had been bare when we left so everything relevant to our holiday had been swept into our holiday holdall. I rummaged in the bag. There was my address book, map, camera, pills, cheque book, some bills still to pay, my scribbling book, Harry's shaving tackle and Gaviscon. There was no red wallet.

I began to panic. I have a sixth sense about emptiness. I know if there is no-one in a room. There were two, middle-aged ladies who phoned twice and booked a twin-bedded room for the weekend. They had arrived and had tea and asked me where the Chinese restaurant was. Next morning when they had not come for breakfast at 9.30 a.m. I went to their room but I knew already that they were not there. They had never returned that evening. I always know if a cattle shed is empty and if all cars have gone from the yard. I knew that wallet was not in the bag. There is a definite, intuitive feel about absence. In that Ullapool cottage I knew I had no tickets for the boat.

I was surrounded by accusing faces. £240-worth of tickets, driving licences, documents and pension books were all missing. There was only one place where they could be and that was amongst my accounts in the brief-case at The Currer. I used the lady's phone and asked Dorothy to look and sure enough that is where they were. She promised to send them on but of course there was no way we would be allowed to board without tickets. The delay had used most of our spare time. I dashed into the ticket office explaining our predicament. The details of our sailing and the record of our payment were all there but I had no identification. I paid £67 for Ullapool to Stornaway tickets and ran back to the car to drive onto the ferry. I was crying inside.

'It isn't fair! We've too much to do! How dare all those people come at the last minute? Why does a bullock which has seemed quiet all winter suddenly change? Why won't the lady in the booking office believe me? Do I look like a thief who has stolen an old Range Rover, with a disabled man in a wheel-chair, an old Auntie and two dogs to sail out to an island? Never! Like Mother, I declare it's the last time we go on holiday!'

But the sail was glorious. We sat on deck the whole voyage. The morning sun was warm and the benches were astern and sheltered from the head wind which was only a breeze anyway. When I sent the unused tickets to MacBraynes they refunded most of the £67 and only rapped my knuckles for negligence. Times have changed since I took children to the islands but my unfair share of ticket responsibility has not. It wasn't the first time I had lost the tickets. Returning from Tiree, with a crowd of children, I could not find them. In those days each ticket was thick cardboard and to lose a bulky packet seemed impossible. Today's flimsy piece of paper would fly away as easily as a Cattle Identification Document in the cattle market. In 1959 the purser had just laughed at my loss and we had been allowed on board.

Lovely Stornoway greeted us in sunlight. Before crossing the mountains into Harris we stopped for lunch and high up on the Tarbert-Luskentyre road, we pulled onto a lay-by and all of us slept. On Harris it is 'to dream and not to dream'. It is beautiful in a way peculiar only to the Hebrides. It is a barren country, a moonscape of rock and peat surrounded by bays on the east and beaches on the west and a sea so green one could be in the Arctic Circle.

We were to stay two nights at Luskentyre with Katie and Angus. Lusky, who was born there, sat up in the back of the Range Rover looking out in amazement. I would love to share his thoughts. What goes on in his head we will never know but I am sure he remembers that these people, with whom he has since lived, took him away from the sight and smell of the sea twelve years ago. Does he know where he is going when cases are thrown on the roof rack, when the car is driven on board? Does he wonder to which island he is bound? I think he does. Does he always anticipate that it will be Harris where he spent the first six months of being a puppy? Is he disappointed if we go to Tiree or is he worried in case we leave him behind on Harris?

Perhaps he is so used to our holidays that he does not really mind where we go as long as it is to the sea and there is a ball or a plastic box to play with. Poor old man his legs are so bad he needs a helping hand to get up and play.

Lusky has always been our fun dog. His back legs were a problem when he was born without hip sockets. He has never been able to jump into the Range Rover without help, and now, in old age he is crippled with arthritis but he laughs at his disablement.

'Where are you, Lusky?' we called to him as with sudden joy we began to drive along the Luskentyre road to the place so many children thought was paradise. 'You're going to see Angus.'

Angus had given us Lusky in exchange for a jacket. Until then he had only heard the Gaelic. It was always his excuse when he did not want to obey. 'I have no English,' he would bark. Now old and decrepit we still thought it was too soon to get another dog. 'One day,' we said, 'We'll get another Harris dog, but not yet. There's still life in our old one. Perhaps when we go to Harris, next, in two years, we'll ask Angus to have another dog ready for us.'

Sometimes, when Lusky appeared very old and his pills seemed not to be keeping his arthritis at bay, we feared two years might be too long, that we might lose him and that Jess would be a long time alone. However the look in Lusky's eye, as he sat bolt upright gazing at the Luskentyre Sands, assured us that we were right to wait. He did not need replacing yet, no way!

At the end of that single-track road, through the gate and down to the shore is the little house of friends we have known for thirty years. They were waiting for us. Katie, Angus, their daughter Kathleen and a brown Collie called Danny. To shake hands with Angus is almost to lose a limb. Margaret risked amputation and then bent to fondle his dog. Most island collies are black and white. This one was almost ginger.

'He's no good with the sheep,' Angus said. 'He's a wonderful cow dog!' Angus is now the only crofter with cows and has no more than two. Isn't that sad? When we were taking children every crofter had milk for us.

Angus continued to talk about his dog, repeating, 'He is chust no good for sheep. Chust for cows.'

We had not even got into the house and we knew Angus was warming up to giving Margaret a dog.

'Now, now Angus,' we chorused. 'We've got two. We don't want another dog yet!'

Nevertheless the crafty man loosed his dog and it wrapped itself round Margaret. 'He's Kathleen's dog,' he said.

Katie was welcoming us indoors but Angus was still talking about the dog. 'She brought him from Bunavoneadar a year ago. He's chust two and he's no good for sheep but he's a ferry good cow dog, Margaret. He's ferry good with cows but no good for sheep.'

The dog clung to Margaret like Sellotape. It is difficult to ignore Angus. Like Bernard he has a way with him of not hearing what he does not want to hear. He was perfectly relaxed in the knowledge that he had a whole fortnight. Och, it was plenty of time, chust!

We went indoors. We had been their friends for a long time but I am sure none of us looked any different and there is no place I would rather sit, for the window, under which Katie lays the cloth, looks over a daisied bank to the sea only feet away. That is turquoise because of the white sand but a peaty river flows that-a-way and makes a golden ribbon all the way down to the bar where a line of breakers marks the estuary, no matter how calm the sea.

The weather was no good for doing anything but sit round the peat fire but there was nothing else we wanted to do. Angus's mother, whom we called Granny, had died and her old house had been dismantled and taken to strengthen the coastline where the erosion comes too close to the Luskentyre road. She was good company to be with but Angus and Katie are no less and, when we had been excellently fed and the dishes washed, we gathered round the peat.

It is at this point, and only when the night is late, the peat aglow and the conversation good, that I am glad there are children no longer on the machair to whom I must return, for whom I must leave the friendly hearth. Any other time I am in mourning for those twenty-five years of children under canvas.

When we went to bed the wind and the sea were noisy neighbours. It was the first really bad gale we had had on Harris for years and the poor weather lasted all weekend which was a blessing for it held us to the peat fires of the numerous people we know in Luskentyre. We had two days of sheer joy, gossiping, close to the heat and drinking far too many strupaks.

When we went to Mary Ann MacSween's lovely self-catering cottage at Scarista Mhòr on Monday morning, the sun appeared. On the mainland the long, hot summer began its relentless burning up of lawns and pastures and its drying up of Yorkshire's reservoirs. There, record temperatures were recorded but, though dry and sunny, we did not experience blistering heat to blight our holiday. We heard repeated warnings to folk on the mainland not to expose themselves to the sun for fear of skin cancer but the Hebridean one caressed and browned us only in a friendly way.

We had not recently been out to Huishnish so, on a bonny day, we drove out. The precipitous road is no wider than it was half a century ago. Single track all the way one is grateful if no other car approaches and one has to use a left-hand passing place. We parked on grass perpetually whitened by blown sand and walked the short distance, Margaret and I, to the jetty. We were talking to two young walkers when Margaret saw a sheep roll off the cliff and somersault through the air to land on the rocky shore below. Harry, ever observant in the car, saw it too.

Margaret interrupted our conversation, 'Hey! A sheep has just rolled off the cliff!' she said and began running to the shore. The young girl ran after her and I, less agile these days, hurried behind. The young man stayed where he was!

The sheep was lying on its back wedged between rocks and bleeding slightly. It would never have been able to free itself. If it had it might have scrambled back to safety but, because Margaret put it on its feet again, it would not move. We dare not leave it bleeding. We had to see it walk, make sure its back was not broken, and that its legs would support it. But it would not try and we had to lift it and drag it with wool and horns, step by step over the boulders all the way we had come. Seeing us three women struggling, the young girl's boyfriend walked over to demand that she hurry, leave it and come with him. But she would not and reluctantly he gave the sheep a feeble lift onto the grass.

'Oh, for goodness sake, come on,' he scolded but the girl waited until the rescued sheep took its first bewildered steps, then she followed her sulky friend.

'That would be the end of him for me!' Margaret said. 'I wouldn't have him given.'

There is nowhere more reverently silent than Harris on a Sunday. We met some holiday-makers on the beach at Horgabost, on a Sabbath so peaceful they were in awe.

'Where is everyone?' they said in whispers.

I think I would like our mainland Sunday to be so. It would do more for the present generation to experience that tranquillity, that supernatural peace, than any amount of hymn singing and psalm chanting. It would do infinitely more for today's malaise than the programmes put out on television.

Having spent the lovely morning at Horgabost we drove to Leverburgh pier. The holiday-makers we had met on the beach passed us and walked to the end of the jetty. Returning they paused once again to remark on the silence. 'We like it,' we said.

Jess was playing around among the rocks alongside the pier. When the amazed tourists left, Margaret went down to join the silly dog and discovered she had found a very dead and smelly sheep. Jess is an ordinary dog really, for all Harry insists she is human, and she did what all ordinary dogs do. She rolled over and over in the decaying, stinking mess.

'Aw, Jess!' Margaret groaned. 'What are you doing?' The dog came immediately but she smelt disgusting.

'You can't get in our car,' Margaret said. 'We couldn't stand it.' The tide was out and there was no pool deep enough to dunk her. Margaret ran into the nearby toilets for water and soap and washed her as well as she could and tried, with little success, to eradicate the smell. The trouble with Jess is that when she knows she has done wrong she wants to climb all over us to say she is sorry. She won't go away and sulk. Nor will she go into the sea. Lusky heads for the waves every time he has a ball in his mouth but Jess had to be dragged in for a final cleansing. Harry thought it was hilarious.

The question of Danny was discussed when we were alone but the idea of taking him home was dismissed. Lusky would never accept him in the Range Rover. A bitch he might have condoned but not a dog and Jess does not like any other of her kind, bitch or dog, except Lusky. The mere sight of another sends her into a frenzy of ear-splitting barking whether she be in the car, or the yard, or on the kitchen window seat. In the car she is public nuisance number one. Margaret and Harry and I can put up with almost anything but she is a most uncomfortable travelling companion. Even with sickness pills she dribbles all the time and she is up on the seat and down on the floor constantly. We decided we could not take another dog home with us as long as Lusky lived.

Angus and Danny thought otherwise. The crofter pretended deafness and the dog clung to Margaret for survival, the one claiming he could not keep a dog which was no good with sheep and the other acknowledging Margaret every time she visited. In May Harris seems to be heaving with sheep. They are all down from the mountains for the lambing and there is a busyness everywhere. All the islanders are doing something and there is a unique air of industry far more positive than in the Highlands. Activity is comparable to that in the Yorkshire Dales. Everybody is doing something.

The fields and parks along the shoreline and the roads in front of you, and behind, are full of sheep and gambolling lambs. Hundreds of them. It is impossible to ignore them. They are everywhere. Far more lambs than seagulls. Because of this Danny was kept on a chain like all young working collies and his kennel was close to the sea. The chain was long enough to enable him to leap onto the wall and watch the mighty ocean and he was near enough to Angus's working dogs for company. It was not a bad life but feeding dogs is expensive and one which makes chaos out of a round-up is not worth having. A collie needs to run miles each day and if he cannot get that working he must get it following his master.

As the fortnight's holiday progressed the problem of Danny became a topic of conversation between Margaret and me whenever we were alone. There were many arguments for acquiring a dog and not a puppy. Of course our guests would just love a puppy. Visiting children love babies of any kind and our aim is always to please. We want them to come again and again for they are the friends and neighbours we never have time to make in the village. They are not just a source of income they are genuinely welcome and we will put up with geese who make a mess in the yard because animal lovers want to feed them from the doorstep, and the donkeys who bray all night because they attract more visitors than cattle.

Life is easier by far when guests have been many times. If none have been before, I have to create the atmosphere in the dining-room, I have to be hostess in the sitting-room and Margaret has to be instructive and available in the yard. Frequent guests can, and like to do this job for us. Only by guests returning do we ever get to know them. They would love a pup, we knew.

We were willing to provide them with a kid but the responsibility of a pup in the yard could only be entertained in winter. We had reared Jess in the off-season. A two-year-old dog was a possibility if it was good tempered, which Angus assured us Danny was. Kathleen, who had brought him from Bunavoneadar a year ago, was attached to him but we did not know the dog at all and, in any case there was no way we could get him home. We said this with authority because Margaret secretly phoned the railways, the airlines and a couple of carriers and not one would entertain a dog.

'It's not that I don't want the dog, Angus,' Margaret said. 'I just don't know how to take him or cope at home. Jess hates dogs and Lusky still has it in him to fight.'

Angus ignored her. 'I could make a box for him and you could put it in the back, with Lusky. They would be alright.'

'Huh, who are you kidding?' Margaret said.

Our dogs, in the car, bark every time they see another even if it is unrecognisable as a dog. When it is a Pekinese, or a poodle Auntie Mary lies to them saying, 'It isn't a dog.' But they know.

We were sailing home from Tarbert to Uig, from Uig to Lochmaddy and driving down the Uists. We were sailing from Lochboisdale to Oban overnight. Hours and hours of our journey we would be on deck and the dogs in the Range Rover in the hold. We had Auntie Mary with us and she cannot bear barking so there was no way we could take a dog we had never handled all the way home with us to Yorkshire. We were not going to ruin a lovely holiday with a lousy return journey.

Nevertheless we were still talking dog when we visited Morag at Lachlea and she, unexpectedly, said she had a BEA dog box, 100% secure. If we wanted we could have it.

'No, no, no,' we both said. 'We're not taking the dog home!'

'Well,' said Morag, 'It is there if you change your mind,' which is why, of course, we were driving over to Lachlea on our last afternoon to collect a BEA box and why Danny became a member of our family.

The airways box was another of those miniature miracles with which life continues to amaze us. Obviously we were meant to be Danny's third family. I think we will be his last for Angus was right about him. He is a good cow dog. He is more than that. There is something special about Danny!

As Angus squeezed him into the perfect-fit box he had the resigned look of a refugee. He seemed to know he must not make a sound. If we did not know anything about dogs we would have said Lusky and Jess did not know what was in the box as we lifted it into the other half of the space we call Lusky's. Danny did not take part in the volley of barking when Margaret started up the engine and we waved good-bye, for two years, to our friends. Subdued and imprisoned he never even whimpered. We drove to the pier and onto the boat and left the vehicle. We imagined all hell would be let loose when Jess smelt dog and discovered we had a stowaway.

Two hours later at Uig, Margaret went to the car deck and all was quiet and similarly, another two hours more, when we disembarked at Lochmaddy there was peace in the Range Rover. Perhaps, we thought, Danny had suffocated in spite of the adequate square of netting. We hastened to a lay-by on the road south, let out our two dogs and ate our sandwiches. Having got the dogs back on board, we lifted out the BEA box and carried it out of sight. A very quiet Danny crawled out.

Margaret walked him a good way on a lead hoping he would relieve himself but he never did all the way home no matter how many times we stopped. I am basically afraid of dogs having been bitten a few times. I do not handle or fondle other people's dogs but Danny did not seem to know the meaning of aggression. He walked so close to Margaret's legs she was in danger of tripping. He did not bark, he would not eat, he just seemed resigned to whatever fate awaited him. His tail did not wag but he did not cower; no joy emanated from him only a willingness to follow Margaret.

Out of sight of our own dogs he was squeezed back into the box. We expected him to resist for he could not turn round in it but he didn't and there was no chorus of barking when we lifted the box back into the car. Nor did we have any trouble on the many occasions we repeated the performance all the way home. Danny meekly obeyed and Lusky and Jess made no comment. They must have been aware of the presence of the new dog. They have acute hearing and smell. We offer no explanation of why there was peace. Danny is not a deaf mute and our dogs continued to bark at every other dog we passed on the pavement. We can only suspect that doggy language enabled them to have a long conversation whilst we were on deck. Danny had most certainly accepted a place in the pecking order and had promised to keep it. In exchange, somehow, he had been told that the new life was a good one and the new owners provided the best that life could offer.

Danny's faith in Margaret was not equalled by our faith in Jess. Danny would be able to keep out of Lusky's way for the old man was too decrepit to give chase. Jess, being a bitch, behaved like one in that she snapped and snarled at every other dog or bitch we met. We would have trouble, we were sure!

But we were wrong. Margaret is usually right about animals. This time she was hopelessly wrong. I suppose the reason was that she did not know Danny. Danny is like Denis. An experience! He is the loveliest of dogs. Margaret, believing she was facing the inevitable, took him and Jess down the Five Acre together and they were both absolutely compatible. They sniffed their way down the field as happily as if they were from the same litter.

'Danny is a barn dog,' Margaret said and she spent the first day of our homecoming introducing him to the empty shed. He was petrified. Maybe it was the size, or the echo, or the distance between him and Margaret. He howled so much we had to leave Jess with him, which was cruel for Jess is a house dog.

The lady currently in the Mistal Cottage christened him Super Glue because he pressed himself so close to Margaret. He folded himself against her so that every part of his brown, lithe body was touching her.

The barn, for him, was impossible. 'I'll have to put him in the yard shed,' she decided and spent all day partitioning off a small area from which he could not escape. We were terribly afraid of losing him. We chained him to an upright so he could sit outside, as he had done in Luskentyre. He came as far towards us as his chain would allow and he howled. And he howled. We could not bear it!

Next day I made him a kennel in the rectangle in front of the kitchen door, our mainly family entrance. The house walls it on three sides so I fenced and gated the fourth side and we put Danny there, loose and he could jump on the seat and look in the kitchen window and was happy. We have never heard him howl since. All the hot summer the kennel was his home but when winter arrived he came indoors and wherever Margaret is he is as close to her as to be a part of the chair.

We bought an identity disc and it tinkles on his collar so that, like the crocodile in Peter Pan, we know where he is. He goes long walks alone, over the moor and is friends with all those out walking their dogs but he never strays and comes immediately he is called.

* * * * *

A few days after our return Olga and Arthur came to up-date themselves on all the Harris news. For twenty years it had been their good fortune to park their caravan at Horgabost. Arthur was a retired headmaster and for many years they had gone to Harris in the spring and stayed three months. Because the journey was so long they had very reluctantly sold the caravan. 'You must go, as we do, to the Scarista Mhòr cottage,' we said.

Arthur was very quiet, letting Olga do all the talking. 'I don't think Arthur is well,' I remarked when they had gone. 'I could almost believe he'd had a slight stroke.'

We did not leave it long before we phoned. 'How are things?' I asked.

'Well,' Olga said. 'Not too bad, considering.'

'Considering what?' I said.

'Well, Arthur's got leukemia. He's having chemotherapy.'

We were shocked. We had known something was wrong but we had not suspected any more than Olga had. The blow had fallen just a few days after they had visited us. Lillian, one of our much-loved guests was having chemotherapy. She had visited us prior to going in for treatment and had brought her wig to show us. She was so brave. Peter was being treated in Berwick and Joan and now Arthur. Whichever way we turned one of our friends had cancer.

We were so proud of them all. Arthur deteriorated quickly. When we visited him he was gentle and welcoming. To visit Joan was a tonic I took as frequently as I dared without alerting her to my nagging concern. She went for hospital treatment every two weeks and, on the alternate week I would make sure I went to her home. She was always cheerful, happy that she could blame the treatment for any discomfort and sincerely believing it was doing good.

My visits to Joan were full of fun for her sense of humour had matched mine throughout our joint teaching career. She did not know how much I was aching inside me. She asked after Peter, always. She was now concerned about Arthur. She was prescribed quite a large dose of morphine and I told her of Brian, our builder, who had been visiting his wife being similarly treated, and had seen a poster in the women's ward encouraging them to 'Grow your own dope. Plant a man.'

* * * * *

In those days of constant anxiety Danny was a necessary member of the family who had to be taken a walk until he could be allowed his freedom. This became my morning task. To me there is nowhere quite so beautiful as The Currer on a perfect summer morning. To rise early is my lot and my good fortune. The scents of my homeland are lost in the heat of the midday.

Early morning, when the valley is lost in mist and late evening when the sky holds only the reflection of the departing sun, the smell of our acres is more intoxicating than anything found in a bottle or barrel. I find it sad that only a few of the hundreds who come, wander away from the farmhouse and view it from below or afar.

I do not do it frequently enough for life is too demanding, but to walk along the fence at the bottom of the Five Acre at dawn or dusk is very rewarding. I do not go down the dry valley of Jimmy's Wood nor down the hollow to 'Jack Fields' for in doing so the house disappears and it is our home, from this angle, which fascinates me.

I find it difficult to explain the sheer joy of looking up so that, gabled and mullioned, it sits on the horizon and is etched against the sky. Guests coming down the road see the house set against a backcloth of the opposite hillside, of walled fields and open moorland. From below, especially whilst dew is evaporating or at night when it is falling, I find the different shape it presents evocative. Against the sky it looks more isolated. It is a sentinel, seen easily from the upper reaches of the other side of the valley and it belongs to us.

Those local people who sometimes exercise their dogs along the fence must see this different side of our sixteenth century home but it cannot pull at their heart strings as it does mine. Looking up to something has a different significance. With silver or gold above its cheerful chimneys and with a smell of woodland behind me and a carpet of wild flowers in front there is an atmosphere of peace.

Modern man rarely smells the countryside these days. Those guests who walk our pastures are in the minority but even if they only walk each morning, from the dining-room door to their cars in the yard, none return home without experiencing the peace which broods, too, within the house. That, we are told by the many, is what ensures their need to return.

One afternoon Margaret insisted it was my turn to check the boundary. It was a hot day early in June, and Danny had not yet been introduced to the herd of big bullocks on the hill. I feel it necessary to point out that our eyes never lift to the pastures we still call moor, without our seeing some person walking a dog. Dog owners everywhere would envy those people who live on the top road and have our one-hundred-and-seventy acres on their doorstep.

Why then did our herd think it was of real concern that I, whom they know, should be walking towards them with a strange brown dog? They never glance sideways at someone with an alsatian or a terrier, a Jack Russell or a spaniel, a poodle or a Rottweiler but all eighty of them made a beeline for Danny and me. We stood our ground as long as we dare, Danny too close to me for my liking. Then the cattle were too close to Danny for his liking and he bolted. Immediately the cattle gave chase and I was in the midst. I turned to face those behind me, shouting and thrashing my arms about and, at the first opportunity, for we were near the boundary, I bolted likewise and scrambled over the wall. I saw Danny streaking across the moor like a disturbed fox. Should our cattle behave like that for other people's dogs they would run riot continually.

* * * * *

We returned from holiday on May 16th and on the 24th the national news reported a light aircraft disaster just a few minutes after taking off from Yeadon Airport. Nine passengers and three crew were killed. Instant disaster and death shocks any community when it happens close at hand and every time it happens shock waves hit those who know the victims for all have friends and relatives, neighbours and colleagues somewhere. Yet one never expects to be one of those to actually know someone lost or hurt, killed or involved. Disasters, one assumes, happen to someone else.

The tragedy, happening so close, stunned the community. The uncommonly hot weather was then thought to blame. We were experiencing sudden, short but quite violent storms.

In subsequent news bulletins those killed were named. Watching a late news, sleepily for I am not at my best at the end of the day, I was suddenly alerted by hearing the name of one of the children I had taught.

'What name was that? What did the man say?' I demanded of the rest of the family. They did not know. Neither did I any more and newscasters do not repeat. Terribly afraid I had heard the name of one of the children I had taught, I strove to remember what the man had said. I was sure he had said Chris Tonkin. The child I had taught we had called Christopher. Desperately I hoped I had heard wrongly.

That night I could not sleep for memories of the Christopher I had admitted to my village school shortly before his fifth birthday. I knew the family well because I had taught his elder brother, and his father was a corn traveller well known in the area. The weather was wet at the time of his admission and, because he had half-a-mile to walk to school each day, he had been kitted out with new oilskins and wellingtons. He was very proud of his fisherman's gear. It was bright yellow and, being new, the buttons were stiff to fasten and the buckle on the belt hard to adjust. One or other of us had to help and to pull on his sou'wester and tie the strings beneath his chin. Having done all this the stiff oilskins prevented him from bending to pull on his wellingtons. Play-time was nearly over before the new pupil was fully dressed.

Joan did not mind this mid-morning palaver because the child was talkative in an old fashioned sort of way and Joan loved to listen to children. She called them her Giles Kids and Christopher was very vocal. Having equipped himself for the weather he was more interested in talking to the teachers than in splashing in the playground puddles with his class mates. They were charging around, coat and hat-less because the rain had stopped.

Play-time over, I took the bell and went into the yard to warn children with skipping ropes and those who had thrown jackets onto the wall, that it was nearly lesson time.

'Come on,' I commanded. 'Pick up your things. I'm going to ring the bell. Pick that up Peter. Fetch it John. Don't leave that out Susan. Bring everything in.'

I felt myself being poked from behind and turned to see Christopher looking like something from the lid of a sardine tin. He gave me an amused and wicked look and said, 'Bossy Boots!'

I dare not look at Joan. I knew she would have the same amused and wicked look she always had if one of her Giles Kids rendered me speechless.

When Rummage Sale Day came there was great excitement in school and our usual relaxed atmosphere was more so. Someone had sent a cane carpet beater for the stall. I asked the children if they knew what it was. They did not for vacuum cleaners had replaced the cane beater we had used when we were children. I should have told them the truth, that we had no fitted carpets in my childhood and that, at spring-cleaning time, we had lifted them and slung them on the clothes line and beaten them until the clouds of dust had stopped flying out.

Instead I offered to demonstrate its alternative use and asked for a volunteer. 'I think I'll buy this, myself,' I said. 'It will be useful for naughty boys. Let's try it out.'

Andrew volunteered and the cane made a swish through the air before landing, gently, on Andrew's proffered rear. An exhibitionist always, Andrew leapt into the air and danced round the classroom in front of an astonished audience.

During the evening sale, Christopher's father came to the stall saying he had been commissioned to buy a carpet beater before Miss Brown did!

When his son had outgrown his oilskins and qualified for my class, he joined the Kildwick and Farnhill Cub Scouts and, because of my life-long Guiding interests, we were often in deep conversation. I loved all my children but I had a real affinity with this one and memories of him kept me awake whilst I worried. Surely I had heard incorrectly.

It could not be, but it was! Joan and I, talking about it soon afterwards, and Margaret and I at the church-packed funeral could scarcely accept it for truth. I had not taught many children in the thirty-one years it had been my daily joy to do so. Village schools are small and children were in my class nearly three years so the turnover was slow. The longer they are with you the more you care. In that beautiful church, called appropriately, The Lang Church of Craven, I mourned others I had taught who had also died. Roy Howker, David McKechnie, Eric Hargreaves, Robert Richardson, Jeffrey Williams, Geoffrey Atkins, Aisla Bailey, Michael Riley, Adrian Rouse, Matthew Holmes, Joe Cunningham, Christopher Tonkin. Where have all the young boys gone? Mostly as a result of road accidents. Writing their names I am appalled that there are so many. Reluctantly I admit there may be more about whom I have not been told.

I do not visit the Lang Church often nowadays. If I am there it is to a funeral. The last time I had been there was to the funeral of a friend's small grandson, a Cub Scout killed on an outing, in a mini-bus with other members of Kildwick and Farnhill Pack. The church had been full, the congregation stunned. Once again hundreds of people filled the church for Christopher and television cameras brought it briefly into the homes of millions that evening. The service rendered many tears. Though he was little more than thirty his record of achievements, his love of the countryside and of music, and his Christian faith and energy made me one of the proudest members of the congregation. As long as my brain can function

I will remember the children I taught and mourn them when they die prematurely.

Joan would have been at the funeral if she had not been having chemotherapy. It seemed to be working. Some day there will be a cure for all cancers.

* * * * *

The heat was intense. It was far too hot to sit out of doors. We wondered how we could celebrate Hilda's ninety-third birthday. Auntie Mary suggested we go to a little tea shop at Hovingham so we did. The coolest thing to do was to drive with all the windows open. When we stopped on a lay-by for coffee we had to draw up into the shade. Even then we could not sit out long.

Driving along one of the maze of country roads on the York plain we passed an open shed and saw the quick, sudden movement of a host of piglets. Periodically we have a nostalgia for pigs. Pigs make us laugh. It is criminal, in our eyes, to keep them intensively. These we passed were running free in a spacious shed strewn with straw and open to the fresh air, what there was of it in a heat wave.

'That's the way to keep pigs,' Margaret remarked and added, 'Can I have some?'

Hovingham really was too far away on such a hot day. Early June is the time to cut hay but the drought prevented the grass from growth. Farmers should already have taken their silage but many had had to turn hungry cows into the meadows. There was anxiety in our profession but we got no sympathy from holiday guests who were revelling in the hot sunshine.

'This is the life,' they chorused every morning leaving without coats or rainwear. The men sported shorts, the ladies followed suit. Everyone treated sunburn. We sweated in the kitchen, grateful for having no fires to tend, relaxed because, connected to the mains, we had plenty of water. In spite of the lack of grass, the cattle were doing well. They eat less in heat but lazing about do not use it to produce energy so continue to get fat. They hovered all morning by the water trough and were easy to count.

Content cattle, happy holiday-makers, Joan, to use her own words, 'Not s' bad,' all added to our serenity so I decided to snatch this moment of calm and fill in those horrendous CID forms which, in our case, must go to the Ministry mid-June. If, at Beeston Castle, we buy two or three calves from the same farmer they have almost identical ear-tag numbers. If I am ever in a mental ward the cause will be ear-tag numbers. We have always one-hundred-and-forty different ones. If we had a suckling herd each would bear our herd number and be relatively easy. Because we buy at market almost every herd number is different and our problem is when we have two calves tagged X 2322 2047 and X 2322 2049. One of these had been the last big animal to suddenly die of coccidiosis in February. Of course we had yellow tagged for easy identification thirty and thirty-three. We think we have a fool-proof system yet, somehow, I had confused the two animals. I had sent in the wrong death certificate and the living one had the wrong birth certificate. Errors have to be corrected or subsidies are lost!

We had to be sure, so I sent Margaret to search amongst the herd for yellow tag thirty. If she found it I had made a mistake and must notify MAFF immediately. She found it!

If I dwell two much on coincidence, I apologise. It fascinates me. Looking for number thirty meant a thorough checking of the herd because it was almost the last to be found. In the process of looking Margaret noticed a bullock, lying down, which did not look all that well. She made it stand and noticed how empty it was. It began to cough and in doing so splattered blood all over the grass.

'That's my guardian angel again,' she muttered, 'Putting me in the right place at the right time again.'

Thinking it might have swallowed a couple of inches of barbed wire she urged it home. Any wire we leave is always far too long to swallow but men repairing pylons tend to ignore minute snippings. We tell them to remove every bit but invariable pick up a bucketful after they have gone. Margaret is fanatical about the danger of losing staples when fencing or spring-head nails when roofing. We put the huge animal in the crush and she opened its mouth but could find no foreign body. 'Aw. I'm sending for the vet,' she said and added 'I'll get him to bring some tools to cut that calf's overgrown hooves at the same time.'

The young vet who had been so reluctant over the bleeding horn had left and his replacement came. We are told there are far more women trainees in the profession these days and perhaps we should be grateful, for female probationers who come to the Practice beat their male counterparts hands down. This new young man could be partly forgiven for being unable to diagnose what was wrong for the animal had not continued to spit out blood. It was a big, fit, apparently healthy animal. Apart from its empty belly when the field provided plenty of grass, there were no outward signs that something was wrong.

There is a reluctance among young professionals to listen to Margaret who was one in her sphere long before they were born. The young man looked in the animal's mouth, as she had done, and seeing no foreign body presumed she was imagining things.

'You'll just have to monitor it and see what happens,' he said patronisingly.

'I did that before I sent for you,' she replied, unsatisfied with his casual examination. 'It has an empty belly. It coughed blood all over the grass in the field. If you listen you'll hear its lungs making a noise they should not. You haven't even taken its temperature. I can't send for you again in half-an-hour when it coughs up blood again!'

'There's nothing I can do,' the young man said, walking back to his car.

This had happened before, too short a time ago for me to forget. Margaret is the farmer but I am the headmistress. 'Oh, no, young man,' I detained him. 'We sent for you and you are not just walking away. Where animals are concerned Margaret knows when one is ill. I have hoped almost all my life that she was wrong, that when she has dragged me out of bed it proved to be a false alarm. She monitors, as you so professionally put it, before she goes to the phone. If she says there is something wrong with that animal there is.'

'Well, I'll take its temperature,' he said. He collected thermometer and a stethoscope from his car. The one recorded a high temperature, the other was not needed because we could hear without its help that its lungs were noisy.

We were tempted to send the huge animal down to the lovely new surgery with its excellent facilities and, because the young vet had not been told to bring the hoof knife and clippers, we bowed to temptation and sent for a cousin who farms neighbouring property, and he brought his Ifor Williams trailer and took both animals down. The hoof trimming was done quickly and the calf was returned but the ailing one was kept. The experienced vet said it would be a lung ulcer and that they would be giving antibiotics and see what happened.

It was Friday afternoon and very hot. Margaret phoned before the end of surgery. The animal was nibbling a bit of hay. If possible it would be sent home in the morning but the first person to investigate it found the bullock dead in a sea of blood and telephoned us.

* * * * *

Saturday, not Monday, is market day at Bingley. We said, 'Aw, let's go and buy some pigs!' We had had enough of cattle.

We had not had any pigs for ten years. We had missed them. Pigs are fun. Sometimes we had real nostalgia for the days when we had free-range breeding sows and scores of baby piglets. During the first five years of bed and breakfast our pigs had entertained our guests. Then Margaret had had an operation and I had been the farmer for a while and we had sent the last pigs to market and not bought more. Free-range pigs are happy souls but they do root a lot and we blamed them for the thistle problem. Rooted earth is an ideal growing space for airborne thistle seeds.

We were fed up to the teeth of cattle. They were spooked. What the heck, we would go to market and buy some pigs! There were seven pens of weaners that morning at Bingley. We had left everything to up-tools and drive over. It must have been a quiet Saturday for, though it is possible to leave the washing-up and do it in the afternoon, it is not remotely possible to leave the bed-making and the cottage cleaning. Guests tend to arrive early if you do. So the changeover must have been negligible for both of us to be at the midday sale.

Whether Margaret is buying calves or pigs she is never satisfied with few. She bought four of the seven pens, thirty-four piglets in all. The guests were delighted. The funny, fat little babies were given the whole of the empty calf shed and they had a beano. Jess was disgusted. She had never seen such things before and she did not approve. She never approves of week-old calves either. Big animals will respond to her herding instinct. Babies will not do anything properly and this squealing variety were intruders. Lusky remembered the days of pigs with real pleasure and spent hours just watching them dashing around the shed every time there was a sudden noise. When not charging about they made a white pile of sleeping bodies, oblivious of watching Lusky and fascinated Danny.

Danny, of course, having just arrived from The Hebrides where there are no pigs, could not believe his eyes or ears. He just could not drag himself away from them and would peep through any hinge hole if the big doors were closed overnight. When the piglets were being fed he leapt over the manger to encourage them to the troughs as gentle as Nanna in Peter Pan. The novelty of the little pigs had a healing quality. 'The slings and arrows of outrageous fortune' had been aimed at us for some time and the little pigs brought us real pleasure. They never ailed anything.

* * * * *

We had not been unconscious of the fact that the Range Rover was being difficult. Returning from holiday we decided it must go to the experts. They checked it over and said that the engine needed a thorough clean. They said we did not use it enough. We complained about the 'learner' start both of us tended to make. Anyway they made a thorough job of the clean-up and the bill was four figures. We have great faith in our Land Rover specialists and less in ourselves for the as-new engine did not solve the kangaroo start which had been our problem in the first place. 'It must be us, after all,' we admitted paying the bill painfully.

Our problem is that we have too much work. When driving we think, 'This isn't right,' but there is something else to do immediately on return, the shopping to unload, the washer to re-fill, the fire to light or new guests to greet. The annoying old Range Rover is forgotten until next time.

The other recurring problem was that of Friesian bullocks inadequately castrated. Margaret was still dubious about the one the vet had returned to just before our holiday. He was still as aggressive and another joined the clan. Margaret sent again for the vet and this time he gave one of them a knockout injection. It is quite a spectacle to see an enormous animal give at the knees and collapse gracefully on the floor. 'Now I can cope,' the vet said. 'That'll fettle him!'

These little tremors in our farming career are never witnessed by guests who leave after breakfast for their wanderings in the Dales and return in the evening when the fun, if that is what you call it, is over.

'It will take a month for the hormones to stop being produced. You shouldn't have any bother after that,' the vet said for the umpteenth time.

'You've said that before,' we reminded him and he smiled.

Anyway, we stopped worrying and devoted ourselves to our paying guests. The phone rang and I admitted I'd a vacancy to an elderly man who wanted to come from the Midlands, without a car. I told him we were half-a-mile from the bus but he said it did not matter. I was still uneasy but accepted the booking and told him to send a deposit, and when everything was settled he said he would also be bringing his six budgies!

'On the bus?' I said. 'Can you do that?' I wondered what the driver would say when an old man appeared ready to board his bus with a large bird cage and six twittering budgies.

'I doubt you could do that,' I said wondering also what I would do with a guest and half-a-dozen birds. 'Pets welcome' really means dogs. We have had people who bring cats. Cottage people sometimes have a budgie and once a family had a hamster. There was the hilarious escapade, many years ago, when professional people from London brought two goats.

'It must be a very big cage,' I said, 'to hold six budgies.'

'They are my friends,' the man said. 'I take them everywhere.'

'What about the other guests?' I parried, ashamed at myself for trying to put him off. 'Will the birds wake everyone up in the morning? We've got cats, too. Is the cage secure?'

But the man was adamant that he wanted to come and I had gone too far with the booking and, anyway, I would not have refused to take such a kind man.

We waited somewhat apprehensively for his arrival and I collected him from the bus, birdcage and all. Life is fun. We found that out long ago. The man was devoted to his birds and took them out into the fresh air every day.

The Vinces came. They are among the few who have been here on holiday annually since we first opened the doors in 1981. We are fond of the Vinces who get more out of their twenty-four hours because they rise so early and have breakfast before everyone else and are in Blackpool, or Whitby, or The Lakes before the resorts are awake. Doreen blames Len who is the early waker and the driver but she complains smilingly. Len has been a good help with the hay and every year he collects wood at home, chops it into enough faggots to last us nearly all year. The Vinces were kind to the budgie man and so were the Johns, farmers from Cornwall. They, too, have been coming for years and Sheila is the lady behind the homeopathy treatment Margaret has found so successful.

Gerald, like all farmers, talks shop all the time with Margaret and, silly man, prefers to stay behind and worm, or tag, or give copper. He was in his element, the year of the budgie man, when only half the herd appeared at the water trough one morning. We had employed John, early in the year, to bring town water to the troughs. We dare not risk another bout of coccidiosis. The weather was building up to another very hot season and and our own supply was already greatly reduced. Cattle stayed round the water troughs in the heat of the day and poached the grass and manured the area well. A wet backend and trouble we hoped would be averted. As the summer wore on we realised that without mains water we would have had none, for drought brought havoc to the region.

When only half the herd came to drink we suspected, correctly, that the other half were astray. Gerald was excited and said he was coming with us. Our neighbours at Transfield reported seeing cattle on the heather and bilberry hillside we call The Rough. Some of it, ten acres or so, belongs to us, the rest to the Council. We only use it in famine but we try to maintain the fence beyond our boundary in case cattle get out. It had not been checked lately and the gate in it was down.

We drove the Range Rover into the far field and I went down to Transfield whilst Margaret and Gerald checked the meadows. They found just one animal. That, too, was a piece of luck for it had a high temperature and needed a dose of antibiotic.

Down at Transfield I found the maze of agitated footprints going every which way. They only led me to the gate onto The Rough, which someone had left open. I closed it and tied it with what string I always have in my pocket. I was angry and disillusioned about ramblers and mountain bikers and joggers. I suspected it would be the cyclists. They are a modern menace to farmers nationwide.

In the meantime Margaret and Gerald had located the herd, once more galloping around the estate and they had to pull down the boundary wall to get them back. When all were safely imprisoned we gently brought home the one with a chill. Gerald had so enjoyed his busman's holiday he would not go out next day until he had helped us give the second worm injection to the seventy eight-month old calves. Farmers are all the same, can't leave their job behind!

On the morning the budgie man was to leave, he accidentally let three birds loose near the dining-room door. This was, for four hundred years, the entrance to the barn. The doors stand back from the curtains so that a reflex action on the part of Len Vince trapped the birds in a four-foot space between curtain and glass doors. The entrance is high and standsteps are barely adequate and an hour later Len was still trying to catch them. Fortunately they, being habitually caged, tired first.

I was to take the man and his budgies to the bus for Birmingham and I said to Margaret, 'No way am I going alone. You're coming with me.'

I had made that request just once before. We had had to turn William out just before Easter one year, because we were fully pre-booked. He had ignored our warnings that he would have to go and we hated turning him out.

'Where can you go?' we said. He was a mystery to us. We did not know what he did all day, from whence he came or where he went when he left.

'I'll be all right,' he said, standing dejectedly with his almost empty hold-all.

'I'll take you to the station, the bus depot or just to town. Whatever you say.'

'Just take me over the moor,' he pointed across the valley.

That was when I said to Margaret, 'I'm not going alone.'

No way was I going to leave the vagrant in the wild.

So we both went and I drove over the moor to Eldwick where he indicated he would alight. He then took us by surprise saying, 'This is my aunt's house. Come in and meet her.'

It was a lovely detached house in a pleasant, cared-for garden. Having introduced us to his aunt he volunteered to make us a cup of coffee.

Whilst he was in the kitchen she said, 'William is his own worst enemy. He comes every day to do my garden. He'll stay for a few days then he'll go. My dog will miss him.' There was no need to worry about William again!

'I'm not going alone,' I said a second time when the budgie man had to be put on the Birmingham coach. The driver of the bus which had brought him had indeed been funny about taking his aviary.

I left Margaret keeping the man company whilst I went into the office. 'I've got a passenger for Birmingham,' I said. 'He's got a cage full of budgies. You brought him. He's your problem and you must take him back.'

I need not have worried. The atmosphere in the office was genial. There was laughter and a comment, 'Oh it's you who booked in a Budgie Man!' but no reluctance to take him back.

'The bus has broken down at Skipton and there's a delay,' the office girl said. 'It'll be about half-an-hour.'

Standing beside our guest in the queue, we learned that if you never want to be lonely, if you always want to have friends, you go around carrying a cage of budgies. Every passer-by stopped to look at them and speak to him. Toddlers had to be dragged away. Old ladies hovered and mums pushing babies crossed over the road. There was no need to worry about this guest either.

* * * * *

Immediately it was the Chapel Fête day. It had previously been held at the home of one of the congregation but a bigger venue was needed so we said, 'Why not?' Auntie Mary was happy making five hundred biscuits which would have been no small task for any cook but was a splendid marathon for such an old lady.

All the days, that summer, were gloriously hot and the Open Day was no exception. Crowds poured in and a good time and a good gossip was enjoyed all day. One old lady said, 'I wish we could have our Christmas Party here.' 'Why not?' we said again. Life is short enough, we may as well do what we can and share what we have whilst we breathe.

Everyone loved the continuous hot sunshine and only the farmers prayed for rain. The drought was severest in West Yorkshire. Leeds frequently had the highest temperatures in the UK. Springs feeding the needs of neighbours dried up completely and they began a long season of carrying water from elsewhere.

The guests couldn't care less. No one could. Everyone took more showers, hosed their gardens and cars, drank more and washed their blankets and woollies. Everything dried.

Only farmers and growers and Yorkshire Water chiefs prayed for rain to green the increasingly burnt fields and fill the reservoirs. As long as water came through the taps the general public did not worry. On scorching Sundays they drove up to the West End reservoir to see the ruined village submerged since the creation of the dam. The enormity of the drought was headline news but no-one took it to heart. When we said we wanted rain those guests listening thought we were crazy, or maybe downright mean.

One guest returning from a day in The Dales said, 'Well, I've certainly seen Yorkshire at its best.'

We contradicted him smartly. Only rain can beautify a lovely landscape and The Dales were crying for it all summer.

The little pigs grew by the hour. We gave them access to the yard and the sunshine but they mostly preferred not to get sunburnt and sought the cool of the calf shed and one bullock shed. Visiting children helped feed them all summer. Guests Paul and Bunty brought another little china pig for the mantlepiece every time they came for an evening meal. Pigs are fun!

The herd of fallow dear were frequently in the Dyke Field. When the stags separated from the hinds and the fawns were born, the menfolk spent most of their time within view of the house. Their antlers were at their very heaviest. We hoped they would be dropped on our land but none were found. Hares we have in plenty.

Because of the heat and drought we began to notice the absence of birds. Swallows had stopped nesting before the barn was converted in 1980. We had tried to provide similar accommodation in the old Dutch barn, nailing bits of wood to the beams should they choose to cement their mud nests there. They never did!

Sparrows had continued to nest in the laithe porch and the dining-room entrance was always bespattered with lime. We noticed, in 1995, that it was no longer so, looked up to the beams and realised there were no sparrows. There were chaffinches and tits in the plantation but no peewits on the moor and fewer skylarks. We had heard the cuckoo on the estate, faintly, but the repeated insult had not been heard coming from Jimmy's Wood. I schooled myself not to be paranoid about that. We believe that, if we survive to see the arrival of the cuckoo, it is an omen of our tenacity. It has been the source of our confidence for as long as we can remember.

We were quite upset about the birds. We use no fertilisers or sprays. We have not drained the moor pond or dried up the dykes. We have not ploughed recently. Only the weather has changed. 'It must be the lack of water,' Margaret said. 'Perhaps the eggs are not hatching. Perhaps the migrants can't make the distance.'

Certainly the goose eggs were not hatching. There was magazine comment on this and our geese experienced the same disappointment. Left to their own devices we would, a few years ago, have had a flock of astronomical proportions. We openly stole eggs and when a goose became broody we made sure it had no more than two eggs in its nest. One goose had been sitting prior to our spring holiday and the defensive bird would not allow even Margaret near. In the pre-holiday chaos the expectant mum had not been seen to leave the nest and we had no idea how many eggs were in the nest and worried about how many goslings would be born in our absence.

Margaret had mislaid a goat foot spray. She, as usual, had accused me of removing it and, in desperation, we had bought another to replace it.

Just before we left for Harris the goose left the nest for water and Margaret hurried to her warm cushion to see how many eggs were incubating there. She found, after all, only one egg, one empty Gaviscon bottle and one lost foot spray. The egg, like all the others, did not hatch. There is considerable, justifiable worry about the present sanity and fertility of the human race. Could it have jumped species and be affecting geese? And goats? Jimmy slept so close to the broody goose his head rested against its wing

* * * *

We had occasion, one Sunday morning, to go to the top of the road and I witnessed the reaction of the bullock the vet had put to sleep. Margaret had continued saying he did not like her, that he bellowed and pawed when she came near. The vet had been confident she would have no more trouble with the last two he had re-nipped. It looked to me as if he was wrong. We could not blame him but I did not like the look of the bullock at all and within days we must send in the CIDs of the twenty one-month-old bullocks. He was one and as soon as the passports were posted he would be in retention and we would not be able to sell him for two months.

I commented on this but Margaret said she thought it was only herself the animal disliked. Indeed we hung around for some time to watch his behaviour with walkers and dog owners. He completely ignored them but came bellowing to the fence aggravated by the presence of Margaret. I did not like it. She had to weave in and out of the herd almost every day. If the bullock objected to her now anything might happen in the next two months.

'Stop worrying,' she said knowing that to sell was to lose money. We were totally sick of this batch of cattle. IBR, coccidiosis, burst lung ulcer and faulty castrations!

After serving the evening meal I took Hilda home and whilst I was away Margaret noticed three isolated animals hovering by the gateway into the road. Animals only rarely isolate themselves from the herd. Among the three was a Charolais heifer and she was a-bulling. With her were the two bullocks which had so defied the vet. There was plenty of evidence that the last attempt had been unsuccessful too.

Margaret was just letting the three animals through the gate into the road when I drove home to help.

'Both of these,' she shouted, 'Could put that heifer in calf. They're going tomorrow. I've decided!'

They did and a few days later I sent off the CIDs and the two months' retention began. Any time in the next eight weeks we could expect a ministry inspection to verify that we had the necessary animals on our declaration form and were not diddling the Government of subsidy. Sure enough the telephone call came but the animals to be checked were the younger seventy only.

'Can't you do them all and have done with it?' we pleaded for our animals were running as a hundred-and-forty head herd and to check just the eight-month-old animals meant a difficult separation. 'I'm sorry,' the caller said. 'It was the younger ones that came out of the hat. If the older ones come out next week you'll have to do the whole thing all over again.' So, unfortunately, works the Min of Ag!

One after another the miracles happen. Perhaps they happen to all and are only noticed if you are looking for them. Two days before the ministry visit we had a thunderstorm. One heavy summer downpour! We could not believe it! The water bounced off the concrete and an orchestra played. There was a beating, as of drums, as rain pelted on the corrugated tin roofs and water music as it gushed down the spouting. A river flowed down Currer Lane bringing with it a sediment and leaving it inches deep at the cattle grid. The empty-for-weeks McDonald Ditch became a lively, leaping snake of muddy water and the long empty duck pond filled and overflowed into the stream. 'Thank God,' we cried. 'Blessed, blessed rain.'

Hilda stayed overnight for she was too old to be alone in such a violent storm and the cattle on the hill took fright and the thunder of their hooves echoed that of the heavens as they poured in one black and white stream to the safety of home. Quick as the lightning that brought them, regardless of rain, we let them into the roadway and contained them overnight in the fields we rent from George.

Behind us are forty years of adapting buildings and fences and gates and sheds to make collection and separation easy. Next day we guided little ones into one shed and big ones into another, possible but not easy. Always some have dived into the wrong shed but they are few enough to re-sort. Having done so we let the big ones back up the hill and the little ones we counted twice and a third time as they went through the shed door to spend the night down the hill. Confident of success we slept untroubled.

There is no water down hill so the seventy smaller steers came up to the shed around breakfast time to drink. Margaret left the bacon frying just long enough to shut the door on them. 'Phew,' she breathed. 'Call that nowt?'

Our reputation as honest citizens and one year's subsidy depended on the number being exact and the ear-tag digits faultless.

The man from the ministry and his assistant came. We were bloated with success. The new crush was a dream. The animals behaved, the tag numbers were correct. We got down to the last five before Margaret, the same double, treble checking Margaret I have always lived with, suddenly announced to a shocked and unbelieving sister, 'I never counted them in!'

She never presumes they all come to drink together. She never shuts the door finally without counting, ever. She is fanatical about one being missing, always. It had just been the fear of the burning bacon, surely. The enormity of the predicament we would be in, had a small group of animals come to drink early and returned to the pasture, left us speechless.

Checked calves had been released back into the field. Should there be one number not crossed off on the Ministry list every one of them would have to be brought back. They should have been counted before the door was shut. Suppose one had leapt a dividing wall between hill and low pasture and joined the senior herd and never come to drink at all!

The last five were checked and we watched, with rising fear, the lady assistant run her finger down seventy multi-digit numbers to see if there was one which had not been crossed off. She raised her head, smiling and said, 'That's it.'

The relief was unimaginable. Once again that narrow line between harmony and chaos had not been crossed.

The deluge only briefly dampened the scorched grass. We worried about what the cattle would find to eat but they seemed content and continued to put on weight. Pigs grew like mushrooms and tourists thought they were in the tropics. Only Margaret and I and the Range Rover struggled to survive the summer.

Chapter Sixteen

All night, all day Angels watching over me, my Lord

Negro Spiritual

The success of the two new corn bins purchased the previous autumn tempted us to order two more and be fully equipped for every manger and reduce the carrying distance to a minimum. The corn place had to be cleaned out to accommodate one. The other we planned to erect outside, at the manger end of the bottom shed, and to tin its sides and roof. The corn place had not been properly cleaned out for years. Hens had roosted there and everyone knows the mess they make. The job of cleaning out often falls to me, the odd-jobman in our farming enterprise. It was unbearably hot and a very dirty job to do between guests leaving for the Dales and returning for the evening meal they would not have eaten had they seen the midday cook.

For a week I worked filling more than a dozen sacks of old, mouldy corn and we heaved them onto the roof rack ready to take to the local tip. We had a major throw out of paper sacks, binder twine, tins and bottles and goodness knows what and filled the rack with the most horrible array of debris. Lastly we piled on top a mess of rusty barbed wire. Margaret, the more agile of the pair tied the whole thing securely and we washed thoroughly and, with no time to spare, bundled the two- and four-legged members of the family into the car Auntie Mary had a hair appointment and Harry wanted to go the baths.

Our lives are such a mixture of filth and cleanliness. No wonder friends eye us with amusement. Having dropped Auntie Mary at the hairdresser, Margaret, who was driving said, 'This Range Rover's going to die on me.' The green fuel light was flashing. 'Perhaps we're out of petrol.'

We limped to the filling station for more and, pulling away, Margaret said, 'I think it's a bit better,' but she did not sound convinced.

'Let's get to the tip,' I begged. 'We can't break down with all this rubbish on top,' but we did. We had avoided the busy main road to the waste dump fortunately, so when the vehicle refused to go further we were conveniently out of the way. There was, however, no telephone kiosk in sight.

We had to have a phone. We had to get the AA. We had to phone the hairdresser. We had to get a taxi to take Harry and me home. 'Which way shall I run?' I asked, floundering like a fish out of water in the middle of the road. Either was some distance. There was a pub. They'd let me use their phone even if I never use their services or drink their beer. In the other direction there was, some distance away, the army surplus dealer from whom we buy corrugated tin.

'Which way?' I repeated.

'Go to the pub,' Margaret said. 'They don't know you.'

A man was coming out of a bungalow opposite. 'Are yer 'aving trouble?' he said. He did not live there. He lived on the top road, three hundred yards from our cattle grid. It was his car standing on the roadside. 'My friend lives here. Yer can use 'is phone,' he said.

Life continues to amaze me. The AA was called, Harry was transferred to the man's car, Auntie Mary was collected and we were taken home. Margaret stayed with the Range Rover for which we had recently paid a four-figure bill for an engine clean-up which might have been necessary but which had not solved the problem one bit.

Margaret prepared to wait for the AA man wondering what she was going to do about the dogs. Jess had been left at home. Danny could be walked home but Lusky couldn't. He was becoming increasingly, but happily disabled. 'Hold his tail when he's walking,' the vet said. We did and the grateful dog would wag his tail and the hand holding it would wave too. There was no question of us terminating Lusky's quite happy existence.

Within the hour the AA man came and within minutes he found the fault. He had the advantage of having the engine cease to function altogether. The mechanics only had our word for it, which they did not believe. He said 'There are two cracks in your ignition coil.' He drove off to the nearest garage and bought a new one which cost £8 and fitted it free because we pay our membership. Margaret drove home on cloud nine.

'That coil has been at fault since the minute we bought the Range Rover,' she was sure. 'It mucked up the engine. It caused the bad starts. Dear knows what it has done to our clutch.'

She showed me the cracked coil. I needed my glasses to see the two hairline fractures. From then onwards we had no trouble with the engine. Only with the tyres and eventually, as expected, with the clutch.

Margaret wanted to go to the library a few days later, to look up some problem in Black's Veterinary Dictionary and Harry needed a prescription collecting from the doctor's. I left Margaret on the steps of the town library and pulled away from the kerb. A lady, in a passing car, seemed to be pointing at me and waving but the car did not stop. I was on a busy road and, whilst the front door of his surgery is on the main thoroughfare the back door has easier parking. I got out of the driver's seat still wondering why the lady had pointed and waved and saw that my front tyre was slowly but unmistakably going down. I do not consider such an event to be of recordable value in the general run of things. I knew that on this occasion, I would have to walk to the library to fetch Margaret because she was expecting to be collected.

She needed time to peruse the heavy dictionary so I thought, 'First things first. I'll get the prescription and collect the Gaviscon'. The chemist is on the main road just a few yards from the doctor's front door so I entered at the back, picked up the bit of paper and went out at the front. I went to the chemist and waited my turn and then I cheekily took a short cut back through the doctor's surgery to my car parked at the back door. I had never done that before. I had always walked right round the block. I ignored the receptionist and the patients waiting and hurried through the backyard and out of the gate so quickly I collided with a pedestrian walking up the pavement.

'I'm so sorry,' I said.

He was an extremely well-dressed man in an immaculate dark suit, white shirt and tie and he put his arm round me and steadied me.

'How are you, Jean?' he said. How, I wondered could I know such a well groomed man? 'Pat Clark's husband,' he jogged my memory.

'Of course, Duncan.' I remembered Pat, a teacher and a Guider who was with me on the last Harris Camp in 1980, fifteen years ago. She had shared the fun, the song, the poignancy of that last wonderful fortnight of thirty years of camping with children, twenty-five of them in The Hebrides. It had been her first experience of the islands and of camping and her wonder of it all had sustained me through the difficult awareness that it must be my last. The bed and breakfast venture would steal from me my summers and make Guiding, as I had known it all my adult life, impossible.

'Of course I know you, Duncan. You look as if you've been somewhere special.'

He worked at the Citizen's Advice Bureau and in that capacity he had had to attend an industrial tribunal but it had been cancelled.

'How are you?' he asked.

'Fine,' I laughed, 'if it wasn't for that!' I pointed to the very flat tyre.

'I'll change it for you,' volunteered the very immaculately dressed man.

'I couldn't let you do that.' We live on a farm and the Range Rover tyres are cow-mucky. Even the spare in the back is none too clean. All wheel bolts are greasy. Mechanics wear dungarees and use Swarfega on their hands.

The good man was taking off his jacket and revealing a spotlessly white shirt. 'Of course I'll change your tyre for you. I've time.'

'It's not that,' I remonstrated. 'It's too dirty a job. The wheel is filthy, so is the spare. The jack is on the roof rack!'

The incredible man would not be said nay so the retired headmistress climbed onto the car bonnet and onto the wire rack and handed down the weighty jack. Thank goodness we had one on wheels. Why hadn't the man calculated that he might have to crawl under the chassis? Knowing that Lusky, his head resting comfortably against the spare tyre, would bite if a strange man put his bulk into the rear, I insisted that I unscrewed the tyre and lifted it down but everything else that wonderful man did for me.

Everything had happened so quickly Margaret had not noticed any delay. I repeat, life really is just a strange encounter. A series of encounters. As Denis says, it continues to amaze.

* * * * *

Another strange thing happened about the same time. The novelty of deer in the vicinity made us focus our eyes more often across our Rough to the two green fields belonging to Marley Hall. The deer were often there. We never saw them cross the Rough but they would appear in the Dyke Field and all our guests would run out with cameras. When we scanned the distance with binoculars one lone goat would sometimes be seen. We did not think it could be one of a pair which had escaped from owners under the hill. They had alerted us to their loss of two angora goats over a year ago. This speck of white had survived a bitter winter in the open and become completely wild. Unlike the deer it never came anywhere within a half-a-mile of us but we saw it now and again and Margaret always worried about it for goats seldom survive alone. They are herd animals. They need companions of their own kind. Our own goats are home lovers who never stray much further away than the paddock. She even reported it to the police but the speck kept appearing.

One August week, children in the Loft Cottage who had helped to rear Jimmy and doted on all our animals, spotted the speck of white and asked Margaret what it was.

'It's a wild goat,' she said. 'It's as wild as the deer and it never comes near.'

These annual holiday children spend most of their time on our acres, collecting flowers, scrambling on the rocks, feeding the donkeys and stroking the goats. A wild goat was magic. Their imagination was fired. Before the day was out the three children were off across the Rough to see the wild, lonely creature. We thought it would be alerted to their approach long before they got there and we forgot about them. Some time later they walked back into the yard followed closely by this long-haired white beauty. It was very wary. Its head was always on the move, its eyes wide and staring but it did not bolt. It followed at a distance until it could see our goats in the shed. Perhaps it had smelled them some way off. Indeed the children probably reeked of goat.

The wild goat was very small and its long, silky hair nearly touched the ground. It had dangerously erect horns and was much quicker than our nanny. It bounced amongst the bales like quicksilver. The children called it Snowybob and begged it could stay. We had enough goats to feed in winter and cope with in the summer without one which was wildly out of control but really we could not have rid ourselves of that goat if we had tried. It decided overnight that it was staying. We could not have succeeded in chasing it away. It was all over the place but it was not leaving. It was up the loft steps wall. It sampled every shed and it got into the garden and ate the standard rose down to six inches of the earth. Its favourite place was on top of the mow. Had village children still been able to get into the Dutch barn I am sure it would have succeeded in keeping them out for it leapt from on high and if we met it in the passage we were unskittled. No way could we have chased that agile thing away. So it is still here, still wild, still incredibly beautiful. Perhaps these children will bring home the deer one day!

* * * * *

The whole of Yorkshire seemed to be burning dry. Our green, pleasant land became arid and tinder dry. Columns of smoke appeared on moorland horizons. There was no water anywhere. The one thunderstorm had little lasting effect.

We had just finished serving the evening meal on Sunday, August 20th and I was preparing to take Hilda home. She had been with us all day but it had been too hot for her and Auntie Mary to sit in the garden. Walking across the yard we saw a smoke cloud rising from the horizon over where The Altar lies. We were amazed that the crowd of children we had in both cottages had not run to tell us. Later on in the week they would not have hesitated but their new arrival the previous day left them somewhat shy.

We did not know how long it had been burning for the evening meal had occupied us for the last three hours. I ran to the phone to call the fire brigade but Margaret's friend Barbara had already alerted them. Appliances were already there, the man said and no, we did not need to go. This advice we ignored. If there was moorland fire adjacent to our hundred-and-seventy acres of tinder dry landscape we felt we had to know what the danger was. So Margaret came, too, and we dropped Hilda home, in town, and hurried back to the hilltop, directly to Harden and through the estate and to the back door of The Altar.

We were flabbergasted at the extent of the fire. Because the precipitous rocky heather and bilberry hillside was also nursery for scores of silver birch saplings, flames twelve and fourteen feet high were leaping sky ward. The changing climate seemed to bring a prevailing wind from the east all summer. The several appliances were lined up on the east of the fire. The wind was sweeping it away from them and it was completely out of control. Two firemen with beaters were making a valiant effort to halt its very rapid progress.

The men were admitting defeat. The nearest hydrant was in the estate. The wind was determined and they did not know how to get in front of it. We left Harry entertained with the unusual activity and Margaret and I ran across our fields and looked over the northern boundary and were horrified. It was just one wall of flame only yards away.

We ran back to the firemen in near panic. 'You'll have to get in front of it,' we yelled at them. 'If it mounts the wall it will sweep across our land. We have a hundred-and-forty cattle!'

'Aye, but we can't do that,' a fireman said. 'This is the only access.'

'We'll pull down the wall and get you into the fields and then you'll get in front,' we offered.

'We'd need a JCB for that,' they said.

'Nonsense,' we almost shouted. For thirty years car drivers and motor-cyclists had found no difficulty in pulling down our dry-stone walls to get access. Hadn't we hastily pulled down walls many a time to return straying cattle? But men rarely listen to us. They would not hear of our pulling down an entry.

'Then you must go round to our farm entrance and get in front of the inferno that way,' we advised. There were appliances from all over the area. One local fireman recognised us and came to hear what we were saying.

'Follow the lass,' he said. 'We're losing here.'

Slowly they began to reverse the enormous machines up the narrow lane. They did not move fast enough for me. It needed little imagination to anticipate what might happen if the fire ignited the pasture. All our land was west of the flames and the peevish wind would consume the lot.

With what then would we feed our cattle? We could not sell because of the enforced retention period. Sadly, we would have to deplete the stacked bales in the Dutch barn. With what then would we feed animals in winter? The hay and straw harvest was over. There would be no second crops. No-one had sufficient silage. At all costs the fire must be held back from the sun-scorched pasture.

'Stop panicking,' the helmeted men said. It was not their livelihood, their animals, their lovely landscape which were in danger!

Margaret leapt into the Range Rover calling, 'Follow me'.

One engine needed to fill up its tank at the estate hydrant. I begged to accompany it for the driver, if he were not following, would not know the way.

The speed at which the vehicle bounced along the potholed cinder track in the estate frightened not only me but several startled hares. 'Don't kill anything,' I called. Behind the wheel men travel fast. Parked beside the hydrant one would have thought they had all the time in the world. I suppose they could go no faster than the water would fill the tank but, to me, it seemed to take hours and when it was full there seemed no urgency amongst the men to scramble back into their engine. Just wait until they get onto the road, I thought. I bet they go like mad!

There is a through road from east to west of the estate. The Mansion House is central and the hydrant in front of it. I was mortified to see the driver prepare to turn east. 'Not that way!' I called. 'Go right!'

'We know the other way best,' the driver said. It was not a local appliance and that was the way they had come and nothing I said would make them turn left so they made a two mile detour and time was flying. Reaching the B6429 I yelled, 'Turn right, for goodness sake.' I was petrified in case they went left into town. 'I'm the only one who knows where we are going,' I reminded them.

The vehicle gathered speed and I saw temporary traffic lights ahead. There must be road works. The light was at red and oncoming traffic was approaching. Nevertheless the little boys on the fire engine put on their siren and the cars ahead all leapt onto the grass verge whilst we belted ahead. That I was shocked is an understatement!

Of course it is a long way to our farm entrance especially if you insist on adding two miles. The driver obeyed my instructions but the officer in charge argued that I was leading them astray, was going the wrong way. I stood my ground forcibly. The only woman in the vehicle, the only one who knew the way, I shouted down the senior officer in a most unladylike manner as we headed for the moor gate and the cattle grid. Ahead of us, half-a-mile away we could see the flashing blue light of the appliance in front of us and the great wall of flame sweeping towards us. 'There's a hydrant just in front of the gate!' I said. 'Remember that when you want to fill up.'

It was at this point my confidence began to desert me. After all my arrogant display of authority, what would we do if our field gateways were not wide enough to admit the fire engines? Could be! Margaret was having the same misgivings as she headed the procession. She reached the first gateway and it looked impossible mainly because there was a secondary wooden gate-post leaning inwards. The driver following her was yelling his abuse. It is one thing pulling down a dry-stone wall and quite another pulling up a stone gate-post. The one is possible. To do the other a JCB is needed.

Margaret jumped down and ran to the hefty wooden gate-post leaning inward, praying that it had rotted below ground and would give. No firemen leapt down to help her. The silly woman had brought them on a wild goose chase! The fire on the other side of the boundary wall was three fields nearer, creeping at a frightening rate, only half-a-mile from the house. Margaret, less than five feet tall, put both arms round the wooden stake and tried to rock it, stared at sarcastically by the helmeted men. She could feel it begin to give. She pulled and pushed and huffed and puffed enjoying the male audience and anticipating their amazement when this little woman was successful. It did not break immediately but she knew that it would for it creaked in its hole and was easily loosened. When the rotted wood below the surface splintered she pulled out the stake and said with a smile, 'See, I eat my Weetabix!'

Men do not comment. Another appliance crossed the moor and both went easily through the gap and continued to within feet of the fire. I alighted at the cattle grid to direct other vehicles following us. Five appliances had been called to the scene. Now that they were all ahead of the fire, not impossibly far from the hydrant, they could cope. In the right place they were professionals.

Having deposited his dozen men and pumped out his water over the flames the driver of the vehicle I had joined came back on the road to refill. Seeing me still directing the traffic he leaned out of the cab and winked at me. 'Sorry about that, luv,' he said. 'My boss is a reight arrogant sod.'

'You're forgiven,' I returned. It is a lovely feeling when you are proved right.

Without a shadow of doubt we, Margaret and I, saved the hillside that night. The local press had a field day but never mentioned us! All night blue lights flickered across the moor back and forth to the hydrant and all the following week men dowsed the area for the peat below the heather kept erupting. I went into town and bought official looking 'No Smoking' notices and nailed them to the boundary gates. We had had a quite nerve-racking experience.

* * * * *

The more people you know the more specialists you can call on in an emergency. So often one of the guests is a friend in need. Lusky became far less mobile and an ulcer formed on his leg. It was very sore and nothing the vet offered did any good. A lady came with her disabled husband and she gave Margaret some homeopathic dressings which worked wonders. The trouble was the leg was painful and Lusky is not the easiest to handle. I am the one which usually gets bitten but this time it was Margaret whose arm was in the wrong place when he snapped. It was not a bad bite and she washed it with salt and thought nothing of it for days. Then on the following Sunday evening it began to swell and harden and we recognised trouble. Late that night we went to the casualty department of the local hospital. The next Sunday, whilst taking home some disabled friends, Margaret had another puncture and this time there was no Duncan. We were becoming fearful of Sunday evenings.

Lusky's ulcer healed but his arthritis invalided him so that he could no longer rise without help. For most dogs this would have been the end of the line for mobility is necessary. Old dogs at The Currer seem to be happy dogs and whilst an animal is happy his disability, however restricting, can be coped with. Just as Angie and Ian, Joe and Pauline, Tom and Judy and a hundred other disabled guests radiate happiness from their wheelchairs so does Lusky and life with him goes on as usual. Lusky needs Margaret in the same way that Harry needs us but I am essential to him in a different way. I always have been and just because he cannot rise alone makes no difference. If I am going out, Lusky is.

Unlike Danny who would follow Margaret to the ends of the earth, glued to her side so inseparably to earn the name Super Glue, Lusky is only interested in excitement. Margaret can go in and out a score of times and only if the car keys jangle does Lusky bark to go. If I put on my coat that is a different matter. He deems that if I am going out a bit of fun lies ahead and for the past dozen years Lusky has been our fun dog. If I am going out with Harry he knows we are going in the car which he will still want to do until he dies. The only place we do not take him is to the Cash and Carry warehouse. 'Cash and Carry! Cash and Carry!' we repeat to him over and over again. If Harry is not putting his coat on Lusky deems the excitement is going to be at the crush, or among the calves and he is not going to miss that either. If Margaret needs me the old dog knows something interesting is afoot and he is not going to be left behind. Not he! Jess can find employment taking animals where they want to go, unaware that they need no persuasion to run to mangers for food or dash into the field after an injection. Lusky is only interested if animals are being forced to do what they do not wish and I am necessary to Margaret in this activity Lusky knows, and arthritis or no he's coming. Though the summer heat did not help, Lusky looked far from dying and was obviously intent on coming with us to Aberporth once more.

* * * * *

We still had had no rain since that storm which had brought home the cattle for the Ministry man but the season was ageing and morning dews which had been absent all summer, drenched the burnt grass and resurrected it. An incredible new greenness appeared and we could almost believe it was spring again. Later rising and earlier setting, the less powerful sun allowed a rebirth of the grass but did nothing whatsoever to fill the almost empty reservoirs. Yorkshire Water was in a mess and imposed every restriction possible short of cutting off water altogether.

The only private source of water we had was the perpetual flow from the spring. Whilst neighbour's springs ran dry ours faltered but never stopped running away to waste. We grieved that this should be and we began to make enquiries about a bore and a pump and we ordered two huge plastic tanks, a 500 gallon and a 250 gallon, determined to be prepared to do something quickly should the need arise. The not-very-big delivery van would not come down our hill the driver said, walking all the way down to tell us so.

'Of course it will,' we laughed. 'Articulated lorries bring hay and straw. Big eight-wheeled BOCM corn wagons come and enormous cattle wagons!' But the man was adamant. He was not coming down.

'Good heavens,' we said. 'Fire engines are always coming down. Coaches bring the handicapped. What use are you as a haulier if you can't deliver?' But he would not come so Margaret had to get out the Range Rover and collect the tanks, one at a time, on the roof rack. Privatisation of delivery services is a fiasco. Once of a day the post van brought everything. The goods were collected from the station by the local man who knew every address in the district. We have recently known three different vans come in one day. If no-one answers the door no driver calls a farmer from the sheds. The van just turns round and goes. This has happened innumerable times.

Towards the end of the summer a man came to ask me for permission to use the land for an ATV demonstration day. I said sorry but our land was grazing, we had a lot of cattle we could not move away for the day.

He put on considerable pressure offering £600 and I think he might have raised even that but I insisted we could not, for any price, move cattle, provide a car park and admit crowds. Did he not see we had less than enough grass as it was and such an event would poach it more?

'If you change your mind,' he said, 'Let me know.'

A mountain-bike rally was to be held, starting in The Estate and the organisers rang to ask if the cyclists could come through our land.

'The Council has just spent £15,000 on the road along the boundary,' we said. 'Use that.' No, the committee insisted they needed to to use ours. 'Sorry,' we said with finality. Walkers, ramblers, joggers and orienteers we can cope with, providing they do not run through the farmyard for the same reason as it would be unwise to leap the garden wall of a dog owner and run past his front door. With all the goodwill in the world, we just cannot cope with cyclists, ATVs and horse riders.

It was slightly embarrassing that, after saying, 'No,' several of the cyclists booked in for bed and breakfast but this did not compensate or allay our anger when we saw scores of competitors using our footpath. The path is single file, anyway, and almost every non-farmer has a pony and wants grazing or galloping space. Many want somewhere for under-age sons to ride a motor-cycle, but those following leisure pursuits must go elsewhere.

It transpired that it had been the Council officer who had given all and sundry permission to use our land. With a holiday a few days away we had deeds to obtain and an ownership dispute to correct. It meant we had two journeys to make to City Hall. 'All we require of you,' we said, 'Is that you tell those you've misled that that hillside is ours.' They never did.

* * * * *

We could not close the door on family and friends and relatives. We see none for months and the ten days before holiday I think all we had came. Maureen and her daughter came from the village and Jack and Betty brought boxes of apples from their Green Hammerton orchard.

The phone rang. When the phone rings in our house almost anyone can be at the other end. A lady booked a twin for her husband and son who were coming to the neighbourhood for the weekend. I prepared to explain how to get here and asked if he, by any chance, knew the locality.

'Oh, yes,' she said. 'He's coming to see his father who lives at Haworth.'

I asked for the name and she said, 'Lorimer.'

'And his Christian name?' I never ask for this but I had taught a Barry Lorimer, over forty years ago.

'Barry,' she said. For me the past invades the present all the time. What a pleasant weekend we had. The middle-aged man was still the boy I had taught. Bigger of course, but I'd have gone straight to him in a crowd and said, not 'Are you?' but 'You are.' He enveloped me in a bear hug.

But the workman who knocked at the door on the Saturday afternoon baffled me. He was a builder with cement dust on his overalls and on his roughened hands and weather-beaten face. 'D'yer know me?' he said.

I always want to be able to say, 'Yes'. It embarrasses me if I do not immediately recognise. 'Robert,' he said. 'Robert Rishworth.' I must be getting old. The last sixteen years are nothing.

'Come in, come in,' I said. 'Of course, of course. You filled a whole exercise book with joy after we came back from Harris!'

'That's why I've come,' he said. 'I've just read your book. I'm doing a job in the village and wondered if I could still find the farm.' He talked all the hour we took to make the evening meal. Sitting there like Robin, who is an airline pilot and Nicholas who was such a nuisance when he was five, and Andrew who came to say his wife was having twins, and Stephen who broke his arm because he fell off the chair when we were having a school photograph taken, and Kate and Lorraine, on and on, ad infinitum.

With just a few days left before Aberporth, our Otley friend lost his fight with leukemia. The struggle had lasted barely five months. We all went to the funeral and so did John and Mamie who came all the way from Horgabost on Harris. Their presence was very moving.

Since the end of her chemotherapy treatment, which had given us all such hope and had seemed to improve Joan's quality of life, she had picked up an infection which was a real setback and a disappointment to us all. Life requires us, so often, to be actors. We hid our real fears. She and we kept up the quality of our performance, determined not to draw the curtain until the very last minute. Life, packed as ours is with comedy, drama and tragedy exhausts us, amuses, amazes and pains us and we do not know which of the masks to expect next.

It was imperative that I must have time to see Joan before going to Wales. I must sit with her to talk about Robert who had won the Rose Bowl for diligence, who was a builder who remembered the 'building' we had tried to do together. It is insufficient merely to teach the three Rs. It is necessary to teach obedience and compassion, loyalty and cheerfulness, endurance and moral integrity. We had tried.

If it gives me such great pleasure to meet the grown children that once filled my life it is illogical that, on the day I went to see Joan prior to leaving for Wales, that I should take such pains to avoid any encounter that day. I had promised to take tinned pineapple to Joan who was having such a dry mouth. 'My mother lived on it,' I had said.

And in the turmoil of holiday preparation I had forgotten. Harry, in the Range Rover as always, said, 'You'll get tinned pineapple in Crosshills.'

'I might meet someone I know, someone I taught,' I said. I hadn't time so I stopped at a shop on the roadside long before I reached the area in which I had been teacher. The store sold almost everything and I picked up two tins of pineapple and hurried to the counter. The middle-aged postmistress looked at me and said, 'It's Miss Brown, isn't it?'

Perhaps I taught her child, I thought.

'You taught me when I was six, remember me, Adele?'

I had only ever taught one Adele. 'Adele Nutter,' I said. 'You had a birthday and brought cake for every child in class. Anthony could not eat his because he had just had his appendix out.'

The woman gasped. 'You remember that? So do I.' It must have been all of forty years ago when I was first out of college and teaching at Lees close to Haworth.

'Do you remember Barry Lorimer?' I asked.

'I do,' she said.

'He and his son were staying with me last weekend.' We laughed together and other shoppers had to wait. I do like meeting my children. I do really.

No way could my visit to Joan be shortened because of my unexpected encounter with Adele. 'Take care of yourself,' I commanded as I left much later than I should considering we were going away on holiday and the week had had its customary diversions. 'Remember I am going on holiday. Don't dare be poorly while I'm away.'

She was becoming very important to me. She was the only link with whom to talk about all these children. I phoned her twice from Wales. I had never phoned her from holiday before. I hoped she would not think I was beginning to lose faith in her recovery.

* * * * *

With all the week's trauma and pleasure is it any wonder that we were throwing up the luggage, haphazardly packed, in darkness?

We seldom felt less like a holiday as we pulled out of the yard leaving Dorothy and George with last minute hoovering and washing-up. The gentle moving of the Range Rover is like a lullaby. We pulled in for coffee and had a sleep. After that we began to feel rather proud of ourselves. We had escaped in one piece leaving reasonable order behind. Just before we left I had said to Margaret, 'Did you order your calf milk?' I had expected ridicule. Did I think she was a twit? Of course she had ordered the milk!

Instead I saw amazement flood her face. 'I've forgotten!' she gasped. She had ordered a bag of glucose, 56 lb of it. She had ginger for colic and medications from McMillan and Streptopen from the vet and no calf milk ordered. 'I'm going senile,' she said picking up the phone to rectify her mistake.

All told we were quite pleased with ourselves. What we had nearly left undone had been remembered at the last minute. I couldn't think of anything we had forgotten. Wales here we come! But approaching Wrexham in declining sunshine for the great ball which had accompanied us all day was setting as we reached the Travelodge, I suddenly remembered, 'I haven't brought one article of rainwear!' Only Auntie Mary had her pack-a-mac. Surely we would get rain. The drought could not continue. But it did, throughout our holiday, and Auntie Mary never took her rainwear from its plastic bag.

Aberporth is magic for us in the same way that Grange was before Mother died. More so, perhaps. It is so totally different from normal life which is, after all, what a holiday should be. There is nothing whatsoever there to remind us of The Currer.

The Cedars is a normal, comfortable quite spacious bungalow except for the fact that one whole wall is a window which overlooks the sea. In front is the private road down to the little snicket which goes to the shore and there are well-cut lawns and beautiful gardens on either side. There is not a farm in sight. The Hebrides are alive with animals and busy people doing the kind of professional work we inherited. They are alive, too, with memories of what was my profession also. Wherever we go among the islands, I used to be there with children. People tell me they are different nowadays but every generation has said that. I prefer to believe that children are an echo. What you are, so will they be also. 'Which one of you does he take after?' I used to say to parents. When my Guides come to see me now, at least for the hours they are here, I think I see a reflection of us. Big-headed we may be but when they go, Margaret and I are warm and happy and, rightly or wrongly, we say, 'Didn't we do well?'

In Aberporth we completely lose our identity and it is a tonic for a fortnight. We drift happily along in a sort of timelessness. The people are friendly but their conversation is not about cattle and that, for the moment, is soul restoring. One of the few October holiday-makers appears from the beach at the right moment to help us hoist up the wheel-chairs onto the rack. In Aberporth we do the normal sort of things retired people do and enjoy every minute. That we do, does not make us wish to lead that life always. No way. The empty hours are thought-provoking and, for a short while idyllic, but we are not tempted from the path we stubbornly follow. Did we choose it or was it thrust upon us? We do not know but we are not veering from it. We would be at Beeston Castle when the holiday was done. The Currer is where we belong. Life is short and how much of it we will get we never know so it is wise to make the most of it.

We had thought Lusky might bid us farewell before we went to Wales but he was not ready, yet, to meet his Maker. He is a beach dog and he was determined to have another smell of the sea. Margaret had collected some pills for him from the vet but wished she had not for he was better without. It was Danny's first experience of being with us and he didn't quite know what to make of it. How loyally a dog will follow its master anywhere. Lusky caused us a problem by barking from the Range Rover several times in the night and Margaret crawled out of bed. 'It's practice,' she said, 'for rearing calves when we get back.' Jess is just a pest on the back seat, licking Harry, slobbering over everyone and barking at every dog she sees. She knows we will never part with her however badly she behaves for she has no experience other than of being with us. Danny has and he makes sure his behaviour is impeccable.

Shortly before the holiday ended we went to Castell Henlys, an Iron Age fort dating back to 400BC. It was an experience for a host of reasons and the memory of it will remain with us whilst others fade. When the man at reception saw the wheel-chair and the old lady he was pessimistic about our ability to get to the top, for the path was very steep and rough.' 'I'll give you your money back if you do not get there,' he promised.

'Do you really think we will admit failure?' we laughed.

'Nowt o't sort,' Margaret said, even as Alison had when the struggle to the top of Ben Luskentyre had seemed too great. I had praised her for getting so far and offered her the opt out of going to the summit.

'Ah'm off t'top,' she had added and so were we. The project took three hours.

Slowly, very slowly we climbed the steep path, pushing and pulling on the wheel-chair, pausing every few steps, acknowledging our stubbornness but still climbing upwards as if heaven were our goal and the path to it, life itself. There were a few who passed us and some we met coming away but when we reached the fort and throughout our hour-long occupation we were alone and it was wonderful. It was warm in an uncanny way for hilltops are usually draughty. It was flooded with sunshine, warm October rays cast only small shadows for it was midday and the arc of the sun's daily journey was still quite high above the horizon.

"'S lovely,' Margaret said. 'I'm off for the picnic basket. Let's eat up here!'

Harry was fine in the wheel-chair. I looked around for somewhere for Auntie Mary to sit. Benches had been plentiful all the way up. There wasn't one in sight at the fort. Also I hadn't my camera. 'Hang on,' I said. 'I'll go for the deck-chair.' I leapt easily after Margaret. The descent was quick without family. Margaret was exercising Lusky in the car park. I collected a chair and my camera and set off back at a trot.

'I'm coming,' she shouted. I couldn't wait, for Auntie Mary had no seat. I was hot. I paused to pull off my thin sweater and I was halfway up the hillside when I realised I had taken my glasses off to do so and left them on the grass verge. They were new, I hadn't had them for more than a fortnight and they had cost the earth. I galloped back down the hill but could not remember where I had stopped. In a panic I scoured the grass verge. The sun caught something shining but it was only plastic. I went back nearly to the car park thinking I would have to go to reception and report my loss. And then I almost trod on them. I heaved a sigh of relief and sped up the path with Margaret hurrying up behind.

We were soon all picnicking, Margaret and I were comfortably seated on the turf and we could have been in many places we have prematurely called heaven. It was warm, bright, silent, utterly peaceful, very, very beautiful.

'What does this remind you of?' I questioned Margaret. We were sitting close to the fire ring amid a circle of bell-shaped, thatched houses.

She responded as I knew she would. 'Camp,' she said. Camp with all our children on the beach or among the dunes and we, the Guiders, snatching a cup of tea before stoking up the fire and joining them. Each thatch could have been canvas for under each dome were gadgets made of twine and wood. We could believe mankind had barely progressed. Close by the fort were the pigs and sheep and goats whose ancestors primitive man had domesticated. Some people were living in high-rise flats and using microwaves and computers but what were we the Browns doing, 2500 years hence that Iron Age man had not?

'I could live here,' Margaret said.

She has experienced that life as I have and many of our children which puts us, as I've always said, in a different category from the majority. We segregate people not by colour or class or creed but by one unique opportunity. Those who have camped and those who have not. There is a sub-category for those fortunate children who camped in the Hebrides with us.

The life we now lead at The Currer is not all that far removed from that of our forebears except, if we are to believe history, we live in peace and they were perpetually at war. Castell Henlys is so beautiful I cannot believe other than that they were mostly benign; just getting on with the job of living as we are. If they really were aggressive their weapons were relatively harmless compared to the horrific inventions of modern man. They killed to eat and they, we are told, frequently killed each other but they were not killing the earth. Early man lived as part of the natural world but we are altering its environment and that frightens us.

Sitting, just the four of us, in a prehistoric settlement, suited me fine I could relate. After a struggle we had reached tranquillity. We had a glimpse of paradise. We were experiencing that most magical thing of all, time. When God made it, say the Hebrideans, He made plenty.

Of course we had to go back down the hill to the moment of God's Time allotted to us. Back to calf-rearing and holiday-makers, flood and drought, sickness and health but, like the man who smelled the Harris tweed we felt refreshed indeed, renewed because we had been a while on that prehistoric hilltop and felt part of timelessness and the near presence of the Great God of The Open Air.

Chapter Seventeen

And best of all along life's way
Friendship and Mirth

Rev. Henry Van Dyke

The 1995 calves were so small each of us feared trouble. For a little while a market bug shook them. The vet took blood samples and sent them labelled with ear-tag numbers to the laboratory. In the meantime the older vet came with the newest antibiotic. We blanket-injected and our troubles were nearly over for another year. The first weeks are always nerve-racking and when the road through the wood seems clear you are never finally out of it.

Whilst many may appreciate the difficulty of being farmers and caterers during the summer not one can really know how difficult it is to be both in November. To rear calves one must be totally dedicated. It is not just a matter that they must come first and guests second. They must occupy all Margaret's time and much of mine and guests be 95% neglected. We are glad, therefore, to have mostly self-catering ones. We do not mind Bernard and Emilie joining the family because they understand and to use a term they would not, they 'muck in'. Their pleasant company could, of course, be a distraction but Margaret refuses to let it be so. She dare not. If she looks away a calf might die.

The 1995 ones were initially frail. It may have been the hot dry summer which had left their mothers less able to produce a strong calf. We were on tenterhooks for some time. When Bernard and Emilie left we had an unexpected and delightful visit from Effie MacInnes and two of her daughters. We had first met her forty-four years ago on Tiree and had visited the island at least twenty times since but we had never been able to return her hospitality. When Ellen rang to say she and Christine were bringing their mother to The Currer, they were more welcome than royalty.

The new antibiotic was working so we were able to enjoy their visit and I could be released for the afternoon to take them round our quite different countryside from the flat island which is Effie's home. I can't remember when we enjoyed a visit more. They were fascinated with the calves. That one died suddenly with a twisted bowel they accepted. They know all about these things happening. Little cattle, little care, remember! We took the unfortunate animal to that place which is as necessary to the farmer as the undertaker is to the human family. We found the owner, just back from holiday, really making the dust fly trying to clear up a backlog of dead animals and grumbling in a masculine way that a staff without a master is less competent. We felt very sorry for him.

'I'm not going to grumble about coming home to a yard full of cars and people ever again,' I said. 'Just pray we never come home to a yard full of dead animals.'

The long hot summer was over when Effie came. Mist obscured the view whichever way you looked and I was disappointed that our friends could not see The Dales as they are, nor Ilkley Moor nor distant Ingleborough. It did not rain. It was just incredibly dour. We scarcely saw the hillside all winter. Guests receiving brochures claiming panoramic views thought we must be imagining things, but it did not rain. I had to put up a lovely photograph in the dining-room and say to guests, 'This is where you are'.

We did not need to describe our beautiful absent view to Joe and Lynne Robinson. They had been coming to The Currer almost since we started providing holiday accommodation. It is their practice to come for a fortnight's peace in November. They occupy the Mistal Cottage and we see little of them being almost entirely occupied with calves. They always have their arrival cup of tea in the farmhouse sitting-room with us. On the weekend they came, a group of Guiders were self-catering and the yard was full of cars. One was drawn right up, almost to our front door. Dusk was falling early for the day was grey and the evening mist was taking away what little light there was. A cheerful log fire danced in the sitting-room hearth and we gathered round happy to have friends to share it and Guiders who were looking after themselves. We tossed a handful of coal on the fire and threw on another log. Fire, in its proper place is uniquely comforting. Anywhere else it is the most frightening of the elements.

Only the roof could be seen of the car which had been drawn closest to the front door. Lynne said, 'I'm sure there's smoke coming out of that car!' Only Joe moved. He got up and went to the window.

'Why, man,' he gasped, filled with Geordie amazement. 'Yer boot's on fire!'

A Guider was rooting in the open boot and flames were dancing round her and smoke curling up into the fog. I jumped out of my seat and ran to where Joe was battering on the window, screaming at the women to get away to safety. Sure enough flames were leaping high out of the boot and the silly woman did not seem to notice. Any moment and the whole thing was going to explode and our windows were going to be shattered and our house in danger. The woman would be killed. That was a certainty!

She heard the battering on the window but did not know what was ailing Joe to scream and shout and wave his arms about. I didn't wait. I ran like mad for the fire extinguisher in the dining-room doorway shouting to the seated Guiders that a car, close to the door, was on fire and would explode any minute but looking through the big windows of the doors, which had replaced the barn ones sixteen years ago, whatever flames there had been were no longer. The puzzled Guider was calmly closing the boot. Joe was still waving frantically from the window. Had I not seen the flames myself I would have believed he was hallucinating.

I ran back to the sitting-room to say there was no fire but joining Joe at the window, I saw flames leaping out of the boot. They were neither spreading nor consuming. They were only a reflection of the cheerful fire in the sitting-room hearth. We were so shocked there was no spontaneous laughter. Just a flooding out of relief. There was almost an embarrassment at being so taken in.

* * * * *

The Ministry of Agriculture penalises farmers when they make mistakes but shows no embarrassment when they are in default. 'It warn't me Mrs Armstrong,' is a saying dating back to my teaching days with Joan. The phone rang and I took the call. The MAFF vet was ringing to say that the results of the blood test on the half-a-dozen calves showed we had an incurable form of salmonella. I told him we had dosed all the calves with Bayatol and not one had died since the tests were taken. He said he needed to check and came forthwith and, whilst still in the farmyard, he donned waterproof trousers and jacket and wellingtons and gloves and mask, and gave me an ear-tag number of one calf whose blood test had showed an incurable form of salmonella. It is not easy to find one calf among seventy but we had already checked numbers and we should have been able to narrow the field down to ten. All we had to do was to see which pen the good man's number lived in.

I went through the ten numbers in each of the seven pens whilst he stood sweating in all his regalia. There was no such number anywhere.

'It can't be a calf that died before the blood test,' we said but just to make sure we checked and then we checked the Beeston Castle bill of sale and announced that we had never bought a calf of that ear-tag number. It was obvious that there had been a Ministry mistake but the waterproofed man was also embarrassment-proof. Had it been our mistake I am sure he would have complained. He shrugged his shoulders and we all went into the calf shed.

'I'll take a few blood samples,' he said in a matter-of-fact voice. The healthy calves insisted on trying to eat his oilskins whilst he very carefully took the ear-tag numbers. He made no apology or comment on the error. He completely ignored it. Leaving the shed he asked for a bucket of water and poured in disinfectant. He calmly sponged himself down and employed Margaret to do the parts he could not reach.

Sometime later we received a letter saying the results of the blood tests were negative. We had no salmonella, least of all the incurable kind.

* * * * *

The wind veered to the north-east and became the prevailing dry wind of winter. Yorkshire Water began transporting millions of gallons of water by hundreds of tankers. They were a menace on the highways and a nightmare for those living in towns and villages through which they had to pass. Water was taken from the River Wharfe putting wildlife at risk and still the reservoirs were almost dry. A mist hung permanently round the buildings making us grateful for the air system and our only cheer was the log fire in our sitting-room.

For Joan the winter was drear but she always kept cheerful. She went slowly, slowly downhill physically and step by step upward spiritually. We all admired her. Peter, in Northumberland, was putting up a fight and the one was always asking after the other.

One small calf got an abscess in its throat which no antibiotic sorted. Margaret sent for the vet and Claudine came. Women vets are a welcome asset to the profession. Together we decided to transport the little fellow down to the surgery for a tracheostomy. We lifted him into the back of the Range Rover where there was just enough space for him and for Margaret who insisted on travelling with him. All the short journey abuse was directed at me from behind. 'Go slow! Don't brake like that! Do you have to go over bumps? Too fast. Don't stop!'

We left the calf for its operation and returned for it in the evening with a tube in its throat and instructions to keep it clear and clean. Hygiene is well-nigh impossible in a calf shed but Margaret did not neglect it and the abscess should have healed, but the antibiotic did not work and the tube could not be removed but had to be replaced. We took the calf back down to the surgery. We left it in their capable hands for several days but specialists cannot always work miracles either. We brought the patient home and a few days later it died. We would have so much liked it to survive, for its sake, Claudine's sake and for ours. We like difficult problems to be solved.

The rest of the baby herd caused us no more trouble, bounced into weight and health and bumped us around like nobody's business.

The two new water tanks remained disconnected. Water was still coming through the tap and we were complacent.

* * * * *

We had promised the chapel ladies that they could have a Christmas dinner at The Currer. We bought a 25 lb turkey from John and collected it ourselves from his remote Dob Hill farm. The season's plucking was in progress. He and his wife, his father and his eldest son were all in the warm, feather-proofed plucking area of the barn. A Calor Gas heater and their pre-Christmas activity kept them warm and a playing cassette kept them entertained as they plucked away at one-and-a-half thousand turkeys. The last thing in the world we want to do is to pluck turkeys but the cosy, friendly atmosphere drew us like a magnet. Given the opportunity we would have sat there chatting all afternoon. We seldom have a chance to talk to farmers and Margaret is deprived. Far more teachers come for holidays to The Currer than do farmers. We all like to talk shop now and again. We can even enjoy talking to other bed and breakfast landladies, as is proven when we break our journey to the islands.

Few people think we are isolated. The very presence of themselves suggests to others that we are surrounded with people. It is true that at breakfast-time, and when the evening meal is being served there are plenty of people at The Currer but they are transient. We, however, recognise our isolation from the community and the extended family and sometimes it saddens. We know far less about what is going on in our locality than those do who live on remote islands or outlying villages, where everyone knows what the other is doing, who go to church and WI meetings, buy at the village post office or gossip at the travelling shop. We are deprived of knowing what is going on along the top lane even to the extent of not knowing if someone dies unless George remembers to phone.

We have no social life other than the 'ships that pass in the night'. We are not unsociable, by any means, just terribly, terribly busy a quarter-of-a-mile from the top road and five fields away from the village. Since Kay went to Florida, we do not know who rents 'Jack Fields' and I had a cheerful conversation in the supermarket car park without recognising our nearest neighbours over the eastern boundary.

To all intents and purposes, we live alone and wouldn't we just have loved to pull up a chair in the fuggy atmosphere of the plucking shed to ramble on with people similarly isolated from the outside world. Village people know us. They pass behind the farmhouse exercising their dogs.

A few, with toddlers, wander in the farmyard to stroke the goats and feed the donkeys. We do not recognise them again for we have often thirty new faces a week and all have two eyes, a nose and a mouth.

We have no near neighbours and have to make a very special effort to keep our friends for we cannot accept wedding invitations or go to evening anniversaries or coffee mornings. When I was teaching and Guiding we had a whole circle of friends and were group members. My retirement isolated us in a way we would never have imagined. When a friend who shared our childhood and has lived in the village all her life, brings her aged mother to see us we realise we do not know her married name, that we have not only never seen her grown-up children but that we didn't even know that she had any. Isn't that a sorry state of affairs? On such an occasion we feel really remote. Our isolation is not only physical. It is social. Guests say, 'You would be lonely without tourists.' Not so. It is they who actually create our loneliness for their invasion excludes the immediate community. It is wonderfully normal in winter, to walk along the top road to the post office, chat to those who live along it, dawdle outside the village shop and catch up on local gossip.

We could have stayed all afternoon at Dob Hill, talking farm talk. Farmers so rarely see each other, but we needed a Christmas tree for the chapel party and we had hungry cattle and young calves at home so we collected our 25 lb turkey and headed for the hilltop centre where more than a thousand trees are for sale. The next day the dogs and I went down to the bottom pasture to bring up berried holly, an annual magical excursion I have made for forty years.

On three occasions I made clouted Christmas puddings. Mother used to make so many in the days when we gave serviced accommodation at Christmas-time. Having turned the place over to self-catering families, one pudding mixing had recently been enough. With a party of thirty-five ladies I made puddings on three days and hanging from the beams they looked impressive.

I continue to marvel how much even old ladies can eat.

I wish I could teach the church membership some of the lovely graces used by the Guide Movement. I am sure He above must be weary of, 'Praise God from whom all blessings flow.'

Everyone was afraid snow would fall and cancel the party but the bad weather, though it kept the cattle indoors, did not close the road. It was followed by a mild spell and when the cattle shed doors were opened the animals not only decided it was fit to venture out but a joy to do so. In one black and white torrent they leapt and cavorted downhill. A wise farmer builds his house below his land for cattle always run downhill.

There is land below us which is ours, but not much. Our builder, some five hundred years ago, knew the ease at which animals can be driven on a down slope and the difficulty one finds trying to drive them up. This is, no doubt, why he walled Currer Lane on both sides so that he could contain his herd or flock and drive them to the summit. When we took away one wall to pitch the road, we lost this facility and, years later we had to fence the naked side and reinstate it.

We witnessed the herd galloping downhill, put our dinner on plates and prepared to eat in comfort. The phone rang and a man on the top road said we had cattle on the crag. 'Impossible,' Margaret said but not one of our animals was in sight. 'I have let them out. They ran downhill. They can't have had time to stray.'

'Well,' said the caller, 'There are cattle on the crag!'

I had made a proper dinner, with roast beef and Yorkshire puddings. Margaret and I left ours 'stawkening' on the table. (That is a very descriptive dialect word which means gravy and beef fat will cool and congeal.) We neglected to clear the pans from the Aga top or wipe down the kitchen table. We left the house a right mess and Harry and Auntie Mary were still eating and we fairly ran up the hill to the crag on the top road. There were cattle there all right, but they weren't ours.

'I knew they wouldn't be!' Margaret said. 'They must be David's.'

'He's gone t'market,' said one of the gathered old men.

The animals were in a frenzy. Cattle can be silly when they get out but these were desperate. 'They're thirsty,' Margaret said.

The mains supply to their drinking trough appeared to have been turned off and, no doubt, the animals had been without water for hours. David arrived back from the market and said the tap was in the house which stands at a corner of the field. As the people were not at home the mystery as to why it had been turned off could not be solved. Obviously the cattle must have water and I was sent to shut some field gates so that the thirsty animals could be let onto our land to drink.

We did not mind that it was not our problem, that we had been alerted when the strays were not ours, that we had missed a very good meal and forfeited our brief after-dinner sit before the fire. What worried us was that we had left such a mess at home, two untouched meals on the plate, unwashed dishes, unstoked fire, pans near the Aga. People are quick to use the phone to tell us 'our cattle' are out and we would prefer them to be wrong than not to do so, but our friends and relatives never think to pick up the phone, that wonderful means of communication, to tell us they are just about to descend on us. Two car-loads of Christmas-calling relatives reached the house before me and I was mortified. 'Living above the shop' has so many disadvantages. Those who would never dream of invading a workplace, visiting you in an office or factory or studio or shop, never expect us to be busy. Even Mother used to think I went out to work and Margaret never did but as Auntie Mary says, life at The Currer has its advantages and its disadvantages and the former far outweigh the latter. I wish I was not so embarrassed, though, to be caught in a mess. We put the uneaten meal into the dog dishes, swept chaos into order and washed the dishes whilst our well-behaved cattle grazed contentedly down the field and Auntie Mary entertained. Chaos in a farmhouse kitchen can be forgiven, I am sure. Dogs and cats and wellingtons and wet coats, emergencies which call the cook outside, animal medicines, newly gathered eggs, nails and staples, hammers and saws all disorganise the kitchen. I am never critical of turmoil in other people's houses and I wish I could be more relaxed when it is mine.

* * * * *

The self-catering guests who spent their Christmas at The Currer did not need us. Hilda came to stay and we took our ham sandwiches and ate them in Nidderdale on the edge of Gouthwaite reservoir, in the snow. The small roads we always use had not been cleared and occasionally ours were the only tyre marks. Gouthwaite tugs at many nostalgic strings. Twice I took my Guide Company camping there before 'the din of the Atlantic surf' began calling us to the Hebrides. I could still name many of the girls who had had their first, exciting taste of camping on the lake shore and had sung, 'There where the blue lake lies, I'll set my wigwam, Close by the water's edge, silent and still.'

There seemed to be a permanent need in us to camp by water, in The Dales or The Lake District, or on the Northumbrian coast or by the great Atlantic. The sun shone on the white foreshore, that lovely Christmas Day, bringing back a host of memories of summer.

* * * * *

One of the Christmas guests said, 'How long have you had a parrot?'

'We haven't got a parrot,' I said, amused to be asked such a silly question.

'Yes you have,' the guest insisted. 'It was in the lounge yesterday.' A pigeon, perhaps! We have plenty of those but the guest insisted a parrot had been seen yesterday. I was taken into the sitting-room. Father Christmas could get down the inglenook chimney so a bird could do so with ease but it was mid-winter and we were not in the jungle or wherever it is that parrots live in the wild.

'See, it was sitting just there. Yesterday. We all saw it. We thought it lived here.'

I had to make a pretence that I believed them. Animals do get in. Jimmy not only eats doors he enters them if they are open. The donkeys frequently have two feet in the Mistal kitchen. I did not believe there had been a parrot. If a guest can mistake a goat for a donkey almost anything could be a parrot. Nevertheless the guest and I looked all over the room and up the wide sixteenth century chimney and behind all the furniture. We opened the door into the passage which leads to the low loo and looked behind the toilet and pulled back the shower curtain and there, on the tiles was a small, brilliantly red and green, dead parrot.

We were filled with a great sadness. Why hadn't the guests told us whilst the beautiful creature was alive. How terribly, terribly sad to find a warm, safe haven at The Currer and then to die of hunger or thirst. Why had these guests not been, as most guests are, shower fanatics. Why was the shower well completely devoid of water? Poor beautiful thing. Had it lived, I suppose we would have had another animal friend to care for. No, we wouldn't. We would have taken it to some aviary where the cages were big and freedom was not limited unacceptably.

* * * * *

The Wilsons had decided not to come for Christmas and we missed them. The new guests were pleasant but they were frightfully cold. Living in London they found the cosy warmth of The Currer far too cold for them. They had no thick jumpers and seldom dared to go outside for they had no suitable footwear. Consequently they never got warm. We felt sorry for them. They had come so ill-prepared for a Dales Farmhouse Winter. They said they had enjoyed themselves. The Currer is a lovely place to be if the hillside is white and the sky is blue. Christmas children invariably build snowmen and some bring a sledge. These from London had no wellingtons or waterproofs and hardly put their noses out of doors except to bring in more logs from the wood pile. Admittedly it was very cold, at least according to the television. At a thousand feet it seemed normal for us.

The scattering of snow at The Currer did not last but the frost did. It is a blessing that a hundred-and-forty cattle mean one at least is drinking and water is more or less continually coming through the mains. As running water does not freeze there is less chance now than there was in 1963, of our pipe freezing. Then we had no water for seven weeks. Now it is highly unlikely. So, whilst all and sundry had frozen pipes we did not.

Having equipped ourselves with four lovely corn bins it was essential that they be full as the new year began. Snow was forecast for the end of the week. 'Deliver before it comes,' Margaret told the office girl in Newcastle-under-Lyme. It is such a crazy system, this ferrying corn long distances in winter, the only season we need it.

'It will come either Friday, Saturday or Sunday,' she was told.

'Sunday?' Margaret said. 'We've never had a delivery on a Sunday before!'

'It's New Year's Day on Monday. A holiday,' explained the girl. I know we have to work on Sunday but we really don't approve of the rest of the work-force doing so. It is illogical but Sunday should be as peaceful as possible.

'Just as long as it comes before the snow,' Margaret agreed.

It did not arrive on Friday or Saturday and that night the snow fell. There was less than two inches. Even so a local firm might have phoned to ask the condition of the road down our hillside. Someone in Newcastle-under-Lyme didn't even know we had snow. The driver must have left in the dark for he was at the top of our hill by 8.30 a.m. We did not see him for we were still feeding calves. He hesitated, parked a while, then decided to come.

Having come halfway down the hill he began to wonder whether he should continue. Perhaps he had better walk down the steep bit first, round the bend; perhaps he had better ask the farmers what he should do.

We have complained before about men leaving their engines running and out of gear. They have more faith than we have. A friend left her car on the house drive and it hurtled through her garden whilst she went inside for some forgotten article. A lorry driver left his huge vehicle on a steep street in town and it ran into a garage showroom and killed two people!

It wasn't until the man walked into the farmyard that we realised we were about to have an early delivery.

'I've stopped halfway,' the man told us. 'D'yer think I should come down?'

We never advise. 'Should I reverse back t'cattle grid?'

'You know best,' Margaret said. 'We can ash it for you to come down or to go back up. Whichever.'

'I don't know what to do,' the man parried. 'I'll get down all right. Will I get back?'

'We'll have to ash it all the way,' she said.

The man still could not decide. We had left calves half fed. It was so early to make big decisions.

Suddenly there was no decision to make. The lorry was beginning to move of its own accord. Margaret saw it first. 'It's coming,' she gasped.

The driver started to run towards the cattle grid. 'Don't,' we shouted. We all three stood transfixed as this mighty eight-wheeled, fully-laden monster gathered momentum and hurtled towards the bend. The vibration on the snow-covered tarmac had set it moving and nothing either in heaven nor on earth was going to stop its downward plunge. It was heading straight for the cattle sheds. If it hit the top shed, which projected westward, it would first demolish the new corn bin. Whichever it hit was full of bullocks. The noise of the impact would stampede them into a heap at the east end of the shed. There was bound to be more trouble than just one very big hole in the wall. If the moving torpedo careered into the full Dutch barn and caught fire, for the engine was running and the petrol tank laden, then we would have a full-scale blaze on our hands. And there was nothing we could do but wait. We hadn't time to do anything. We just had to stand transfixed.

Without driver, of course, there was no way the lorry was going to negotiate the bend. The banking, merely a foot high, was not going to pose a problem to it either or even brake its descent. The living machine enjoyed the bump and leapt joyfully into the air, just like my children used to do at play in the sand dunes of Luskentyre. It seemed airborne for ages before the front wheels touched terra firma once more and flattened the first fence. The take-off had been uneven and direction had been lost in flight and when the landing was successfully accomplished the vehicle was no longer heading for the cattle sheds but veering to the right where water from McDonald's ditch joined the underground overflow from the pond. When the Dutch barn had been built the gulley alongside of it had been deepened and cleared and fenced against cattle. The lorry was going to have to flatten another fence and then leap a six foot dyke. We prayed it could not. The gradient was too steep. The front wheels were six feet below the back and a nose dive into the gulley was inevitable. We breathed again. Still with its engine running the BOCM monster was nearly on end. We waited for it to burst into flames but it did not and after a few moments the driver dared to open the door and switch off the ignition.

'Can I use your phone?' he said.

We went hurriedly to the calf shed to feed the rest of the hungry herd. We had lost all ability to talk. The driver arranged for salvage men to come from Skipton to assess the damage to the vehicle, and to the ripped-up fencing posts, and the mutilated netting and barbed wire, for insurance purposes. They took photographs and decided three cranes would be required to lift the lorry out of its hole. They agreed nothing could be done until the road was clear of snow enough to allow the JCBs to come safely.

*****

That was how the old year ended. It was a great entertainment and excitement for our New Year self-catering guests who numbered twenty-three; I remember this because of what happened on New Year's Day.

I don't think it was the BOCM lorry which had thought it could take a running jump which brought so many family friends on the 1st of January. Some came to wish us a Happy New Year. Dolores and Raymond came to fill up their water storage containers for their supply had frozen. We were making coffee all morning and then Raymond put his plastic carriers in the big sink and filled them full. When everyone had gone we turned on the tap for water to boil the potatoes for lunch and there wasn't a drop.

'It can't be frozen,' we were sure. 'Someone has turned it off. Yorkshire Water must have a problem on the top road.'

But our neighbour had a supply. The office girls were harassed and we put more pressure on them. Yes, they had turned off the water at an empty house. No they had not turned off ours. We complained we had a hundred-and-forty cattle wanting water and twenty-three guests using loos and wanting cuppas even if they neglected to wash. Every time I phoned the Water Board a different girl answered. No-one made any attempt to solve our problem. They had burst pipes all over the place, everyone was ringing, everyone was persistent.

If I say I phoned almost continually all day it would not be much of an exaggeration. If we had a burst, if our pipe had been frozen we would not have pestered the suppliers. We would have sorted out the problem but we were absolutely convinced a workman had inadvertently just turned off our supply. The cattle began to bawl and guests took containers down into town to get drinking water. I carried bucket after bucket from the spring for us to drink and for our guests to flush the toilets. I twisted a hamstring in my leg and had no end of bother for weeks. Margaret emptied the tank which catches the rainwater from the sheds, bucket by bucket into the calf mangers.

At two o'clock in the morning Yorkshire Water brought us a two-hundred-and-fifty gallon bowser and long before dawn we filled some buckets from it and then attached a hose and flooded the bullock mangers. The bowser was emptied in no time. Our nearest water was in the trough in one of the fields we rent from George a quarter-of-a-mile away. It would have been impossible to drive the cattle to it. Our spring was still running but far too slowly. We deeply regretted neglecting to do something positive with the two new tanks. The girl in the office said, 'All businesses were advised to have a five hundred gallon emergency supply,' implying that it was my fault for not reading the small print on the last bill.

Our nearest neighbour walked the length of our buried pipe with Margaret, looking for rising water. There was none. He went down to George's to impart the latest bit of gossip.

'Browns er 'aving no end o'bother,' he said. 'It's a reight carry on. They've 'ad no watter fer twenty-four hours. Loads o' guests, umpteen cattle and no watter!'

'Aw 'eck,' said George. 'Ah got a lad t'turn off me trough yesterday morning. D'yer think 'e's turned wrong un?'

The two friends hurried up the road, George carrying his tool for turning off well-sunken stop taps and sure enough they found the lad had lifted the wrong iron lid. They quickly turned on our tap and hurried down our hill. George's tail was well and truly between his legs and his face became redder than ever when he saw the empty bowser, heard the thirsty cattle, saw the line of buckets and the uncoiled hosepipe. No doubt he was very grateful for his friend's support.

The relief! The ability to turn on the tap and for water to come prevented us from lynching him! It had not been funny but suddenly it was remarkably so. George was so contrite, so apologetic, so red about the ears we had to laugh. He, acutely embarrassed, changed the subject by saying, 'Hey, what's that BOCM lorry doing head first in yon ditch?'

A few days later three heavy cranes came to lift it and the tons of our corn it contained, onto the level yard. The men tried the engine and it worked. They must have reported this for at about six o'clock the phone rang. The driver was being brought over to collect it. He would be here about seven.

'But he can't.' I protested. 'It's very foggy here. That lorry has been upside down for days. It's never been checked. The bonnet's bust. The petrol sump might leak. Who knows whether the brakes will work? He can't venture on a motorway. It's suicidal!'

'He's already on his way,' the girl said.

'Well, call him back on the mobile,' I suggested but no one would listen to me in Newcastle-under-Lyme.

Around seven another equally big corn lorry ventured down our lane in the fog. The driver climbed into his disabled cab and switched on the ignition and the engine leapt into song but the lights wouldn't work. We heaved a sigh of relief. No way could the man drive home in the dark. They were, however, able to use the auger to empty the tons of corn into our bins, a job the recovery firm which collected the lorry next day may not have known how to do.

That is not quite the end of story. When we needed a further delivery a few weeks later a fall of snow sent us to the garden centre for expensive bags of rock salt. We cleared and salted the road down to the tarmac but when the lorry arrived from its distant mill, its driver refused to come down the hill. He turned the vehicle round and went back to Newcastle. We were indignant. Clearing the road had been backbreaking. We sent for the traveller, a more local man, to come and see for himself that the road had been cleared to the tarmac. Men!

* * * * *

For days, until the fencing had all been renewed, we could not let out the cattle. We were pretty fed up with corn lorries. Even the snow was not the good, clean, deep variety which is so beautiful and crisp. Cattle can be let out in that and dogs are clean and the sun shines and no-one comes. The snow of 1996 was wet and dirty and fog was ever a blanket. We could not see the top of the hill or anything in the valley. It was a nuisance we could well have done without.

A group of handicapped pensioners came from Hawick and we had to ask them to leave their cars along the top road and we ferried them back and forth in the Range Rover. We wondered what we would do if snow isolated us further for they would never have been able to walk out.

We had had a warning to be complacent no longer about water. The five hundred gallon tank must be connected quickly, before any friend of George's should again turn off the wrong tap. I remembered once being without water in the village where I taught so long, because Nicholas, a five-year-old had turned off the main village supply. All the housewives in the houses opposite the school had been without water. Nicholas came to see me not so long ago. It is a wonder he and I lived to tell this tale for the infant had been preoccupied with turning on every switch within reach and if it was already on, he would turn it off. He twiddled every available knob and left us totally bewildered until we found the culprit, a quiet little fellow, absent-mindedly pressing switches. The most serious was the day he strolled round the minute school kitchen and turned on the empty sterilising sink. That was when he nearly blew up the school.

Hopefully George would, in future, instruct his friends against any human turning off our water but any day frost can do that just as efficiently and the Water Board is prepared to turn off water in drought situations, or they would not advise us in small print to make sure we had a temporary reservoir of at least five hundred gallons. We sent for John and he came to inspect the tank and list the fittings he would need. He stood on a bale and looked inside.

'Someone's put a loaf of bread in 'ere,' he said.

'Never,' we scoffed. 'Who'd do that?'

'There's a loaf of bread in there,' John repeated so Margaret the smallest and most agile, climbed into the tank and lifted out the parcel. It wasn't bread, it was a brown, heavy packet addressed to a Bradford firm and, presumably it should have been delivered in October. Thinking it held fittings we did not look at the address and only when we had opened it did we realise the contents were for someone else.

We have no doubt that the lost part would have caused quite a panic the previous October and knuckles, somewhere, would have been well and truly rapped. We reported our find to the makers of the tank who knew the haulier they had employed and a representative collected the packet.

John had brought the necessary fittings and connected the tank to the mains and an air-locking problem we had had for years was unexpectedly solved.

* * * * *

In Northumberland my college friend's husband, Peter, lost his fight with cancer. I did not tell Joan for she no longer asked and I could not bring myself to depress her. Joan's cancer had been temporarily halted and Peter's had gone apace. He had been the healthiest of men, revered headmaster, golfer, gardener and countryman. His illness had been unexpected, sudden and incurable. He left his lovely family and many, many friends totally bereft. They mourned a parting which twelve months ago would have seemed impossible. Life and death are mysteries modern science tries to tamper with. Perhaps we should leave those decisions to the Almighty. At all costs we should have our bags packed and our passport ready just in case we are called away early. Joan's had been ready for some time. Her house was in order and she was cheerfully relaxed. She would have taken Peter's death philosophically. It was I who could not find the words. I was not as ready to lose Joan as she was ready to go.

For the second year running we had a remarkable young couple in The Loft for the first two months of the year. William is small and slight and does not look strong enough to carry his disabled young wife up the barn steps or dig out the road every time isolation threatens. That, in moderation, is a blessing to us but not to a young couple wanting to do their shopping and visit their friends.

For the other ten months of the year this extraordinary couple live in a caravan on the moorland above Baildon. Green belt regulations, or some such stipulation, decree that the caravans must not be occupied twelve months so they come to The Currer for January and February. Margaret and I were feeding calves and did not hear the fire alarm go. The Loft smoke detector is the most sensitive and it is the one which always gives the false alarm. William went to the farmhouse. Auntie Mary, being deaf, had not heard it. It would be the top cottage, she said. No it wasn't, William said. Then he must fetch the girls, she replied. We will always be girls to the last member left of the generation before us.

William came to alert us and we ran to find the light which told us which alarm was quickly bringing the fire brigade. Sure enough the Loft passage detector was red. So were our faces and so were the appliances hurtling down the road.

'Are you cooking?' said the fireman.

'No.'

'Have you had a shower with the bathroom door open?'

'No.'

'Have you been using hairspray?'

'No.'

'Air freshener?'

'No.'

'I can smell scent!'

'That was Shake n'Vac,' said William.

It is a bonny state of things when the use of such will activate the smoke detector and bring two fire engines from town.

Chapter Eighteen

> Teach me to live that I may dread,
> The grave as little as my bed.
> Teach me to die that so I may
> Rise glorious at the awful day.
>
> *Thomas Ken 1637–1711*

Amongst the pile of anxieties and pleasures, the routine and the unexpected, the musts and the needs of everyday life there is always that particular something which surfaces, sits on top, occupies ones thoughts and will not go away.

Increasingly worried about Joan I did not want it to go away. She sat proudly and courageously on top of my thoughts challenging me, and all who knew her, to greater patience and awareness of what is, what may be and what cannot be changed. I visited her more frequently, fearful she might suspect that I knew what she knew, that time was short. I like to think she needed me, my matching sense of humour, our shared memories, our similar faith. I needed her. I couldn't bear to miss one moment of the limited time left.

It was thirty-six years since I had walked into the village school where she had already been teaching twelve years or so, and found I had the most competent infant teacher for miles around. Thirty-six seemed too few, faced with a life without the sunshine she brought. To no other could I turn to talk about those twenty-one magic years of my headship of that village school. Every fortnight I would collect Auntie Mary from the hairdresser's, bring her home and then take Harry to Cononley.

Joan's cheerfulness will remain in my memory as long as I am capable of remembering. That she was really ill did not escape me. I knew her too well and Margaret and I had nursed both Mother and Father with cancer. Faced with terminal illness in another one's soul cries out to be able to do something positive to ease the pain.

Six years ago I had found that the best advice and reassurance had come from another Joan, a pupil-Guide-friend who had trained to be a nurse and who was then working at Arden Lea, the Marie Curie Hospice in Ilkley. She knew not only how to control symptoms but how to advise the doctor and our excellent district nurses.

I took her to see Joan and the two, who shared the same name, chatted happily together whilst I sat in the comfortable knowledge that my engineering was a success. The younger Joan explained how capable was the team at Arden Lea, how those specially trained were willing to make outside visits taking their knowledge and their comfort cheerfully into homes to people like Joan, who could manage if given a little help; who did not need residential care, just information on how to minimise the discomfort of the symptoms. The younger Joan's reassurance led to a male nurse coming regularly with incomparable support, which made Joan feel in control and safe. Just a phone call would bring help.

The winter of 1995/96 was too long. It imprisoned cattle in sheds. Joan's snowdrops appeared and then disappeared under a carpet of snow. She said, 'As soon as spring comes, and some good weather, nothing is going to keep me in.'

'Phone me any time at all and I'll take you out,' I promised.

Denis came at the end of February for two, too short, weeks. I had promised to bring Joan to The Currer to see him or take him to Cononley but winter bugs were thriving. We had flu jabs but our guests had not and I was petrified of bringing her into an atmosphere created by the constant turnover of people. The Guide Reunion on the first Saturday in March brought the usual pleasure. Many complained of persistent coughs. Their visit was followed with the early contingent from White Windows. They brought their germs along with them and Denis had travelled in a packed train from Lyon and through the Channel Tunnel. We decided it was ridiculous to bring Joan just then so, the weather being sunny, albeit cold, I rang and said, 'Let's go for a run,' and she said, 'Let's go to Burnsall.'

Both Margaret and Denis thought one tyre of the Range Rover was slowly softening. It was an illusion for there proved to be no puncture, but with Harry in the back seat and Joan in the front I could take no risks. I called at a garage and asked them to change the wheel and check the other.

We had a lovely drive around the empty side roads. There are no tourists in The Dales in early March. The snowdrops were lovely. We chatted away like we always did. 'Remember taking children to see the snowdrops in The Folly? Remember going to see Hagar's lambs?'

The hedgerows were leafless but the hazel catkins were dancing in profusion and the lamb's tail display told us spring was imminent. The pussy willows, too, were dusted with yellow. 'We've got all summer to do this,' we agreed. There is no greater beauty than that presented by The Yorkshire Dales. 'Do hurry,' I prayed of the spring. 'Do be quick and come.' Arriving at Burnsall we parked on the roadside outside the small cafe. In the summer, parking there is impossible but then the road was almost empty.

'I should have brought a flask,' I apologised.

'I could murder a coffee,' Joan said.

'Let's go into the café,' I suggested. Harry was enthusiastic. We hadn't done anything like that for ages.

'I didn't bring any money,' Joan said.

'I did.'

'OK. It will be my turn next time,' she promised.

The café was almost empty. The cakes were home-made and lovely. On the wall, as in my kitchen, was the certificate saying the proprietress had taken a hygiene course. Millions of pounds had been spent by caterers to conform to European regulations. Stainless steel had appeared everywhere. Under government pressure all wooden surfaces had been replaced. Enforced hygiene rules had put people out of business, closed down long-established firms and made the comfortably-off bankrupt before it was realised that a scratch on steel was more of a germ collector than on wood.

We enjoyed the coffee and the cakes and the afternoon sunshine. I put Harry back into the car and Joan and I walked among the mallards on the riverside. She looked so very ill and I was reminded of how Mother looked, on her trip to Bolton Abbey, on the Maundy Thursday of the Easter she died. Joan was just as cheerful. Just as happy. When was she not? I cannot, for the moment remember one occasion when her lovely humour deserted her. Occasionally, when admitting to being a bit low she would say, 'I don't like myself like this.'

We walked slowly by the water among the ducks. Stray animals were often brought into school. I was the farmer and the trained biologist but it was always Joan who assumed responsibility. 'Remember the swan which wouldn't go away from the school kitchen door?' I reminded her.

She linked her arm in mine, laughing at her weakness. 'I've never done this before,' she said. I felt privileged to have worked beside her for so many years. She had always been surrounded by children. They had leaned on the back of her chair, rested on the arms, grouped themselves in front of her on the hearth-rug, tugged at her skirts and prodded her for a turn to speak. She had attracted them even as the Galilean Teacher she strove to follow. She had never tired of their company or wearied of their need to communicate. Everyone had admired her ability to cope with children, to quietly discipline and to teach, not only the three Rs we still believed in, but how to behave socially, to live happily and to have fun. She had taught them how to live with each other, to be kind, to wait their turn and to clear up their own mess.

No parent had ever found fault with her and had only ever complained if a child thought Mrs Armstrong knew best and was prettier! One mother said, 'She never noticed I was fat until she met you!' Children had climbed on her knee when she heard them read. I had never heard one of them answer her back. Her's had been the ability just to look and command obedience, like my mother whose look, Margaret said, should have been bottled and sold in Mothercare shops.

It was nearly fifteen years since we had both retired. Where had they gone? It seemed only yesterday. Joan had stopped saying, 'Don't be silly,' even as I opened my mouth to say something nice about her so, with her arm in mine, I said, 'I want you to know how proud we all are of you'.

She replied, 'That's nice to know'.

When I took her home I repeated, 'We've all summer to do that,' but I knew we had hardly any time at all.

* * * * *

We had had a disappointment over arrangements to go to Tiree in the spring. Since Charlie's death extensive alterations were being made and it wasn't ever going to be a holiday cottage again. That was the bad news. The good news was that Effie's son, Hughie, had plans to convert a steading at Coales. Coales is utterly beautiful. A three-bedroomed cottage there, close to the white sand, the turquoise sea and the playing seals would suit us fine but we would have to wait. 'Then we'll go to the Isle of Man this year,' we agreed. At the back of my mind was that it wasn't all that far away. Mamie and John had come all the way from Harris to be at Arthur's funeral. I thought I would feel easier not being quite so distant, with Joan so ill. Harry was as pleased as punch. He loves new places. Auntie Mary was over the moon. She had been there with Auntie Janie thirty years ago. It was lovely she informed us. So we booked a cottage at East Bradda, a mile-and-a-half from Port Erin.

On the 20th of March catastrophy hit the farming community, and brought down abattoirs, hauliers, meat processors, steak houses and butchers in its trail. The morning news reader alerted us to an announcement which would be made in Parliament that afternoon. The Health Minister, Stephen Dorrell, had important information to impart re BSE (Bovine Spongiform Encephalopathy). It had not been our problem for we never have cattle old enough for it to manifest itself. Dairy herds had known the disease for a great number of years. It had reared its ugly head in thousands of herds but over 90% had had less than ten cases in as many years. Although it made headline news it killed far fewer animals than pneumonia or liver-fluke, or accident, or black leg, or John's disease; milk fever, or calving, or IBR or coccidiosis, or BVD or salmonella or twisted gut, or bloat or scour. I could go on all day. Cattle die of copper deficiency, lack of sufficient colostrum, untreated lung worm and farmers take it in their stride. Slowly but surely they have eradicated brucellosis, anthrax and TB. They were used to treating staggers. A cure for this was found and, caught soon enough, a magnesium injection can be given and recovery is immediate.

BSE looks like staggers but is incurable. Brain damage, bovine or human, distresses those who have to cope with it more than most illnesses and farmers do not like BSE any more than people like Alzheimers or CJD. Bovine spongiform encephalopathy proved to be so similar to scrapie found in sheep, and well-known for over two-hundred-and-fifty years, that fears were raised as to whether it could have jumped species. Diseases rarely do however.

Meat meal from animals was generally fed to cattle, pigs and poultry and thousands of tons sold abroad. If the disease had jumped species it may have come by this route. The fear was that perhaps CJD, human spongiform encephalopathy may have come in a similar way from infected beef.

Lest this be so, six years previously, the use of meat meal protein in cattle food had been banned. As a further precaution some councils had taken beef from their school meals menu. Having taken away the suspect source of protein the Government expected the incidence of mad cow disease, as it was more easily referred to, would rapidly decline and, each year, there were fewer cases but eradication was not total and many began to think the meat meal theory, unproven, was a red herring and that the problem would have to be solved some other way.

Regulation in the beef industry became so tight as to strangle. Abattoirs were constantly inspected. No part of brain or spinal cord was allowed into the food chain even from young cattle. Everyone, under threat of closure, exposure to media headlines and exorbitant fines, got his house in order and still there was incidence of the disease. In the case of TB, brucellosis and foot and mouth, all contracted by infection, a policy of slaughter was effective. Because BSE is not contagious, such a method was pointless. There was literally no proof of anything, only wild, unsubstantiated guessing.

We invariably do not get midday lunch until mid-afternoon and we were just finishing it then on 20th of March. We switched on 'Westminster Live' in order to hear this statement from Stephen Dorrell. What we heard shocked the nation and the world and put every farmer's livelihood in jeopardy.

Scientists, Stephen Dorrell said, true to BSE (Blame Someone Else) strategy, were alerting the country to ten cases of a new form of CJD in much younger people, very similar to mad cow disease, which for want of a better explanation, may have come from contaminated beef. If it had, perhaps 500,000 people would have already contracted it. Nevertheless, he was sure that the precautions which had been in place since 1989 would be sufficient protection and to the best interpretation of the word, beef was safe to eat.

Suddenly there was media mania, people panic and consumer concern. Europe and the rest of the world banned all meat from Britain. Recently-imported British calves were immediately slaughtered on the Continent. No more calves crossed the Channel. No animals could be sold, markets were closed, workers at meat processing firms lost their jobs, fat animals remained on the farm costing farmers millions of pounds. There circulated scare stories of mass slaughter, mass bankruptcy, despair and suicide. There was no other topic of conversation, nothing else was watched on television. The only ray of hope lay in the fact that all our British guests asked for beef to be on our menu.

Of course the price plummeted and everyone stocked up freezers. 'I've eaten more beef since the scare than ever before,' guests said.

I went to the Cash and Carry and all the huge parcels of beef were stamped IMP. 'Hey,' I said, 'Where's the British?'

'Don't panic,' the butcher reassured me. 'I've some in the fridge.' I filled up my trolley.

'Where does IMP come from?' I asked.

'That comes from Botswana,' he replied. It said it all!

Our annual sales are never until September so whatever the disaster meant to us would not be known until then, but our hearts ached for those with cattle ready for sale and calves which could only be sent for slaughter. It was soul destroying. Would there be a market for ours in

September? Would there be calves to buy in October? No-one could tell. Farmers were in deep shock, blaming Government for creating panic with no scientific proof of anything. Only 'might', 'could', 'the nearest we can guess' and a falsely-conceived idea that mass slaughter might appease Europe and satisfy the consumer. There was no consideration for the farmer and, more especially, for the animals he loved and the herds built up caringly over many years. Farmers had not asked for meat meal in their corn and, indeed, many of them did not believe that it was the true, sole cause. Initially they were too stunned even to think.

Then they began to put two and two together and each came up with a different set of facts. Most had had so few cases it was difficult to find a pattern. In any case, until he could let out his fat cattle to grass, his chief worry was could he pay the feed bill? Then he began to talk to neighbours whose experience never matched.

The mystery disease had everyone baffled. 'All my cases were bought in,' a local farmer said.

Some thought it was the calf milk. Since milk quotas farmers had fed surplus milk to calves instead of using milk powder. 'That's why there are fewer cases,' they suggested.

'It's the pour-on stuff,' others believed.

'It's that warble fly eradication we were all forced into.'

'It's o.ps. Organo phosphates.'

'It's hay fleas.'

No farmer thought it had jumped species. Most said their experience did not point to the meat meal which had also been fed in Europe. The truth was that no-one knew. Even those who had researched it for years disagreed with everything except that nothing was proven.

What distressed and angered us most was the total disrespect for that time-honoured species, the cow.

> 'The friendly cow, all red and white,
> I love with all my heart.'

Who cared? No-one in Westminster. Where were the animal rights lobbyists? They who objected to calves being reared for beef remained silent when calves were killed in their hundreds and thousands at birth. Awful words were used which had never entered the farmers' vocabulary such as slaughter and incinerator, consumer and compensation. Money, even if enough and on time, could not replace a carefully-bred herd. I am sure a farmer does not work all hours of the day to feed an ungrateful, critical consumer. On the contrary he has a tremendous respect and love for his herd or flock. To hear politicians callously list the tough measures needed to regain consumer confidence, they themselves had shattered with a few ill-chosen words, was soul destroying.

*****

We had to seek our entertainment other than on television. The news terrified us and boxed comedy was an insult. So we laughed at Chintzy. It seems strange that I so seldom mention the cats we continually fall over and feed. Chintzy has been with us since 1987 when our neighbours, Matt and Betty Watts left 'Jack Fields'. The new people did not want the two cats and, as the Watts stayed with us until their new house was ready, their feline friends were left to fend for themselves in the farm outbuildings which had long been their home. The new people sent for the RSPCA to catch them and take them away but cats have to be tame and restrained if you wish them to be caught. Betty continued to feed them on the hillside between the two homesteads and, though wild, they came nearer and nearer to The Currer. When Betty left to occupy her new suburban house, Margaret continued feeding and the cats came up to the buildings. Children in the cottages tamed Chintzy. She had still-born kittens in the Mistal Cottage and Margaret spoke to Betty who said all recent litters had been dead so the cat was taken to the vet's and neutered. The black tom, called Tom, for we are unimaginative where names are concerned, was withdrawn and never came near the house but, little by little, Chintzy wormed her way in and assumed she was welcome. Years later Tommy came indoors too. He is far from handsome but Chintzy is beautiful. Her coat is multicoloured and softer than silk and she lives in the front room and is always waiting. Whoever enters she abandons her more comfortable seat for the less secure and only temporary knee. Why she bothers at all is amazing given that we stay seated only minutes at a time.

She prefers a lofty seat which we applaud for she is invisible on the hearth-rug. She insists on choosing her perch. Sometimes it is on the mantleshelf, sometimes on the back of a chair or settee. From this high location sooner or later she will fall, suddenly, with no warning. If Margaret is sitting on the hearth this weightless ball of fluff will fall like a stone, believe it or not, onto the marble flagstone. If the cat is asleep on the chair back she falls either behind or suddenly on anyone seated. Should she choose the ironing board she is in disgrace and she gets short shrift if she leaps with youthful agility onto the table.

Chintzy must be a very old cat but she is as light as a feather. Her sight is poor so, though her leap is unimpaired, it often misfires and she crash lands on the carpet. She is a naughty cat who will sneak into the dining-room or sitting-room and be missing for hours. We never leave the staircase door open for she is like Goldilocks and will try anyone's bed. She makes us laugh and to her we are grateful.

Tommy does, too. He has managed to get into the house, after many years of being fed outside. He has far more problems than any cat in the country and, had he sought shelter with anyone else but Margaret, he would have been put down long ago. A less strong and determined cat would have died a decade past but Tommy lives on. He has survived a road accident which has left his back-half appear to be unattached to the front-half. His rear follows his shoulders nearer to the floor and does not continue the same direct line. His ears are badly torn, or eaten and he suffers from eczema which is controlled by infrequent visits to the vet. The pain this suddenly causes turns him into a firework. Without warning, he begins to scratch, furiously, overtaken by a form of madness and like a rip-rap, lit on Bonfire Night, the scratching monster is fired every which way. We quickly get out of his way Margaret says, 'I'll have to take him to the vet!' and any holiday-maker currently in our kitchen has hysterics.

Only Margaret will tolerate this scruffy animal on her knee but we all have the greatest respect for him. So do the veterinary nurses who entered him in their computer as Winston when his name is a simple Tom. During the period of suspense over BSE the laughter in our front room was almost solely triggered by the sight of Tommy gaining entry. Chintzy is in the room, full stop. Tommy is allowed when Margaret is there to propel him out if the need arises. Neither he nor Danny can open the door from our living kitchen into the room we call 'front' and in which we briefly repose. Jess can. She knows that if she takes a running jump at the door, hitting it just below the handle, it will fly open and we know that she will hurtle through it and Danny, who has been urging her to just that, will follow her so closely they are almost siamese twins. After a pause, for he is disabled, comes the black cat. He wobbles across the hearth rug to find Margaret. We know the exact details of this parade but we always sit in suspense for the third member of the comedy act and we always laugh.

Danny, too, makes a beeline for Margaret and between him and her there is not a hair's-breadth. Wherever she is Danny is there also. In the front room he is just a rag-doll attached to Margaret. He seems to be sleeping soundly but if she rises so does he. He cannot bear separation. If she goes upstairs he waits at the staircase door until she returns and he spends the night there for he, too, has gained entry.

Unlike Danny, Jess is a problem in the front room because she is not a rag doll sticking like super glue to Margaret's armchair. Danny has no nose and no paw. Just adoration in his worshipful eyes. Jess has all three and unfortunately she wants Auntie Mary's attention and pokes her with a clawed paw and a cold nose. Mother used to say, 'Not up!' to Jess who will, now, never respond to the word 'down'. She will not be ignored and a cross word excites her. She likes funny noises and particularly she likes it when Auntie Mary says, 'Down.'

To remove her from Auntie Mary we say, 'Go to Harry,' whereupon she climbs heavily onto Harry's knee and licks his face. Tired of this she then sits on Auntie Mary's feet. Never on mine or Margaret's so she is in the dog-house and turned out of the room. She knows, of course, how to get back again. The new door handle I bought to repair the worn one remains in its cellophane pack for our fun lies in her ability to open the door. Our fun, not our peace!

* * * * *

Fun, at the beginning of April was at a premium but it is essential in our diet. Had the BSE fiasco come in February we might have been starved of its remedial benefits. April brought Easter guests in plenty and we could turn to them for relaxation. For sixteen years they, too, have been the comedy on which we have relied. In general we find the human race a deteriorating species but of our guests we have only praise. They fill the dining-room with laughter, making lasting friends, visit each other in all corners of the world and their tolerance of each other's shortcomings is remarkable.

They feed the donkeys at the door, most of them laugh when goats knock off car side mirrors and geese leave unhygienic droppings. They love to tell us a bullock we have let into the lane, is there. They are happier still if cattle we have not, are galloping all over the yard.

Richard entertained us, as he always does at Easter. Tom, the retired headmaster, who likes to serve the meals, showered all the carrots onto the table-cloth. If, as I frequently say, bed and breakfast and farming are incompatible then perhaps it is a good thing for when the one distresses the other acts as balm.

The holiday was approaching. We were quite relieved not to be going to stay among farmers. On the Isle of Man perhaps we would hibernate from our problems.

Joan caught a chill and the doctor gave her antibiotics which made her very sick and upset her routine medication. BSE took second place and Joan became my biggest worry. I hastened to Cononley, left the others in the car and went in alone. Joan was up and dressed and well-nigh back to normal. She smilingly said, 'I'm not s'bad. It's passed.'

'Every time I'm going on holiday,' I moaned. 'You give me such a scare.'

'Well. That was it. S'over. I didn't like it but it's gone.'

Because evening meal numbers had dwindled to five I felt suddenly more relaxed than I had for weeks. It was Friday and the day we normally have fish for the evening meal. It was easy. We had served the main course and the apple pies were golden and had been lifted from the oven. 'Come on,' Margaret tempted me. 'Come on, let's sit by the fire for five minutes. Let's put our feet up.' It was the last Friday of an Easter week when frequently there had been twenty at the table. Five was so very easy.

She went into the sitting-room first and, as she sat down, a few burning balls of soot fell from the chimney and bounced off the hearth and on to the rug. She jumped up to retrieve them and suddenly more fell and she called me and grabbed the fireguard. Soot continued to hurtle onto the burning logs in an alarming way. It fell in an increasingly enormous pile, glowing red and spilling onto the hearth. We removed the rug, ran for more fireguards and brought baking tins to contain the red hot heap.

We ran for the fire extinguishers. Guests gathered. The heat was fantastic. We rang for the fire brigade and this time our helmeted friends had a job to do. They brought in dust covers to protect the carpet and a hosepipe to put out the pyramid of glowing soot. They took a ladder up onto the roof and directed water down.

They would not believe us when we said the one chimney served four rooms. They continued to disbelieve until jet black water began to pour out of the Aga in the kitchen and soak the rug in front of it. Opening the oven door I found the plaice destined for the family meal swimming in a sea of ink. Where is the peace we are always looking for? We brought an empty water cistern in and the men filled it with smouldering logs and soot and carried them away from the house.

Having had their floor show for the night the guests went back to the Snug to gossip about the evening's excitement and drink their belated coffee from their armchairs. Next day we sent for the chimney sweep to put his brush up five chimneys. We did not tell anyone how long it had been since last we had sent for him!

* * * * *

On Sunday morning Joan chatted on the phone for half-an-hour in high spirits. Told of our recent escapade with the chimney fire she vowed to call in the sweep herself before such a catastrophe happened. 'Aw, I'd just die if that happened to me!' she declared.

I commented to the family, 'Joan's talked better today than she has for a year.'

Encouraged by her cheerfulness I rang again on the Wednesday to suggest we went for a run in The Dales that afternoon. It was her friend, Eileen, who answered. She had visited early and found her not at all well. 'Just a minute,' Eileen said, 'I'll see what she says.' After a moment she returned with the message, yes, Joan would like to go. My depressed spirits rose. She can't be all that bad if she wants a day trip, I assured myself.

I took Auntie Mary to the hairdresser and Harry to the barber, for only a few days remained before we left for the Isle of Man. Back home I made some open ham sandwiches and covered them with clingfilm. I whipped up cream for a sponge. Joan loved cream cakes and I had kept her freezer supplied all winter. I heaved Harry into the Range Rover and we went to Cononley.

Eileen was there and Joan was still in her dressing-gown. She looked far too ill to be gadding off for the afternoon.

'It's alright,' I said. 'We don't have to go,' but she wanted to. Her brain did but her body made no effort.

'I'll go and get ready in a bit,' she said.

I remembered a child asking her how old she was and she had replied, 'Twenty-one and a bit.' The child had gone back to her friends and said, 'Mrs Armstrong is twenty.'

'She isn't,' they had said. 'She's more'n that.'

'I asked her,' said the child. 'She said she was twenty-one in a bit.'

That was more than twenty years ago. It seemed like yesterday.

'There's plenty of time,' I said. Lord, I thought, there's no time at all. I am going away on holiday and all the calves, nearly grown, are to ear-tag and to worm. There are a hundred-and-one things to do and Joan so ill. I always want to go on holiday but I did not want to go the Isle of Man one bit.

She kept dozing. I don't think she knew when she slept. Harry was left in the car. I hoped he was dozing, too. An hour later Joan decided to go and get dressed so that we could leave. She went slowly up the stairs as Auntie Janie did, twenty years ago, the night before she died. Eileen was still with us. 'Come too,' I begged. 'She's so poorly and I've got Harry on the back seat.'

So Eileen came and we went to the deserted car park above beautiful Wycoller. It was late afternoon and the winter sun quite weak. Suddenly there was a radiance behind the clouds, silvering them across the heavens. Joan noticed it too.

'There's someone up there,' I said.

'I know,' she answered. She was not afraid of dying any more than she had been afraid of living. She had told me so on the day we both went to the funeral of the school secretary's husband, two years ago when cancer had reappeared in both her lungs.

She did not like the pain, though, and I believed her drowsiness was pain-related and I was afraid. Nevertheless she ate more sandwiches than Eileen and had a slice of cream cake. She kept dropping asleep but in the in-between times she was having a good time. She was talking about camping with my Guides shortly after her husband Jimmy died. She had stayed in Charlie MacLean's cottage. Had he lived we would have been going there in a fortnight's time.

'I was the friend with the civilised loo. I was easily the most popular,' she remembered. 'Didn't we have fun? Can you remember Bella, in the old folk's home at Scarinsh, who kept saying I had such lovely teeth?'

Harry laughed. 'Every time she saw me she kept saying, "Oh the poor thing!"'

We meandered over to Oakworth and back through Slippery Ford and down the precipitous road to Sutton.

'I'd ring Arden Lea, if I were you,' I suggested. 'They'd know what to do about your sleepiness.' She did not like being told what to do but I took advantage, now and again, of once being her boss.

'I'll do that in the morning,' she promised.

She got out of the car and walked slowly to the gate of her home before she realised she had not said, 'Good-bye' to Harry and she turned and came laboriously back to do so. When she was safely back I hurried to the car. I said to Harry, 'We might never see Joan again.'

I looked at the car clock. It was twenty to seven. A new guest, from Zimbabwe was expected. She wanted an evening meal and I realised Margaret would be in a panic. Reaching home I rang Joan's son, Ian, to alert him to his mother's extreme weariness but Eileen had already rung and Ian had gone to Cononley.

Miraculously, when I phoned next day Joan answered, 'I'm not s'bad. I've phoned Arden Lea.' She said Eileen was coming to take her the short distance to the hairdresser. When I phoned in the evening she was cheery. 'George is taking me to Skipton tomorrow'. On Saturday I phoned Ian. It was not my practice to phone Joan every day and I could not let her know how worried I was, and on Sunday morning Ian phoned me before coffee to say his mother had died peacefully in her sleep. Three days ago I had expected it. Then I had begun to wonder if I was mistaken and the Sunday news came as a shock. Through my tears came the relief that she had 'gone home' peacefully. 'No more stress or pain, No more waiting for the day.'

Wherever had the fifteen years gone since our retirement? Sixteen years of bed and breakfast, all those people, all those cattle, all those evening meals, beds to make, sheets and dishes to wash? Joan was only my senior by a few short years. We had been a working, teaching team for twenty-one years and after our retirement we had kept in touch. Her illness had brought us closer together. I hoped it had been a comfort to her. It certainly was to me. It left no recriminations, no guilt feelings of having done less than I could or more than I should for independence was very necessary to her.

Then the nicest thing happened to me. It is important to be able to do something practical when a friend is ill. Through the other Joan, I had been able to reassure her that Marie Curie nurses are not the last resort but very good friends along the way. It is equally important, and a privilege, to be able to say something when a courageous, laughing friend goes through the open door into the next room.

On Monday evening her family phoned to ask me to do that, in church, at her funeral. I could hear her say, 'Call me by my old familiar name, speak of me in the easy way you always used. Put no difference into your tone; wear no forced air of solemnity or sorrow. Laugh as we always laughed at the little jokes we enjoyed together. All is well.'

To prepare for such an opportunity one must have space and silence and time and I had none of these. We were heading hell-for-leather towards holiday. On the run-up to such there is no time to wield a pen, no slot in the common day to think clearly, to prepare, to remember, to differentiate between what is important and what is trivial.

* * * * *

With only days left on the April calendar we had to ear-tag seventy-one calves. Monsters they were at six months old. We had delayed the job because Margaret had been walling on the boundary, fearful lest they breach it, and had badly crushed her finger. 'I'll never grip the punch,' she had moaned, days later, when the swelling was severe. We belatedly called at Casualty to make sure it wasn't broken. The calves never even tried the walled boundary. They had headed straight for a wonky piece of fencing which encloses the woodland and that was a far bigger problem to solve.

When a guest comes purposefully towards me my heart wavers. If the problem is only 'How do we get to Haworth?' I heave a sigh of relief. If Margaret walks towards me in a similar manner my heart plunges to my feet.

'The calves are in the wood!' she said. Our antics are comedy. We shout at the calves, at the dogs and at each other. Those we force back into the Five Acre must be prevented from returning. Three people are needed and there are only two. The wood is on the precipitous hillside and the wonky fence is above the quarry. All told it was a miracle we managed to get the giddy calves back and a blessing they had found its weakness before the problem had been Dorothy's.

Still with swollen, black finger and barbed wire scratches, we decided to ear-tag on the Tuesday and leave the worming until after the funeral. The robust animals looked quite attractive with their massive red ear-rings and that evening, when Margaret counted, one calf was missing. In the afternoon there had been a tremendous clap of thunder to frighten the herd. Perhaps it had jumped the stone wall boundary Margaret had been repairing and joined our neighbour's very similar cattle. The only reason Margaret knew it was ours was the big red ear-tag clipped in that morning. 'Phew,' she said on the night air, 'That was only just in time!'

The phone was ringing frequently. Our school-children had heard of Mrs Armstrong's death. The school secretary rang and also the eighty-year-old school caretaker, retired before us and living miles away. A pupil who had become a Leicester vicar phoned. All had loved the long-serving teacher and were shocked to hear she had died. Few had known how very ill she was. Somehow, or other, I had to collect my thoughts and my busy life provided no time. On the Wednesday afternoon, before the Thursday service, the phone rang. Margaret answered it and booked in five adults, four from Australia, and Gladys, neé Spencer, their relative who grew up, in our village. Our evening meal total rose to twelve. I could not pause in the bed-making, cooking and phoning, and time was running out.

Margaret, who has many late night jobs, came to bed at 2.15 a.m. and her entry disturbed me. I'd had two hours sleep and, on an impulse, I got up and crept downstairs, stoked up the fire and made a cup of coffee. No-one, I hoped, would disturb me until dawn.

I let my mind wander back to Morning Assembly. Now and again, Joan used to say, 'Give them a little homily,

Miss Brown. Go on.' She used to enjoy my being serious. She would sit in the wings watching the children's faces as they waited to see whether I was pleased or cross. I never dare look at her because she was wicked and should I catch her eye I would lose my concentration and give the good news too soon or mess up the cross bit.

OK, I thought, it's your turn Mrs Armstrong. This homily concerns you and I am not going to reprimand, I'm going to commend.

I cannot remember when I have had so long, so totally, to myself. As I write this I pause trying to recollect ever being awake and not on call. Dawn came, but until I wakened Margaret at 7.45 a.m. the only interruption was the visit of the dustbin men who do not need help and the postman who is safe as long as the dogs are inside.

I remembered the day I first met Joan when I took up my village school headship in 1960. She was waiting in the staffroom surrounded by children. That is how I will always remember her. She had an extraordinary ability to communicate yet she was a very private person. She listened to others but rarely burdened them with her own troubles. I was still thinking and remembering when the cock began to crow. My vigil had to end but I was prepared.

*****

Margaret and I made breakfast for twelve. One of the Australians did not come to the table. 'She's got 'flu,' her husband said. 'It has kept bothering her in Europe. She'll be OK after a morning's rest in bed.'

They decided to go out and leave her to sleep. She and her sister were Keighley-born but had emigrated whilst still school girls. Their time was limited to three days and they did not want to miss a minute of it.

'You know we will be out at a funeral?' we warned.

'She's alright. She's only got a headache. It's 'flu.'

We decided to leave the dining-room door open so that, if they returned to collect her, and she felt better, they would be able to get in. I put my head round the bedroom door and said we were leaving and she answered.

For weeks the weather had been bad but Thursday was a beautiful day and daffodils seemed to have bloomed overnight. Joan had been a dedicated gardener. Her backyard had been a blaze of colour from the amazing collection of flower tubs. I envied her skill. We forget to water and are ashamed.

It was sunny but still windy and Ian advised us to park on a postage stamp of tarmac behind the church so that Harry and Auntie Mary did not have to climb the multiple steps from the road.

It was a lovely funeral. It is a mockery of belief if, though we mourn, we despair and a tragedy if lovely people do not have a lovely funeral. I was nervous and not always composed but I was grateful for the opportunity to commend.

When numbers of school-children had nearly doubled, Joan and I had been joined by two other members of staff, Winnie Annan and Liz Wheatstone. They and Mr Paul, the elderly caretaker, whose son-in-law had brought him from Barlby, Mrs Wass who used to be dinner lady, and our school secretary, Gladys Crossley, were all present. All were retired. Dotted among the congregation were our mature school-children. Ian and Jacqui wanted us to go to Stirton, to Tarn House. I declined. It was already late in the afternoon and we had twelve for evening meal. Life has to go on. 'I'll take Mr Paul with us,' I said. 'His son-in-law is anxious to get home but he must have a cup of tea first.' They understood.

Margaret had to pick up something we needed from the supermarket so the Pauls took me back to The Currer and the thought came to me that if the Australians had returned for their sick relative, and taken her out, they might have dropped the latch on the door. Margaret had our key. Until she returned we might be locked out. The weather had not encouraged us to open windows lately. My guess appeared to be right for when we reached home the latch had been dropped on the dining-room door.

The weak, April sun illuminated the landscape and reflected on the myriad windows. Not one was open. I tried nine of the ten doors into The Currer. All were locked. I wondered if it was worth trying to climb into the garden. To do so is not easy for we have made determined efforts to keep out the goats and if they can't enter it is presumptuous for a lady in skirts and nylons to do so. However there was a possibility that Margaret, having fed the birds, might not have bolted the door. It responded when I pulled. Although the sky was blue, the wind was chill and the old man had been ill. I was immensely grateful for her negligence as I let the two men in.

It never occurred to me to see if the sick lady was still in her bedroom. The latch had been dropped. We still do not know who did so. Half-an-hour later the Australians' car drew into the yard but we did not count the passengers. It was fifteen years since I had seen Mr Paul. Guests had to take second place. We poked the fire into a cheerful blaze and warmed ourselves with the cup that cheers and our faith in a hereafter. Mr Paul is a devout and God-fearing man. We were happy for Joan that her suffering was over but sad for ourselves having lost an irreplaceable friend.

I was weary, of the winter, of the worry over Joan and BSE, of work, of watching and waiting and writing. My eyes were closing with lack of sleep. We desperately needed a holiday.

Margaret went into the Snug to adjust the radiator and found the Australians in a worried, conference circle. 'How's the lady who was sick?' she queried.

'We can't make her out,' the husband said. 'She doesn't seem to know us. She doesn't seem to be able to speak.'

'Didn't she go with you?' Margaret told them the door latch had been dropped.

'We've only just come back a short while ago. We thought she was asleep but she won't wake up.'

Margaret went with them into the bedroom and she genuinely believed the woman was dying. She ran to the phone and described what she had seen to the doctor who told her to ring 999 immediately and get an ambulance. It appeared within minutes and the paramedics had whisked her away before we could even explain to the family what the commotion was.

We teeter on the brink of disaster. By rights I should have accepted Ian's invitation. Mine had been a privileged presence at the funeral. I should not have come home. We should have all been at Tarn House in which case the door would have been locked on the Australians and maybe it would have been too late to dial 999. I do believe the lady might have died. Her relatives followed the ambulance in their car and we said a stunned goodbye to Mr Paul and forced ourselves to make the evening meal.

We thought the woman might have tetanus. Certainly her jaw was well and truly locked. At seven the relatives came back to try to eat their meal. She was paralysed, they said, unconscious, dying. She was in intensive care and all manner of tests were being taken. Immediately after eating they all returned to the hospital and there was silence among the remaining guests.

Nearing midnight the lady's husband phoned to say it might be the early hours before they returned. The results of the tests showed his wife had meningitis. I do not blame those who read this who say it cannot possibly be true. The sheer variety of joy and sorrow, and the speed at which they alternate leaves us spellbound. In the early hours, filled with antibiotics, the lady regained consciousness and her family, equally dosed, returned to creep belatedly into bed.

The next day we, and all our guests, were similarly treated and warned not to be alarmed when we went to the loo and everything was bright orange. I must commend the wonderful way in which the holiday-makers reacted to the quite alarming news that they had been in contact with someone suffering from meningitis.

There was now less than a week before holiday and nothing towards its preparation had been done. We wondered, secretly if we would go at all. The hospital staff reassured us that the meningitis was viral but we did not fully believe them because they were so adamant that contacts should have antibiotics. We had new families coming in, some with very young children. On Monday I called on Eileen, then I visited the lady in hospital and heard that the cultures did prove it not to be the infectious variety. Only then did we relax.

Chapter Nineteen

Spring is here at last,
Weary winter's past.

A favourite choice from BBC Singing Together

The florist brought me a beautiful display of flowers from Ian and Jacqui, a 'thank you' for my part in the funeral, reminding me that a friend had recently died. It seemed a decade ago.

On Saturday we wormed all seventy-one calves and that left too few days to our holiday. This time we will not make it, we prophesied, but we left for the Isle of Man before the cuckoo was heard. The weather for weeks had been too uncommonly cold for him. Its dreariness had matched our sluggishness, our 'feel good' normality was smitten with tiredness. Rain had fallen enough to water the earth but not enough to alter the low level of the reservoirs. Warmth was desperately needed to herald a growth of grass. The daffodils were the only harbingers of spring.

A too-late flush of green has been known to bankrupt farmers. In 1996 its tardiness was disastrous. Farmers with fat animals over thirty months old were not allowed to sell into the food chain. Because there were inadequate facilities for incineration farmers had to hold them. They could scarcely afford to keep them on anything but grass and that was in short supply. No-one knew what was happening other than that they could not sell older cattle and the trade for younger ones was shockingly poor.

The weather, we assured ourselves, would change when we went on holiday and sure enough the forecast became encouraging. We had no time to be apprehensive about two weeks on the Isle of Man. We would enjoy it, we always enjoy holidays, but we require to live in beautiful surroundings. We are fortunate. We have the choice. Most of the people living on this beautiful planet have none. We are used to magic on holiday, peace and space. In the Hebrides these are assured. We knew nothing of the Isle of Man.

I think we appreciated the short distance between home and Heysham. We drove the whole distance on minor roads. We passed through Wray, very slowly, for the village was uniquely decorated for its Gala Day. Almost every garden sported a life-size straw doll enjoying some pastoral activity, a spectacle unequalled anywhere. We were fascinated.

Had we not been so tired we would have been very disappointed with the sail. To board a Caledonian MacBrayne boat is sheer joy. Be it the *Lord of the Isles*, or the *Hebridean Isles*, or the *Isle of Mull*, moored to a West Highland quay it presents its broadside majestically. There is a buzz of nautical activity, a screech of gulls and a rich smell of sea wrack and fish boxes. Those already aboard are not to be found in the lounge. They line the deck rails watching cars being driven onto the car deck and cargo lifted on board. The sun shimmers on the sea all the way past the lighthouse and out to the horizon.

We never saw the *King Ory*. It was hidden behind a facade of ugly buildings whilst we waited nearly two hours for the signal to drive on. There seemed to be some problem with our roof rack. We never quite understood why we had to wait until the last and had to reverse into the gaping hole. Even then there was a problem and Margaret had to drive out again, with four others, whilst the seaman in charge rearranged his cargo. We had asked to be loaded near the lift. The MacBrayne staff always oblige. The *King Ory* crew were equally kind and pleasant but had not put us anywhere near the lift. Squeezing between the cars and lorries with Harry and Auntie Mary was a nightmare and when we eventually found the lift and asked the seaman operating it for us to put us out near the observation lounge, he looked at us blankly.

'What's that?' he said.

We were used to the Hebridean ferries with seascape windows offering a vista of unparalleled grandeur. There was no such thing on the *King Ory*. The kind man managed to find us the last four empty seats, mercifully close and in the non-smoking area. We had to temporarily disturb two sleeping men to squeeze past and once seated we were prisoners. The windows were too high to look out unless we stood which was almost impossible for the seats in front were too near. Our nostalgia for the *Lord of the Isles* was fractionally less than our tiredness. That won the day and we slept most of the way.

There was a brief entertainment when the sleeping man beside Margaret wakened and saw the child sitting in front of him, across the table. The ten-year-old girl was wearing a hideous hat with an old fashioned, floppy brim. The waking man suddenly saw this otherwise pretty girl and gasped, 'What an awful hat. My grandmother had one like that and she threw it away!'

The child's mouth fell open. Perhaps she had never experienced disapproval. Her father said, 'I know but we can't get through to her.' I am sorry for the present generation. Criticism is a necessary part of education.

When we berthed we had this nightmare, again, of threading our way through the cars, in haste for we would be required to drive off first. The dark car deck flummoxed Auntie Mary so that she neither heard nor saw the barking head of a noisy Rottweiler protruding from an open car window. We had a moment of panic when her hand rested on the car, dangerously near. We hastily re-routed, regardless of how many wing mirrors we disturbed.

We were the first to drive off into the maze of roundabouts and traffic lights of Douglas. We never looked behind to see the gaping mouth of the boat. There was no space, no long pier over a green translucent sea, no gaelic greetings, no scent of peat and mercifully no time to do anything but drive with the traffic, following the signs westwards towards Port Erin.

Suddenly it was all behind us. Ahead was an empty road, banked with primroses and celandines and the fields were ablaze with gorse. We let out deep breaths of pleasure. 'It is beautiful,' we said. Within half-an-hour we turned right into East Bradda and found our cottage. The key was under the mat and we let ourselves in.

It was deathly cold but one press of a switch activated the Calor Gas central heating and immediately the radiators warmed the air. The kitchen door opened onto a sun-flooded patio, full of primroses and with an uninterrupted view of sheep and pasture and well-kept golf course. In the distance was Port St Mary and the sea. From the lounge window the green hillside climbed towards the blue of heaven. It was dotted with gorse and only a few homesteads and a track running horizontally towards Freshwick Bay. We were happy. Immeasurably so!

We desperately needed to distance ourselves from The Currer for a host of reasons. We needed calm water and some time in a cathedral with a ceiling of blue. We craved a period without people. I had had no quietness in which to mourn. The Isle of Man provided all those necessary-for-the-moment things to allay our extraordinary tiredness. There was never a ripple in the sea around us. Mother, who loved the great ocean to swell and crash, to boil and foam on the rocks, who wanted white horses to ride ashore one behind the other, would have called the sea 'uneventful'. We, her children, who need a challenge and thrive in a storm were nevertheless willing to exchange its excitement for a taste of its eternal peace.

The island was as silent as Harris on the Sabbath. We could hear our own footsteps in the empty shopping precincts of historic Castletown. We appeared to be in some sort of timelessness. The whole island seemed to be standing still for that fortnight of the year, waiting for the sound of motor-cycles to awaken it briefly. All activity seemed to have ceased and we, strangers on that lovely isle, seemed to move about alone, breathing flower-scented air and creeping along empty roads.

The weather, though not hot, was very blue. The sun was shining from the moment it climbed out of the eastern sea until it dipped behind Bradda Head. All those many people who had crossed on the ferry had disappeared. If we individually walked the pleasant track to Freshwick with the dogs, before mid-morning coffee, we met no-one. If I pushed the wheelchair the mile-and-a-half to Port Erin or the two miles to Port St Mary no-one was going our way. There were few shops and fewer shoppers and the Victorian hotels on the promenade were either sleepily waiting for the TT races to waken their proprietors or were displaying For Sale notices against their dusty, curtained windows.

Everywhere we went it was beautiful. The trees were just beginning to leaf. I have never seen so many primroses. In millions they carpeted every banking and the Sun God illuminated the celandines, and the violets were surely confetti from the eternal blue of heaven.

To be among the Hebrideans is to witness pastoral industry. Crofters are busy people. Hughie and Angus, Duncan and Alistair are all moving cattle or gathering sheep. Donald and Donny and Donald Alick and Donald John are at the peats or the lobster creels. There is a tilling of the land and a fishing of the sea. Mary Seaside is off in her Land Rover on some important task. John is loading sheep onto his little boat to take to Taransay, Donald is off to meet the boat at Scarinish and Donald John is in his weaving shed. Mairi is calving a heifer, Morag Ann is home for the lambing, Rachel is driving the school van and Mamie is attending the petrol pump. Margaret is in the Post Office. Effie is feeding pet lambs, Katie is out at the travelling shop and Lexey is hanging out the washing. Maggie is off to care for the old folks, Calum is sorting fishing tackle and Kathleen the district nurse hurries to a patient. Only we are on holiday and only on a Sabbath does silence and inactivity last all day.

On the Isle of Man it seemed to last all fortnight. We wondered where the farmers were who tilled such well-kept fields, why the pastures appeared empty, if every day was early closing, why so few inhabitants needed to shop and if everyone had taken sleeping pills. We really felt and appreciated our isolation. It was lovely in a different way. We felt far-removed from the whirlwind in which we thrive and everywhere we went there was inland beauty. In the Hebrides we cannot leave the beaches. We cannot turn our back on them to look elsewhere. Their white emptiness haunts us when we come home. On the Isle of Man it was the the inland beauty which attracted us. The beaches could not be compared with those of the Western Isles and, where the sea had, for centuries, spread sand and shingle, men had built hotels and laid concrete promenades. Strangely we did not mind that the shore was less white, the sea less blue because we could take tonic from the wonderful greenness of new grass and bursting buds and the myriad spring flowers. We realised holidays on Tiree and Harris had lately been too soon for the flowers. We had been deprived of sea pinks. The Currer meadows are rich in clover and a botanist could find a hundred different species but we have walls not hedges and we are too high for primroses. We must go down to the River Aire if we want to find the water-loving plants and walk along the canal bank if we want to find the butterburr and water flags.

If, and I hope when we return to the Isle of Man it will be for the flora. Of the domesticated fauna we saw little. Perhaps the cattle were still housed. Perhaps the shepherd had hidden his flock. Perhaps the fairies, whose home is there, had made them invisible so that we could forget the BSE crisis and bury our fears for our profession under the sentence of death.

We found a friendly, almost empty swimming baths on the outskirts of Castletown with a waterside changing room for the disabled. The temperature of the water was cooler than at home but Harry did not let that deter him. Finding a new pool on holiday is important to him.

Behind the Bradda East cottage was a secluded patio which caught the breakfast sunshine and held it until midday, so that Auntie Mary could eat her cereal outside and still be seated at coffee time. The dogs could be restrained there though Lusky preferred to stay in the Range Rover. We had been afraid of the problems taking Lusky with us would cause. He and the Tom cat were geriatric members of the family. Short of being put down, which was unthinkable, Lusky had to come with us and Tommy had to stay. Anyone who is agile enough to get up to open the door for Tommy if he is in a hurry, and tolerant enough to mop up the lino if he doesn't get to the door in time, can handle him. Margaret had taken him to the vet for the injection which stops him being a rip-rap and she told Mr Greaves that, should Dorothy have problems, he must make whatever decision was necessary. Dorothy would not want to be left with such responsibility.

Lusky had to come with us because he was a holiday dog who loved travelling and was at home by the sea. What's more, no one can handle Lusky except Margaret. He would allow no other to lift his aching arthritic bones into the comfortable bed in the Range Rover. The problem we anticipated was incontinence. When Lusky wants to go out he barks. He is a clean dog. Nights on holiday are longer than at home. Should we not hear him and he not be able to wait, his bed would smell. We had taken a pile of old towels thinking we might have to throw them away but the cottage had a spin dryer within and continual sunshine without and Lusky was 'nay bother at all'. The homeopathic pills seemed to alleviate his arthritis. The aloe vera juice Harry had been taking seemed to be making a younger man of him and had begun, with the help of the sun, to bring back the healthy look that winter had stolen. 'Doctor Green', that wonderful agent of spring was doing his rounds.

* * * * *

It was cooler at The Currer but the weather was pleasant for Dorothy, too. Animals and guests behaved for her. The Currer seems to present its problems only when the owners are at home, or near enough their timely return for the responsibility not to be the burden of the holiday staff.

'Tommy is going to die,' Dorothy said when we returned home. She appeared to be right. 'He hasn't eaten for a few days and he's not going to the toilet. If you hadn't been on the point of return I'd have taken him down to the vet. The last time he went in the cat box it seemed all blood.'

'Poor Tommy,' Margaret said. 'I'll see to him.'

It did seem as if Dorothy was right. The old cat looked as if he was ready to meet his Maker but having seen Margaret he decided to stay on this earth a little longer. Other animals too had been waiting for her return before displaying foul, spring scour and lumpy jaw.

The chart for the month following our return had been filled long before the BSE crisis. July and August were noticeably yet to fill. We suspected that the men wanted to be at home for the Euro 1996 football and the ladies for Wimbledon but feared it might also be that tourists were avoiding beef-rearing farms whilst the mad cow disease dispute flared. There might well be a reluctance on the part of the Europeans to take holidays on farms. The British public reacted wisely to the media-mania and housewives took the opportunity to fill their freezers with cheap beef. They turned off the television when the boring subject was discussed and did not know who to believe, so they believed no-one. Farmers could not dismiss the whole thing, just like that! It was far too painful. Our guests had infinite compassion but our obsession with the awful mess must have embarrassed them. We were caught in a political web that had nothing to do with reason or silence. Farmers just wanted to be left alone to sort out the problem without Europe.

Friends, whose cattle sales came in June, were losing thousands of pounds. Our turn would come in September but the cuckoo was doing his level best to cheer us and it was too soon for us to panic. There were plenty of distractions. The deer were seldom seen. I suppose it may be that we have little time to stand and stare. We did see two in the Five Acre being followed by seventy-one very inquisitive calves who mistrusted them enough to stop whenever they did and curious enough to follow whenever the deer led. The stags must have been in a mischievous mood for they did not head for the wall until they had zig-zagged all over the Dyke Field. When they finally leapt gracefully over the boundary, seventy-one calves looked longingly over the wall. We thought they might well decide to attempt the impossible and only when they turned away did we resume work.

The wild has been coming closer to The Currer lately. The fox, a too frequent visitor, dares to come into the yard in daylight. Shortly after our return he managed to take our remaining cockerel. Free-range birds have to take a risk or be imprisoned during the day. The fox must be hungrier or more daring than usual. He runs from the chasing dogs but he does not know which way to turn when attacked by the crows. Margaret witnessed an Alfred Hitchcock episode below the silver birch trees. The fox was being well and truly harassed by a flock of crows. Only when the dogs joined in did Reynard escape. We like the wild things to encroach upon our door but the crows are quite aggressive. Compared with the fox they are far more fearsome. One took to pecking on the window panes, usually at day-break. Guests drawn to the locality in search of The Brontës may well have leapt out of bed to see if the ghost of Catherine Earnshaw was outside. The tapping at the window would set the dogs leaping onto the window seat and bring me scurrying downstairs expecting to have to cope with the early morning problem of some worried guest or the very premature arrival of the postman.

We were glad of any diversion to distract us from the chaos which had erupted when one government minister had ill-advisedly said, yes, if necessary they would slaughter all eleven million cattle in the country.

'Leave it to us,' I heard one farmer on a phone-in programme say. 'As we successfully got rid of TB and brucellosis and anthrax, so will we exterminate BSE.'

'Ah, but this is different,' said the politician answering the calls, 'BSE can be passed to humans in the form of CJD.' Of the four terrible diseases the last one was the only one, as yet, unproven. The others had been eradicated for the simple reason that they could be passed on to man. Did the politician, in such a post of responsibility, really not know why all the herds were tuberculin tested and almost all milk pasteurised? Yet these men could, and would, order the slaughter of large numbers of cattle without proof. It was heartbreaking.

Summer, somehow, must be got through normally. It was foolish to panic for politicians change their minds more often than the British weather. Our eighteen-month-old steers grew like mushrooms and, blissfully unaware of human incompetence, peacefully grazed all day. Out of earshot of our guests we could be found singing stoically, 'Keep right on to the end of the road'. Asked whether the crisis would put us out of business we shrugged our shoulders privately determined that it would do no such thing. Come hell or high water we'd eventually be buying another on Monday!

* * * * *

The profusion of wild flowers on the Isle of Man made us far more conscious of the similar display which awaited our return. The Currer was a garden of bloom. The May blossom on the hawthorn trees seemed heavier and more scented than ever before. When the dandelions were over our lane was a mass of purple clover, bird's foot trefoil, hawkweed, lady's slipper and daisy. Buttercups ignited a blaze of glory across the pastures, orchids flowered round the almost empty moor pond. We spent hours measuring walls and taking photographs and filling in forms with the hope that we might be considered for a Countryside Stewardship Scheme which would give grant-aid in rebuilding more of our traditional stone walls. There could be no hesitancy. Applications had to be in by the end of June. It was a ten year scheme and by the end of it I would have been retired twenty-five years. Also one hundred-and-forty CIDs had to be sent back at the beginning of the two month retention period. I filled in the multiple forms, carefully writing the thirteen-figure ear-tag number of each one, copying it in the order of our herd book which had been entered in the order of the Beeston Castle bill of sale. Each number begins with UK then two letters XY or CF or CK or QR. A new system had been computer devised by MAFF. Previously the checking necessary was in two months time when new documents arrived in the order they had been received from me. Checking there must be, when it is a human which feeds thirteen digits and a date of birth into a computer.

The new system floored us for the numbers had been reshuffled into alphabetical order. This typed list of a hundred-and-forty thirteen-figure numbers was sent by return for check matching and it was a nightmare. Thirty years ago all that was needed to obtain a subsidy was a punched hole in an animal's ear. The system had gone mad. Farmers thought it was the government and not the cows that was diseased.

Our circumstances were different, during this 'summer of discontent' from those of our neighbours in that we had a steady flow of holiday-makers to pay towards the hay and straw we must order in the summer. We could not neglect to buy. If we had fewer cattle the barn would just remain well stocked all winter. We would still have the seventy yearlings. That was a certainty. If we could not sell our seventy older animals we would have to keep them in the hope that there would be a future market. All winter, if we kept them, they would be hungry and enormously big for a five-foot sister and a long-retired headmistress to handle and they would cost a fortune to feed. They would occupy what housing we had and we would not be able to buy calves. That was one possibility. The most likely one would be that we would be able to sell but at a grievous loss. We would have to take it on the chin as we have done so many times and be glad of our diversification.

Because the government was paying farmers handsomely to have their week-old babies incinerated there were fewer calves on the market. There was a grave possibility we would find buying our October ones very difficult. There was nothing we could do but pretend all was normal and order the deliveries of hay and straw to fill the barn.

* * * * *

We discussed whether or not to buy pigs again, but worry had tired us more than work and we decided we had enough to do. June was full to bursting point with the nicest possible people. The schools came, less-able children from Isebrook and Northgate and local children on farm visits. The Longcakes came from Australia for six weeks in The Mistal. They were Bradford people who emigrated in 1981 and return every six years or so. Vera is in a wheel-chair after a road accident way back in their early days in Australia. Colin's parents had joined them years ago and all four came to revisit old haunts and close relatives. They arrived before midday and put us on the spot.

All week The Mistal had been occupied by a truly delightful woman who had the misfortune to have been suffering from chronic ME for ten years. She found it impossible to get up early. I had been taking her breakfast at about 2 p.m. I had warned her she must realise the Longcakes would maybe arrive early depending on their incoming flight, that Vera would be exhausted and the old people shattered. 'The Mistal must be ready and vacated,' I said.

I cleaned the cottage and made up the other beds and took all the lady's cases into the dining-room but I could not force her out of bed. She was still there when the Australians descended on us early and I was most embarrassed. It was lovely to see them again. 'We're just ready to roll into bed,' they said.

'I'm sorry,' I told them. 'Goldilocks is still in yours.'

Colin's father was on the verge of collapse. We lowered him into an armchair and Margaret ran to make a pot of tea. I went to the lady severely handicapped with ME and forced her out of bed. The taxi to collect her wasn't due for two hours. 'I can't sit upright,' she said. 'I'll have to lie down.'

I ran for a camp bed and put a mattress on it among the tables in the dining-room. 'What a crazy life we lead!' I grumbled to myself helping my patient to her makeshift bed and snatching sheets to make up the last bed in the cottage. A car drew into the yard. It was only 12.30 p.m. Our brochure says 'Arrive after 3 p.m.', but Germans were getting out of their car. Imagine welcoming Europeans with a lady flat out on a camp bed! We had such nice guests in June and the Germans were no exception.

* * * * *

Tom rang from Newton Aycliffe. He and his wife of twenty-three years, are both in wheel-chairs. Judy was awaiting a call into hospital to have both legs amputated. 'Can we come for a break?' he asked. We would be prepared to forfeit our beds and reshuffle guests to squeeze in Tom and Judy any day. Valerie, whose husband Eric died five years ago, had met the couple several times for both families had been regular Easter visitors. She went out of her way to give them a good time accompanying them and arranging a trip on the Settle to Carlisle Railway. I think she enjoyed it even more than they did. Do I emphasise frequently enough that the real necessity in life is to be needed?

Guests who come often, frequent our sitting-room. In summer, when there is no fire in the hearth, people in armchairs tend to face the south wall mullions and focus their eyes on the rising landscape. We are forever watching cattle. Visitors tend to follow the activities of the pigeons on the shed roof or the clouds putting illusionary mountains on the horizon.

Valerie came home early one day, before we had begun to prepare the evening meal, a chore we never start before we must. Cooking is a job we have to do and not a pleasure we look forward to and we take as little time over it as possible. The result is that nothing is overcooked and everything is served straight from the heat. Our guests say our cooking is superb and each meal, as Mother would daily say, 'goes down alright' but there is no way we will spend all day in the kitchen and we flatly refuse to watch any cooking programme on television.

Valerie lifted her late afternoon cup of tea from the Aga and we followed her into the sitting-room. We have only a once-a-year opportunity to chat with Valerie and, over the years, she has become a friend rather than a paying guest. We were all sitting comfortably, facing the sunshine pouring through the windows, when Valerie suddenly said, 'A mouse has just run across the carpet. It's gone under the chest!'

Most people would concede that if you live in the country a field mouse will, now and again, get into your house. It is a most infrequent occurrence at ours. We cannot have a straying vagrant, however beautiful the velvety creature is, for our hygiene rules forbid it and guests will not return if they fear a mouse is in their bedroom. We know, for I have counted and Margaret has painted, that we have some fifty-three doors in the house and that all are being perpetually opened and most have half-an-inch clearance that could admit a creature which can squeeze through a crack.

Next morning however, we stopped worrying because I saw the stray animal leap onto the window ledge below the new mullions and disappear down a perfectly obvious and inviting cavity left by the builder when he was laying the stone slabs which make the sill. I sealed the entry and we forgot the incident. That was foolish for, had I reasoned, I should have concluded that a mouse must have some outside entry into the wall.

It was when Ian and Angie Cameron were watching Auntie Mary baking biscuits that we all heard the activity in the wall behind the Welsh dressers. We were too busy to investigate and whilst we were fully occupied with the summer tasks, a mouse was fully occupied nibbling a way under the skirting board. Only when he had succeeded did we exert ourselves. We found a nibbled bag of rice and, pulling the dressers from the wall, we found the hole. Again we blocked the entrance without investigating the outside entry, wherever it might be. There is a limit to what one can do in a twenty-four hour day!

* * * * *

Ramblers came for their seventh consecutive year. Their eighty-five-year-old leader is a Bradford lady we know as 'Auntie' Nora due to her relationship with one of our Guides. Eight years ago she and a friend went on a wild flower expedition to India and met others on the tour who now come for an annual get-together and weekend ramble in the Yorkshire Dales. They entertained our guests in the evening with a slide show of a recent visit to New Zealand and when they left 'Auntie' Nora rebooked for the next, their eighth, year! The oldies put the young people to shame!

Philip Oakes came for his sixteenth year and surely his twentieth time and the Vinces enjoyed their annual week with us. The Johns, farmers from Cornwall, came promising not to talk shop. I threatened to put my every penny on them not keeping that promise but they would not take me up on it. It was their fifth June with us and Margaret has a great deal to thank Sheila for her homeopathic advice. The cooler, damper weather brought such an abundance of grass our donkey, Joe, went down with laminitis. He became very lame and Margaret asked the advice of the blacksmith who frightened the life out of her by saying the animal should be kept indoors and fed only on straw for as long as it took, maybe all summer. She brought the inseparable animals in and Joe proceeded to frighten the life out of her again by lying flat all the time. 'I can't face another donkey dying,' she almost wept. 'I can't cope with keeping him indoors. What'll I do with Jasper? He can't be kept in all the time! Heck I'm fed up!'

'Try McMillan,' I suggested, he is the homeopathic expert Sheila Johns introduced to us a few years ago. Unfortunately he lives in Cheshire and any remedy must come by post.

Margaret phoned and told

him about Joe and the blacksmith's warning and he said, 'Laminitis? That's easy. A ten-day treatment and he'll be as fit as a fiddle.'

The medication came by post van next morning and after only a few doses the donkey was back on his feet and raring to go. More credence should be given to homeopathy!

The Chapel ladies used The Currer again for their annual money-making open day and once again we had the embarrassment of being visited by local people who obviously knew us but who seemed strangers. Each time we rub shoulders with our near neighbours we realise we only recognise the farmers and are shocked to find we do not know the others.

Our family now consists of 'birds of passage', 'ships that pass in the night', and this isolates us from the local community even more than farming does. Thousands of bed and breakfast guests, names and faces and dialects confuse us. We seldom recognise people when they return, unless they have done so frequently. Thank goodness I still know my school-children and the Guides I have taken to camp. It is sad that, whilst we know who lives in every house in the Hebridean townships, we do not know who live in many of the houses on the top road.

We live in a nightmare of being known by everyone and ourselves being completely blank as to who they are and where they come from. What's more our brains are unwilling to learn, knowing we must immediately forget or go round the bend. So we smile and say, 'Hello,' and 'Good-bye' without quite knowing to whom. There are others in similar circumstances I'm sure, at the supermarket checkout, on the wards in hospitals and men driving coach tour buses.

Exactly one week after the Coffee Morning, at 10 a.m., a stranger's car drew into the yard. Four people we did not know climbed out and approached the front door. I hastened into the dining-room to greet them with an outward show of hospitality. Our bed spaces were full and I had barely begun the morning chores.

'We've come to the Coffee Morning,' one said and all were flabbergasted to learn they were seven days late.

'Never mind,' I said, 'I can most certainly make you a cup of coffee.'

'We really came because we think we are distant relatives.' The voice was Australian.

When Auntie Mary was very small, nearly ninety years ago, a grown-up married cousin went out to Australia with her husband and children. Only her immediate family kept in touch. All of Florence's generation, except Auntie Mary, have died. Her ninety-one-year-old son was still alive in Australia. It was his son and daughter-in-law who were visiting the husband of a deceased cousin. He had wanted to visit the birthplace of his grandparents and meet any of his surviving relatives.

How terribly restricting the Coffee Morning would have been. Auntie Mary, taking the entrance fee for the 'Chapel Do', would never have had the joy their unexpected arrival brought. She would have been deprived of hours of happiness had our guests succeeded in coming on the right date. The oldies could not stop talking and Auntie Mary recounted story after story of family fun at the turn of the century. She painted such a lovely picture of life in the hilltop village, where their ancestors had lived, we were all entranced. We, as children, had been nurtured on stories of the village characters and their strange nicknames. We knew, but our visitors heard for the first time, how a local man received the name, Bornso. He had been born with a slight deformity. His head did not sit squarely but was always on one side. Strangers picking him up, one night when he had been drinking too much, tried to straighten it and he had shouted, 'Born so, yer buggers, born so.'

'I remember hearing that,' said the husband of the late cousin who still lived in the locality. Thus encouraged Auntie Mary brought out a narrative poem about the digging up of a buried friend and the taking of the coffin elsewhere to re-bury.

'And,' she concluded, 'there was this man everyone called Tommy 'Offee. He went into the village shop and asked for 'offee because he could not speak properly. The shopkeeper thought he wanted toffee but the man explained, 'No, 'offee, that th' mak tea on' and after that he was called Tommy 'Offee.'

'He was my grandfather,' said the husband of the late cousin. Auntie Mary went red and everyone laughed. It was a great reunion.

* * * * *

Every two years the College of Ripon and York St John holds its annual Reunion at Ripon. I seldom attend on alternative years when it is held in York.

I can never get to Ripon in time but I am known and my late arrival is predicted and understood. I am late principally because we all go. Margaret drops me at the college gate and then takes the family into Studley Royal though to watch deer, nowadays, is often not even to leave the farmyard.

Joan's habit, for some years, had been to go for the full weekend, stay in college accommodation and enjoy more than just the Reunion meal. Being so ill she had hesitated to make the booking. 'Can I go with you?' she had said. 'Don't let's book a meal, even. I couldn't face it but I want to see my year.'

So we'd agreed to go alone. I could leave early if Margaret was not leaving too. We could assemble with everyone in the Bishop Chase common room, well before the meal, then she and I could eat sandwiches in the grounds. It was a good idea but sadly she was to die too soon and I was left debating whether to book a meal or not to bother going at all. I left it to the very last minute and hastily sent my cheque. College Reunion Day is always brilliantly sunny and it was a shame to deprive the family of an outing. Things were made easy because the guests we had in that weekend, were a seventeen-strong rambling club who come every year, make their own beds and tolerate a later evening meal.

So I went to the Reunion. Margaret fed the mallard in Studley Royal Park and I sought out Joan's college year to tell them all how she had lived and how she had died. One of her contemporaries had opened the Book of Remembrance to reveal the page bearing her name. There was a moving silence in the frequently-used chapel. If college wanted proof of its success in training teachers fifty years ago, it need look no further than Joan. I could vouch for that.

Margaret's profession was taking a beating and so was mine. We were taught, in the first half of this changing century, that academic excellence should come second to moral and social behaviour. 'Do not say to children "will you?" our Education Lecturer said, 'Say, "You must". Do not shout. Speak no louder than "lovely". Do not condone but be positive that certain things are not allowed.' We were taught that our example was all important. As closely as possible we must follow Caesar's wife and be beyond reproach. When did it become accepted that what one did in one's private life was unimportant, that it was not necessary to practise what you preach? I only know that standards fell in adults first and when, as Bernard remarked, the television became the second parent, and failed miserably, the present chaos was to be expected. It is quite wrong to think the clean-up must begin with children. It must begin with those who are teaching, guiding and preaching, with those who are parenting and those who are governing.

Our moments of leisure were stolen by under-age children riding motor-cycles on our land. They lifted heavy ramblers' gates off the hinges and could not hang them back. Local farmers rang to say our cattle would be out in no time. We went to lift back the gates and turned one gudgeon so that vandalism was impossible. We caught the boys at their mischief and persuaded them kindly that what they had been doing was illegal and very dangerous. Straying cattle on busy roads cause accidents. Someone, I hope it wasn't the boys we had been lenient with, thereupon took off the wire securing the gate between our land and a neighbour's which was too tempting for seventy huge bullocks and put us in grave danger when getting them back, for our neighbour has strong cattle in that field, too.

* * * * *

We sent for the vet for the animal with lumpy jaw. We feared it might choke and wondered if leaving it to struggle was inhumane. The vet came and said a few doses of streptopen might alleviate the infection but would not cure it. 'That is not your problem anyway,' he said. 'Because of BSE you can no longer send cattle as casualties except in extreme necessity. I would have to be called to witness shooting the animal, on the farm, by the remover. You would get nothing for the animal and have to pay considerable costs. Your best plan is to give it streptopen and see what happens.'

Almost simultaneously with this the Prime Minister went to a European Summit Conference in Florence, and it was agreed that if Britain engaged on a slaughter policy of something in the region of a hundred-and-sixty-seven thousand dairy cattle and all calves normally sold to Europe, which amounted to many thousands every week, the ban on sales might be partially lifted in the Autumn.

'The beef crisis is over,' said the Government and as far as the general public was concerned that seemed to be that. For the farmers it was no such thing. It was the beginning of a tragic waiting game and the immediate killing of countless week-old calves. The Government had made promises it could not immediately keep. It was ill-prepared for execution on such an unnecessary scale. Anticipation of the bereavement which was to come, left the farmers paralysed physically, mentally and financially. They were already suffering from neurological conditions caused by compulsory sheep dipping, disinfecting and spraying with the organophosphorus chemicals that many of them believed were also responsible for the BSE problem. Now many of them were suicidal with worry. How could their overheads be paid and their overdrafts kept in control when cattle they sold brought 25% less than they should?

The dairy-bred calves normally reared on the Continent were sent straight from the market to the abattoir. The Government compensation for this being over £100 dealers were willing to pay even £90 to buy, en mass, every dairy-bred calf in the market and drive straight to the abattoir. 'God help mankind!' These were the calves we hoped to buy in October. They would cost us dearly. We visited the markets more often. It distressed us to see farmers sitting in the ring so depressed. "Ave yer brought some t'give away?' one farmer said.

'Not yet,' Margaret replied. A haulier employed to take a load of calves straight to the abattoir was almost weeping. He was a big, strong man. "Ell,' he said, 'Ah'm sick!'

We had three bullocks over the thirty-month limit. They would not be allowed into the food chain but because the queue to dispose of such was so long they would hopefully have all summer on the hill.

We had kept two of these animals longer than the usual twenty-three months because they were Jerseys and not really beef cattle at all. A Jersey bull calf is mostly slaughtered at birth but Margaret had outbid the butcher because the beautiful babies had looked at her on the ring-side and she could not resist their wide-eyed appeal for the same opportunity of life other breeds get. At twenty-three months they had not been suitable for sale as stores. We had kept them another winter and they were enjoying another summer. The third bullock was a Friesian and he was a year older than they. We ear-tagged him number twenty-two but always called him number seven because he had been Lot Seven at the market and the sticker on his back had remained there long enough for him to be identified. He had not been ill but he was small and had not thrived. We had penned him alone during his second winter because the yearlings were too boisterous for him. Since then he had done well and would not have looked amiss had we been expecting to sell him in October but his age, now, took him right out of the food chain and we had entered him with the Jerseys into the thirty-month scheme. Our cattle love the hill. They pasture in the low fields but choose to be aloft at night. 'Enjoy it,' we told them. I think there must be a hilltop in heaven specially for cows and for all this unforgivable wastage of calves. I noticed more particularly how pink the cattle were on the sky-line, at sunrise. Surely God must be aware of man's inhumanity to animals as well as to his own kind.

Yet, had it not been for the appalling governmental mess, I think we would have been highly satisfied with July. A nation-wide slump in the tourist trade did not affect us and we were frequently full and we had the nicest possible people. A wealth of grass took away all fear of famine and enough rain fell to fill the reservoirs.

And Lumpy Jaw recovered. The antibiotic had done nothing to relieve his symptoms. The expensive visit, the medication and the information re disposal of casualties brought one positive result. The bullock was still alive. The huge swelling in his jawbone prevented easy grazing. He had breathing difficulties and we could hear the peculiar noise he made a field away. We feared a passing walker would report us to the RSPCA.

Something distracted the herd and there was a stampede which is a fearsome spectacle. The only mishap, however, was a tear in one calf's foot. Right from the start this calf had been small and difficult to rear. At ten months old it was fit and healthy but still smaller than the rest. We name only a few calves every year, the ones which cause us problems. This was one of them and was called Tiny Tim. Doctoring it was time-consuming as it moved with the herd and had to be brought home to the crush. I was frequently needed and saw more of the grazing herd than usual and I grieved for poor Lumpy Jaw noisily struggling to eat and breathe. Nothing could be lost by a phone call to the homeopathy man who had so miraculously cured the laminitis and whose medication appeared to have halted the coccidiosis.

The good man had never heard of lumpy jaw so Margaret enquired after the correct medical term whereupon the man was able to prescribe and sent a package in the post. Incredibly the lump began to disappear and breathing and eating problems became things of the past.

Every new day brings some good, some bad. The recovery of Lumpy Jaw was complete but the animal was permanently named, even as was Tiny Tim.

The fruit harvest was good. We went to Strawberry Farm at Birstwith and gathered strawberries, filling almost a score of baskets and emptying our purse of exactly £100. Obviously we expect to be feeding guests in perpetuity and if the season is good we gather for several in advance. Three weeks later July guests Grace, Joan and Ron Latham came with us for the raspberry-pick and a similar weight of them was put into the freezer. I gathered some blackcurrants and we froze some and Auntie Mary made jam with the rest.

Strawberries and currants we pick comfortably from a crawling position. For raspberries the picker must stand. I find that difficult. I tried to remember advice given to Margaret in the supermarket by her over-ninety-year-old teacher. For over half-an-hour they chatted about school-days more than forty-five years ago. The incredibly old lady stood there without fidgeting whilst Margaret stood on one foot and then on the other, leaning first against her trolley and then against the stacked shelves. Eventually she said to her school-teacher, 'Mrs Beaumont, you have been talking to me for ages, standing comfortably. I've had to lean on something all the time and I'm still collapsing.'

'Stand at ease, Margaret,' the wise lady said. 'Stand at ease!'

I tried to do so, feet apart, at the raspberry canes. I am excited and intrigued by the wandering of my mind whilst fruit picking. My thoughts do not follow any logical pattern and are always in the past tense. Their irrelevance interests me. They leap from one vivid memory to another without order or bidding. One minute I am travelling on the coach to School Practice in York. Another I am lighting a hike fire in the grounds of Steeton Hall or sitting, as a child, in a hayfield, my hay rake idle beside me whilst I drink cold tea from a lemonade bottle. Then I am leaning over railings in the school yard and, on seeing a stranger's car parked on the road below, I am remembering saying to Mandy, 'Oh dear! I hope we are not having visitors,' and she replies, 'Aw no Mith Brown. Ve people 'ave gone onto ve canal bank wiv a lavatory door!'

'They've what?'

'A ... lav ... a ... tory ... door.' Mandy is a little uncertain.

'She means one of those big black dogs, Miss Brown,' her friend explains.

And I am eating tomato sandwiches in the Black Cat in Gairloch which has no connection whatsoever with Mandy. There is is just Margaret beside me for when we were in Gairloch we were without family or children. Then my mind jumps to Bradford Street and I'm in my first term at grammar school and Auntie Janie, the family tailoress, and not my sewing teacher, is forcing me to pull out seams she says are too wide on the green check gingham dress I am making in school. Immediately, for no reason, I am leaning over the deck rails watching basking sharks with my camping children. Then, I'm walking down Alice Street in the new, navy school gaberdine bought at Verrall's in Cavendish Street. The smell of it is stronger than that of the raspberries!

Without pause my thoughts lie in the staffroom of my village school and I am reprimanding my children for condemning a boy for being naughty. 'Have you never been naughty?' I am asking, intent on teaching them that people in glass houses should not throw stones. 'Think,' I am saying. 'Think of a day when you were naughty. We must teach John not to be so but we must not be unkind or think we never do anything wrong.'

Paul wants desperately to confess. I try to stop him. 'No, no, Paul. I don't want to hear about naughty boys,' but I cannot silence him. His guilty conscience hurts. He must confess.

'Miss Brown, I must tell. I spilt the Indian ink on our sitting-room carpet and I put the bottle into baby Claire's hand!'

Nothing provokes thoughts when I am fruit picking. Their sequence is without order. Each memory is unconnected and follows one after the other at speed. If we are to believe correctly the only other similar experience is death when, we are told, the whole of our lives flashes before our eyes. If it is as pleasant as fruit picking we need have no fear. The berries fill up the basket as fast as my thoughts fly to the college chapel where I am singing, 'Beloved let us love,' to my mother running with her children too near to the tide line, to Joan telling a child who is looking for me that, 'Miss Brown is in a green jar in the cupboard,' to my father's frown when something on television was vulgar, to Dorothy whom I first met when she was teaching swimming to my school-children at Glusburn Baths and to Andrew who is screaming because he has tied a knot in the string of his swimming trunks and cannot undo it.

Please God, never deprive me of fruit picking!

* * * * *

Sadly the years have flown nearly as quickly as the fleeting memories from that apparently unfillable box. Sending a brochure to Alan and Pat Chew at Brighouse I glanced at the wording. It says that the barn has been recently converted into holiday accommodation. Recently means sixteen years ago. Where have they gone, these busy years since teaching and Guiding?

Pat Chew wonders where the last forty-five years have gone!

Margaret had found her, a stranger, in the yard, apologising profusely for trespassing and for staring so at the house with its mullions and its Yorkshire slated roof. She introduced herself and said her grandfather had been tenant until his death in 1952 and that she and her brother had spent every Saturday here until she was ten. After that she had never returned, but a short while ago she had found two old black and white photographs of the farmhouse and, having just started adult art class, had wondered if she could paint the place she had known as a child. She had come, with her husband, to see if the place still stood and if her painting bore any resemblance.

Margaret invited her in and she stood in the front porch with tears streaming down her face, unashamed of her emotion. It pleased us that she could easily recognise the haunt of her childhood. It seemed we had made no basic change. The Snug she remembered as a cattle shed, but often used for farrowing pigs. She was amazed that such a beautiful inglenook fireplace had been hidden there. She obviously found the encounter with her childhood an evocative experience and kept wiping away tears. We said we would dearly like to see her photographs and even ventured to say we would be pleased to see her painting. We did not seriously expect her to want to display what would be an amateur attempt to recapture something remembered from forty-five years ago, when she was only ten.

Len Connor, a founder member of the Watercolour Association, exhibiting in Ilkley and staying at The Currer, sat at the top of the hill and skilfully painted the road, the farm and the hillside. Such talent is unusual. We did not anticipate anything to equal that from this ordinary lady whose grandfather had lived here without water, or sanitation, without road or electricity. Yet when she appeared the following Saturday with a colour photocopy of her painting, we found it to be an inspired masterpiece. We could not believe the accuracy of her gift. It was not just exact in every detail, it was not just a picture, it had atmosphere, a spiritual quality quite unexpected. We hung it above the desk in the kitchen so that our eyes could frequently stray from our chores, to recapture a past which constantly excites us. We do not regret the conversion of the barn and the multitude of visitors, for sharing what we own has always given us pleasure. Nevertheless the memory of the days when we were alone is very dear to us. This picture, painted and donated by Pat Chew, was the nicest thing that had happened to us for ages.

Looking at it we see what we saw when we first came to look, after Mother had said, 'I don't want to go to yer mother's. We'll go t'T'Currer.'

On the picture there is glass in the sash windows which had all disappeared five years later but essentially we see what we remember. We see the black and white striped door, the ivy-choked pear tree climbing against the wall, the unpitched yard and the flagged terrace. Pat has remembered, accurately, the laithe porch entrance not captured on the photographs, the barn doors with interlocking cross bars and the small inset door to give easier access than swinging back the heavier portals. She has remembered the cobbled entrance and the flagged wheel tracks and the butt to catch roof water and provide a source nearer to the kitchen than the well in the cellar or the spring in the hollow. She has not forgotten the old-style pigsties across the yard or the sycamore which still stands sentinel outside the east gable. Her brush had been guided by an unseen hand.

This unexpected gift seemed to tug at our heart-strings and was a positive balm in an otherwise tortuous summer.

* * * * *

The season wore an old-fashioned garb and gave farmers a good hay harvest and 'the poor the growth of the moor' with a bumper crop of bilberries. An occasional thunderstorm and valuable dews soaked the abundance of grass. A sudden heat wave activated the mowing machines and hay was harvested in quantity. Our first load came on a ten ton articulated lorry which, when reversed into the Dutch barn almost equalled its length. A fit and energetic guest opted to stay behind and help unload. Having been emptied, the cab failed to pull out the trailer. The heavy wheels spun ineffectively and the vehicle threatened to become a permanent feature. Margaret went for the Range Rover to tow it out but it needed the extra pull of the energetic guest's Toyota to heave the monster into the yard. The amazing thing was we had not had a guest with a four-wheel-drive vehicle for years.

Everyone wanted to sell hay. It was annually a rare commodity. Farmers who regularly made silage opted for saleable hay. A near neighbour had too much and ferried his surplus into our barn. The price plummeted and we bought more loads than usual. A news report that BSE could be passed from cow to calf created further chaos. Those who thought chemicals were to blame said, 'I told you so'. Some countries had, even before all this happened, banned the use of certain pour-on treatments on cows-in-calf. Children of those veterans of the Gulf War suffering neurological symptoms, were being born with abnormalities. The future was so dreadfully uncertain some sold their hay. Some sold their cattle. Some sold their farms and land and bought bungalows. Their work-load was too big, their job too lonely, financial stakes too high, paper work too exacting, penalisation too swift, errors too easy, inspection too frequent. Continually harassed by storm without they sought a safe anchorage in a peaceful harbour. We feared for them knowing that a pastoral way of life is none too easy to desert.

Others bought the hay, stocked their barns to capacity determined to 'trace a rainbow through the rain'. No one knew which was wisdom and which was lunacy.

'Little cattle, little care' was an idyllic profession. To calve a few cows to provide milk for a few people; to make enough hay for the winter feed and to provide work for a few hired hands, sent no-one round the bend. A guest remembered, 'My grandfather farmed land near where I live, a century ago, by cutting hay with a scythe along with thirty other men.' Imagine the banter, the cajolery, the harvest supper at the end of the day, the friendliness, the neighbourliness, the young and old together!

'Now,' the guest continued, 'A man in the closed cab of a tractor, ploughs, harrows, drills and seeds his forty-acre field in one day, alone with the diesel smell and continuous radio.' Laughter, we maintain, is the essential ingredient and, because of loneliness, it is missing. Perhaps, at the end of the day, we will have kept on course because we have not become mechanised, because fumes and noise have not obliterated the atmosphere of our countryside and because humour has not emigrated. I think we survive because of comradeship, the compatibility of the partners and our ability to have fun. That is a talent we have and long must we hang on to it.

* * * * *

Somehow those who are to survive must grow accustomed to inspection. The Fire Service inspects every year and tells us we have to add further clauses to the certificate we must hold if we accommodate more than six holiday-makers. Less than that number and we could burn them alive and not be disobeying rules.

'Chain self-closers aren't accepted any more,' said the nice fireman in July. 'You must get some Henderson fittings or suchlike. I'll come back in about ten days.'

'Where am I going to get a joiner immediately for something which isn't an emergency?' I complained. 'I'm ringing my joiner tonight. Ah'll ask 'im if 'es' free and ring straight back.' The fireman was a nice man but he didn't ring back, for all joiners have their work schedule planned and cannot down-tools for a small job on a farm off the beaten track. I was wondering, next morning, if it might be a job I could do myself. Where, I deliberated, was there a door with a heavy self-closer that I could inspect and copy? A red van drew into the yard with Johnny, our plumber and his father a retired joiner-cum-builder-cum able-to-do-every-thing. We had a small two-month outstanding job to do on the shower. I had forgotten it, it was so unimportant.

I grasped my opportunity. 'Hey,' I said. 'You're a joiner! I've got the fire officer on my tail. Could you? Please?'

'No problem, Skipper.' (His brother came to camp.) 'We'll go straight away and get the fittings. We did a similar job the other day.' Our Guardian Angel must still be hovering!

Once a year A1 Extinguisher inspects our appliances and Ellis Electrical checks all the three-pin plugs. By law these must be done. The Tourist Board makes inspections and so does the Hygiene Officer; health and safety and environment being important issues. Regular inspection of dairies and kitchens are routine. Checks are continually being made by bank managers and accountants. The animal welfare representative checks that herd and medication books are kept properly and MAFF sends inspectors for this reason and that.

The phone rang announcing an inspection of the ear-tags of all our seventy twenty-two-month old bullocks. We had just allowed the next generation from the low pastures to join the herd on the hill. 'Why didn't you ring a few days ago?' Margaret moaned, knowing we would have a big task separating adult monsters from adolescent giants. We drove the hundred-and-forty-strong herd into the lane and firmly closed the gate. A growth of grass along the roadside kept them happily occupied whilst we prepared to separate the big bullocks into one shed and their younger brethren into another. Colin from Australia came down the road, thought the animals were trespassing and deliberated whether to open the gate and try to urge them back into the first intake alone or come for help. Fortunately he chose the latter. 'Don't you dare open that gate,' we screamed at him.

It is not continual inspection which harasses farmers for, by and large, all those who come here are pleasant people and they are not responsible for the letter which comes from whatever Government department sent them, if some minor fault is found. Indeed they are embarrassed by the wording of the duplicated letter sent in every case whether the fault be major or minor. 'Don't be upset by the wording,' one said. 'It's the same letter which goes out whatever the error.' Typewriters wrote courteous letters! Word processors can be quite cruel!

Those who are to survive must not only be able to cope with Big Brother forever watching but also with the computer with its blunt statements and criticism relevant only in cases of genuine malpractice.

* * * * *

It was inevitable that many, whose ability to survive was not in doubt, should choose not to try. Some were forced out of the profession because their business was no longer viable but most of those who put farms up for sale were plainly just fed up. Those who had farmed the estate over the wall for nigh on forty years phoned to say their giving-up-sale was at the end of August. Their neighbourliness had been invaluable when cattle had strayed. We could not imagine life without them. George had already sold his milking herd and was renting us land. Our stalwart neighbours on the east had sold Marley Hall to a non-farmer who rented his pasture to a young man living seven miles away. In the Post Office I spoke to a man who owned adjacent land on the north-west and had always kept a suckled herd. 'Ah've sold most of me cattle,' he said. 'An'm BSE free. They fetched the best prices.' Coming home from the Cash and Carry, a big 'For Sale' notice was displayed outside a farm and buildings newly-built within the last twenty years or so. We were devastated. Farmers are tired and lonely and fed up. They have been used by society suffering post-war famine and urged into excess productivity. Now they are tired out!

To survive we must still find life to be fun. We must still recognise a miracle, still laugh over a coincidence, still have a sense of wonder, still believe 'it will pass', that 'something will turn up', 'we'll get used to it', 'we'll sort it out in the morning', and believe the promise that 'the morn will tearless be'.

It is not necessary, yet, for us to recapture a sense of wonder. Ours is well preserved. Whatever the news flashed on television it is never more interesting than the call from Margaret to alert us to baby swallows sitting on the telephone wires. Swallows stopped nesting at The Currer the year Father died. That was seventeen years ago. Why, we do not know but we were heart-broken. They were Father's summer joy. His eyes danced with their arrival. He worried if they came late. He monitored their nesting activity and followed the progress of the fledglings on the washing line, preparing for their marathon flight to Africa. Perhaps it was the severity of the winter which deprived them of food, or some abnormality of the late spring which interfered. A year later and we would have blamed ourselves for converting the barn and stealing their habitat but they had already disappeared. To lose both Father and swallows was double bereavement.

Seventeen years later and there were fledgling swallows on the wire! 'Come quickly,' Margaret called. That swallows were once again nesting at The Currer was supported by the fact that, cleaning the calf shed in anticipation of a 'buy another on Monday' in October, I heard hungry babies yelling for food and found the nest which cradled them.

There was definite evidence that the bird population was increasing. Perhaps they were re-finding our unfertilised pastures. Perhaps it was just the less-hot summer suited them more. Margaret never came down from her daily count without some comment on the diversity of the wildlife. I never opened the door in the morning without noticing the silvery snail trails on the doorstep. It was a major event for guests to look out of their bedroom window, early in the morning, and see the fallow deer close by, or spot the increasingly daring fox slide along the wall-side. Patient guests would keep a late evening vigil hoping the deer would come out of the coppice and up through the hollow. Sometimes they would be lucky and, when darkness made photography impossible, the shy creatures would investigate the lights from the many windows of The Currer and come closer. Their presence was the more exciting because they were elusive.

* * * * *

The damper atmosphere of the summer not only greened the Pennine ridge on which we live, it brought the opposite hillside closer and lifted its summit higher so that Ilkley Moor seemed nearer to heaven. No guest failed to tell us we had a beautiful view. As if we did not know! Similarly we never descended Currer Lane without experiencing great joy in our ownership. There comes a time in mid-August when the sun is setting over Ingleborough and we, as a family, have slumped in the sitting-room armchairs after feeding the multitude. We will do without a second cup of tea, we will ignore barking dogs or children knocking for attention on the window pane; we encourage Auntie Mary to ring up her friend and so use the phone to prevent it ringing for us and if Harry wants something he must wait for we are not getting up for anything. Yet, after four decades of living here, when familiarity could breed contempt, everyone jumps ups and hastens to the eastern mullions at that brief moment when the sun, nearing the western horizon, casts long shadows of house and walls and trees on a scarlet carpet. The moment is brief but we cannot afford to miss it, for though we live as long as Auntie Mary is doing it will not be long enough for us to be at The Currer.

Much of the winter darkness finds us outside feeding cattle beneath a canopy of stars. Because summer darkness comes late a special effort, perhaps to take the dogs down the Five Acre, must be made if we are not to miss the wonderful scent of summer nightfall and hear, and fleetingly see, the active colony of bats which live somewhere we have never discovered. When we go to bed we do not draw the curtains for only the moon can look in. We ought not to feel annoyed when cottage people draw theirs for they are urban dwellers. We have to resist temptation to enter and draw them aside, critical of their ignorance of the greater beauty outside, and we never close our bedroom windows for every time we climb into bed the owl is calling.

Margaret's diligence in cutting the thistles and pulling up the ragwort results in fewer plants every year. Not one stem of the poisonous yellow weed will she allow on the land and we cut most of the thistles before they flower and seed. The spring rain had nurtured fewer of the enormous specimens of the species Scotland proudly chooses for its emblem. Margaret got out the ATV and the flail mower and spent hours on the bumpy seat cutting them down. She left a few, here and there, knowing they would reproduce next year but mindful of the fact that they were home to a multitude of red admiral butterflies showering them with scarlet blossom.

The barn was filled more quickly than usual. Laden lorries coming along the top road told everyone we had no intention of giving up but under our show of bravado we were trembling. Sitting in the cattle market, on one of our frequent observatory visits, we were silent when we overheard old men saying to colleagues, 'Ah've sold up!' and when a weekend group timidly told me none of them were eating beef I found I was shaking.

'Aw, let's go to the baths,' we said, 'and afterwards let's picnic near to The Rough and gather some bilberries!'

So I dashed into town to buy necessary groceries and at the top of Currer Lane there was an unidentified noise in the Range Rover. I stopped and got out and looked underneath but found no fault. I drove into town and collected my vegetables and went home. I didn't even tell Margaret there had been a funny noise. We threw bathing gear into the car and levered Harry in and chose to go the quick way to the Leisure Centre via the cobbled snake which zig-zags steeply and joins a road which feeds the bypass. Approaching the junction a sudden cloud of steam escaped from the bonnet and boiling water overflowed. Margaret braked at the feed road and we both got out. Lifting the bonnet we found the fan belt had gone. That must have been the cause of the strange noise I had heard.

There is always that moment of not knowing what to do. The trail of water attracted Margaret. 'It doesn't go far,' she said. 'I suppose we send for the AA.'

I took my card from the glove pocket wondering where the nearest phone box would be, dreading having to go to the nearest pub. Only seconds had passed. My eyes followed Margaret returning from following the short water trail. I turned and faced the junction and an AA van came over the bridge which spans the railway. I leapt into the air, waving my card and the man's reflexes were as good as mine. He stopped his van and poked out his head. 'I'm just going to ring for help,' I gasped and laughed at the same time. He laughed, Margaret laughed. Harry laughed. The comedy we still call LIFE just goes on and on.

We were too late for the baths but we were able to drive across the moor to the bilberry slopes and we gathered 12 lb, which is millions of berries. We chuckled frequently over the sudden, welcome appearance of a driver's best friend. It was the second time the AA man had magically leapt out of the hat.

Chapter Twenty

> Does the road wind upward all the way?
> Yes, to the very end.
>
> *Christina Georgina Rossetti 1830–1894*

It cannot be that the life one leads mirrors the land on which one lives for, if so, those living on the plain would experience none of the heights and depths I believe we all must expect. Having said that, I can liken our lives to the terrain on which we dwell. I look through the mullioned windows of our sixteenth century homestead and see that we are balanced just short of the hilltop. Not precariously, though the drop into the valley is quite steep. We have shelter and our foundations are firm and we can see far into the distance. I could easily believe that our life is dictated by the pattern of hill and dale, mostly climbing, sometimes falling; frequently in sunshine but uncomplaining if the storm rages, confident it will abate. A pinnacle is to climb but not to live upon. We are fortunate that, though we see the bottom we also see the opposite hill climbing out of it so that it is inbred to see a way out. Though mist may shroud the opposite ridge, so that the way only seems to be down, we know it is an illusion. Providing we keep moving, though we go down we will come up again.

This we believed as we took twelve bullock stores to market and received, as everyone did, at least £120 less per animal than we had last year. Apart from being wounded financially we were hurt by the unfairness of political wrangling. Wounded and hurt but determined not to die! Margaret booked in another fifty-two cattle for the next sale a month ahead and we put the meagre sum into the bank to pay for the hay and straw almost filling the barn.

We had noticed that organisers of the annual mountain bike race had, again, flagged a route through our land and we had phoned the Council. On our return from the market the organiser's car drew into the yard. He denied ever having been told that the land was ours by the Recreation Department who subsequently told us they had never been told by the Property Services who insisted they had passed on the information twelve months ago. The dispute was their problem. We had more worrying ones.

The postman brought a letter from Janet who had already been a Guide in the company when I was appointed Leader in 1951. She had been the first teenager we had taken to the Hebrides in 1953. When she married and went to Leeds we had remained friends. She and her husband, Michael, had been the first to release us from the bed and breakfast so that we could go to Wales for a week in October in 1981. Though we saw them only once a year the bonds between us were strong.

The letter told me she was in hospital, 'Tomorrow I am having an operation for cancer.' The news had been delayed until over the weekend by a Post Office day strike. I rushed through my chores and caught the Leeds train. It was the hottest day of the year and the city was like an oven. My feet are unused to shoes. 'How do you keep going?' a guest had queried.

'I just work barefoot,' I had replied, 'And I can shout!' The latter releases tension extremely efficiently. Margaret had not heard my answer but had repeated the exact formula when the same man asked her the same question. I would have given anything to slip off my shoes and walk up to the Infirmary but the pavement was too hot. The heat was as intense as my fear for one of my children. Life expectancy, for all of us, is a fragile thing. 'Plan as if you will live for ever', I believe Baden Powell had said, 'and live as if you might die tomorrow!' With the operation behind her, Janet was rapidly gaining confidence. An hour with Janet is always a tonic.

* * * * *

That night Veronica McGinty rang. She and Tommy are part of the August Bank Holiday Gang. Tommy had put up our three corn bins. Though his legs necessitated the use of a wheel-chair, if he had distance to go, his arms and shoulders were immensely strong.

Veronica said, 'Tommy's just had an operation for cancer but we're still coming.'

We laugh! We cry! We pray! That same night I had occasion to phone a Tiree friend whose sister-in-law had died of the same malady. The funeral had been on a lovely day: 'A serene day,' the minister had said, 'for a serene lady.'

A hundred-and-twenty-thousand cattle were destined for slaughter because of a suggestion that BSE may have caused CJD in one or two cases whilst one in three of us, we are told, will have cancer. Why? Mike, the boy I took to camp, who is now our electrician, says it is because we are all living in a microwave space. Television rays, mobile telephones, radios, microwaves, fax machines, computers, fridges, exhausts ... The air, he says, is vibrating with the inventions of this new age of technology. He could be right. If there is danger to life on earth I do not think it is in the form of an unproven link between the cow and man! Governments may be willing to slaughter animals but are quite unwilling to do something about pollution and are only beginning to question OPs.

The Gang arrived bringing their wonderful brand of human sunshine. In our profession, names are duplicated many times. The Joan who comes at Bank Holiday had her legs amputated half a century ago. Tin legs had supported her admirably until the heat of the previous summer when metal had burnt flesh and she had discarded them, perhaps for ever. Now she relies on her wheel-chair and on her wit which is incomparable. Bernard, another legless guest, came for the first time and the competition between them entertained everyone.

Bernard said, 'We can be legless without being drunk,' and Joan said, 'I'm 'armless as well but no one will believe me!'

Tommy had a lovely holiday coping exactly as we knew he would. Torrential rain fell for the Bank Holiday, swilling and bleaching the yard.

Harry, who always sees the rainbow first, called us to the window. A perfect arc grew from the moor top, soared into the heavens and bowed into Jimmy's Wood. There would be no moorland fires to frighten us, no water shortage, no headaches for the Water Authority. We had such a happy week with the Gang. Ever increasingly am I aware of the 'Joy that seekest me through pain' and that 'The morn', in this world or the next, 'will tearless be'.

* * * * *

With a future uncertain it was pleasant to be even more frequently visited and comforted by the past. I am sure we have less of the present than anyone. We blink and the day is gone. Yesterday is a lovely memory which jumps unexpectedly into today and stops us worrying about tomorrow. The past is invariably peopled. School-children or Guides; friends or holiday-makers. The postman did not come every day but, more often than usual, he brought post-cards from one or other of the Hebridean Islands, from Barbara camping again at Eoligarry, from Kathryn saying, 'Thank you Jean Brown, if it had not been for you I might have been in Benidorm!', from Sylvia spending her umpteenth year on Tiree.

Mandy came on the afternoon before she left for Luskentyre. Many of my children feel the urge to go back and all are even more appreciative, in middle-age, of the wonder of their childhood holidays. 'Thank you, thank you, thank you, thank you,' echoes from all corners of the compass and from as far as Sandra in Canada and Susan in New Zealand.

Beryl stopped me outside the supermarket, wide-eyed with sudden recognition. I could not return it. 'Guides?' she queried. 'Temple Street? remember me?' I am always expected to do so. It was forty years since Beryl had been a Guide. We had not seen each other since but we could not stop talking. 'Please, please come to Reunion?' I begged, eager to get to know her again though she no longer wore plaits.

Enid, now living in Scotland, came for an hour and stayed until after midnight. Michelle, rearing a family in Wales, saw Margaret in the car park and overflowed with excitement. She came to The Currer that evening and stayed till morning. Jonathan and his wife and son came for Sunday morning coffee and had to be chased away lest the roast his mother was cooking for dinner should be ruined. Susan Flesher and her mother stopped me in Marks and Spencer. Mrs Smales, a school dinner lady from forty-five years ago phoned me to say, 'Hello'.

Jackie, now living in Oxford, brought her new baby to see us and her toddler called Harris. 'So that he will want to go there when he is older,' she explained. 'Every night I sing them to sleep with,

> "Far over the mountains, far over the sea,
> There lies a small island so dear to me.
> It's Harris of splendour, of beauty divine,
> A heaven indeed is that home of mine."'

There is magic in the islands for all of us. Now, many years hence, Margaret and I realise that the job we did so joyfully, for so many years, was even more important than we appreciated. 'I was thinking about you,' Janet said when I phoned on her release from hospital. 'You taught me: "A Guide smiles and sings under all difficulties."'

To teach, I am sure, is the most important job of all, as teacher or parent, Guider or friend, minister or neighbour. Somewhere, someone thought children should learn from experience and should be left to get on with it. That someone was wrong and teaching is the most important job of all.

The phone rang. It was Craig, the handicapped boy who had learned to walk at my village school more than thirty years ago, who was carried pick-a-back to Rhienigidale, whilst in camp and who hugged my mother shortly before she died.

'Are you in Skipton?' I asked. 'Are you coming to see me?'

'I'm bringing my wife and our six-week-old baby,' he said and I cried. She was a lovely wife, they had a beautiful baby. Craig was a happy man.

* * * * *

There were no happy people at the calf sales. The black and white Friesian calves were penned in twenties and sold in bulk to be taken straight to the abattoir. We were only observers but there was no point in buying at Beeston Castle. Our autumn calves could be easily bought nearby providing we bid more than the Government. Thankfully a depression is only a hollow. I am reminded of the two frogs which fell into a bowl of cream and could not get out. One gave up trying and died. The other went on struggling and his efforts churned the cream until it was butter and solid so that he could leap out. Eventually those who go on struggling will find the way out.

There were cherries on the newly-planted trees in our infant orchard, a cupful of blackcurrants and a bowlful of gooseberries; the apple trees gave us a few apples and there was one pear and two plums. If we live to see it mature, the orchard will give us pleasure.

The last Saturday in August was a busy day wholly enjoyed by Harry stepping confidently into his seventieth year. We went to the closing-down sale in the estate because it was a beautiful day, because Harry wanted to go and see everyone and because most of the guests were not moving on. It was a sad social occasion which, ironically, everyone enjoyed. Even the farmer's wife wore a happy smile. I think, at that moment, she was relieved to be seeking retirement. Our last load of straw came the same afternoon. The barn was nearly full and most of the load had to go elsewhere. When there was no more space we shut and bolted the barn doors. The harvest was over and weeks and months would pass before the fields would be bare and every one of the thousands of bales fed to the herd.

The still-laden lorry was drawn up to the calf shed so that bales could line the calf pens should a sale of the rest of the bullocks enable us to re-buy. Following the lorry to the open door of the shed I saw these very bullocks walking through a gate onto the Rough and shouted my dismay. Someone had untied the gate and a dozen bullocks were already through and more heading in that direction. Margaret and Danny were running with speed the four hundred yards to the escape route lest a hundred-and-forty animals galloped every which way.

She returned, abandoning those already in the heather for the lorry must be emptied of its load so that the driver could go home. He and I had put enough bales into the calf shed and when Margaret came, out of breath, the lorry was taken to the entrance of the yard shed and emptied of the remaining bales and a hitch-hiker. I heard the driver having a one-sided conversation with the mouse which had travelled from the corn field on the York Plain. The poor thing was disoriented and leapt from the bare lorry and zig-zagged across the yard away from the shed and towards the house.

It is hitch-hikers we fear. Home-bred field and barn mice generally know their place. Foreigners don't know which way to go. We alerted the cottage dwellers that a mouse was at large and begged them to keep doors shut. The space it found itself in would be too big and it would hide wherever it could. Then we bundled Harry into the Range Rover to retrieve our bullocks suffering from wanderlust.

Perhaps, we thought, the gate had been opened by spectators of the mountain bike rally. Maybe local children were imitating their elders and using the track. The bikers had widened the footpaths into highways and the cattle had used the easy trail downhill and could not be persuaded to re-climb the hill, so, with the help of a neighbour out mushrooming, we pulled down a wall and let the animals back into the Dyke Field. Pulled-down walls must be rebuilt immediately. Too many things were happening and we had many for evening meal.

Stan and Barbara had arrived. They had been coming to the Mistal Cottage for the first week in September for many years. Barbara is in a wheel-chair like so many of our guests. Next day Stan reported a nibbled bag of biscuits in his bedroom and our heart sank to our boots. We gave him a trap but there was no sighting all week so we presumed the stranger had not liked the Mistal as much as Barbara and had departed. It seemed to be an isolated incident but we were uneasy.

A week later, when I was crossing the yard in view of all the guests, I spotted, and Danny spotted, and they spotted a mouse doing a Samson on the concrete. My efforts to divert it in the direction of the barn caused amusement but failed miserably when Danny joined in and chased the mouse behind his kennel, near the front door. I told the laughing guests to shut all doors and Margaret and I moved the kennel but there was nothing there. I have great admiration for mice whose ability to squeeze through a hair's breadth is incredible and whose courage, to enter the dark unknown, is phenomenal. In their place, I could even agree with Rose Fyleman who thinks 'mice are nice'. I read the poem frequently to my children, but I have a fear that, if they see a mouse, guests will leave our house and take with them our reputation. Fortunately most people think it is normal for there to be field mice near farms but it has never been normal for us to have a mouse in the house.

There was a sudden shout from the dining-room. Not a frenzied one. We ran to the door and were told the mouse had just run under the piano. We hastened to disturb it and a guest saw it run under the fire door into the Mistal. Everyone thought it was great entertainment, live 'Tom and Jerry'.

Sylvia and Henry, annual September guests who'd booked the Mistal for a fortnight, were not one bit perturbed. They placed our live mouse trap on the floor and caught the fellow promptly and the 'fun' was over.

* * * * *

The day before Sylvia and Henry left was the day of selling cattle at the monthly sale. We had fifty-two huge bullocks to get to market on time. The three over thirty months were called in, on the same day, which simplified things for us. We have never had such a trouble-free sale. Margaret planned professionally, opened the gate into the road at the right moment and had gates in the right places to divert and to sort. She could not have organised things better. Late at night she let the fifty-five animals into the shed and at 7 a.m., in the stillness of next morning, the two cattle wagons came. Loading posed no problems. Five loads were taken and not one animal behaved badly. Even the Jersey bullocks had lost their stubborn streak.

Only we were tense. We were coming to the end of the waiting game we had been forced to play all summer. We were coming to a T-junction. Which way we went would be determined, not by the size of any cheque we might receive but, more fundamentally, whether we could sell at all. Financial loss we would have to take on the chin but failure to sell at all would be catastrophic. The animals were already big enough for us to handle. Certainly if we brought them all home again we would have to cancel our holiday. They were far too big to leave with Dorothy. There would be no buying calves. Our existing stock would fill all our sheds and eat all and more, of our store of bales. They were such lovely animals, not one of them belonged to the bank. It should have been a bonanza year. For everyone it bordered on disaster.

I had taken a piece of foam into the auction theatre. A pulled ligament in my thigh was causing me sitting discomfort. We sit so seldom, to do so uncomfortably was nearly the last straw. Farmers spotted my bit of foam which caused amusement and banter, sadly lacking in market atmosphere that autumn.

It was a relief to witness that buyers were there and that cattle were changing hands, albeit for a ridiculous price. We had drawn about mid-sale but there were cattle galore and we sat for, perhaps, two hours before showing ours. All the animals looked good. Farmers rear fine animals nowadays, but that day in the market, as on any other day of sale in the autumn of '96, every farmer lost £120 on each of his stores. The depressed fat stock market made dealers cautious but they were still buying and all our bullocks sold. It was a great relief.

Many farmers were thinking it imprudent to re-buy. Beef was still being imported by the ton. Exports were banned. Until the one was stopped or the other restarted things would never be normal. We walked away from the office with a buoyancy which did not reflect the cheque Margaret was carrying.

We could buy another on Monday, but first we must prepare for that event by gating the calf pens, putting up the barriers and taking a much-needed holiday. Our next sale was twelve months away. Anything could have happened by then. The Currer seemed seasonally empty with only yearlings. Empty and silent. We prepared the evening meal and fed our guests. Margaret and Harry finished washing the dishes and we were inexplicably weary. So apparently was No. 7.

The phone rang and Ben, the auctioneer reported that the two thirty-month-old Jersey bullocks had been transported away but that No. 7 would not get up!

'It's never been ill,' Margaret said. 'D'yer want me to come?'

'No, no, no,' Ben said. 'We'll get the vet in the morning and Mitchell.'

Next morning No. 7 was standing and he walked as far as the removal wagon and then he sat down again and nothing would make him stand. He had decided to meet his maker sitting down and we did not blame him though it cost us dearly.

So now, with just six days to go before leaving for holiday we must go hell for leather towards departure.

We went into the calf shed determined to hang gates to divide it into pens. For years we had struggled on with binder twine, the farmer's best friend. Margaret had said she would struggle no longer and ordered strong men to bolt on the gudgeons but they never came so we decided it was just another job we must do ourselves. Lifting on the twelve foot iron gates was child's play compared to screwing bolts into railway sleepers. Margaret walled each pen with straw bales. There was no doubt in either of our minds that when we came home we would buy. But not at Beeston. There was no need. There were plenty of calves locally.

Margaret said, 'Drop me at Skipton and I'll buy there, drive on to Otley and you buy there and on Thursday we'll go to Gisburn,' and in farming she is bossman.

My goodness, we needed a holiday and, heaven forbid, we had another mouse. John said, 'Aye, they're coming in early this year!' Our physiotherapist said, 'I've got one among my plant pots on the window ledge.'

They were so complacent and we were all worked-up and agitated. We tidied every cupboard and drawer in the house and threw out everything we did not need. Friends bring us clothing to wear out on the farm. Some we can use. Some are too big or too small or too warm or too fancy. What we could not keep we took to the Oxfam tank in Sainsbury's. The mouse was not resident. It came in and went out again and we were frantic to find its entry. When we did it was so big Margaret declared a badger could have used it. Auntie Mary's bedroom had been made en-suite and the new drain had been introduced to the existing one and the workman had left without completely finishing.

Before going to the Isle of Man I had filled the uncovered drain, but, in my haste, I had not noticed the open invitation to wandering wildlife in the kitchen wall. On investigation under the sink Margaret discovered that, after the mullions had been put in when we made the new bed and breakfast kitchen, the higher sill should have been bricked. It would have meant taking out the sinks to do the job and builders are not rodent controllers and ours had not anticipated a problem. We made this discovery on the day before leaving for Wales. We had just fed breakfast to twelve guests and Harry and Auntie Mary were still sleeping. It would be more accurate to say Margaret made the discovery for only she can get under the sink and even then it is with difficulty. From her contorted position she said, 'I think I'm going to cry. There's nothing we can do without taking out the sinks!'

The new, raised window ledge was of wood and laminate and was wedged securely and screwed onto a hidden scaffolding. We were amazed to find we could lift it out and, by golly, with a two-by-four from the wood pile and a piece of plywood from the calf shed we were able to fettle that mouse. Margaret cemented up the cave outside and I made order out of the chaos the operation had caused in the kitchen.

* * * * *

We always aim at having the cases on the roof rack the night before we leave. We failed completely. Dorothy had a dog show in Belfast that weekend so it was Sandra and her family who were releasing us. Sandra had not done so for the last ten years. She had been busy getting married and raising a family. Indeed little Sam was not yet two.

We fed the twelve guests. They were so nice and so talkative they would not go. 'What do you do on holiday in Wales?' one lady asked.

What do we do? Perhaps she should have said, 'What don't you do?' I told her that we go to a lovely cottage with an incomparable view. That, if I rise at eight, it is still two hours later than usual and easily two, possibly three hours before the rest of the family begin to stir. What do I do? I write or I read what I have written. I make a cup of coffee, maybe two and I take the dogs to play on the lawn and I retrieve the fallen apples. The birds and squirrels have been before me and they are nibbled. I sit by the huge window and watch the open sea and the mountains of Snowdonia. I write a few more pages, surrounded by the space I am deprived of, and I drink another cup of coffee and have a lovely holiday.

For as long as he lives, and it looks like being for ever, Margaret's first job will be to see to Lusky. Though he cannot stand without help, like all our disabled guests he radiates happiness and looks all set to revisit Luskentyre, next spring.

In the Hebrides no-one else is walking a dog and we and our canine friends have the beaches to ourselves. Everyone in Aberporth has a pet but Danny makes friends everywhere and Jess keeps her distance and snarls if another dog comes near. Lusky, the warrior, would have them all if he were able.

It takes us till midday to pack some sandwiches and fill the flasks, for Harry to be shaved and maybe have a bath, but we are all having a lovely holiday even though the sun may be on the decline before we climb out of the Range Rover onto the first beach of the day. We do not want museums, or shops, or amusements, or people, or cafés or attractions. We just want beaches and we have a lovely holiday. Of course we spend half our time lifting the wheel-chair from the roof rack or the rear space next to Lusky, pushing it miles on hard sand and then heaving it back again. We say that one day we will buy a lightweight one but some day is not yet. We throw balls for the agile dogs and waste too much time looking for ones they lose but even then we are having a lovely holiday. On rare occasions when it rains we have a lovely time watching it stream down the windscreen singing, like the Oxo family, 'Oh, we do like to be beside the seaside'.

We talk to natives, most of whom seem to be retired and have dogs. When dusk is falling we go to our temporary home, cook a few new potatoes and a vegetable and slice the cooked meat we take with us. We stew some of the nibbled windfalls and blackberries gleaned from the hedgerows. We never open the oven door or switch on the grill or use the microwave. We have a lovely holiday. The lady shook her head. She did not understand.

But first we had to send away tardy guests from The Currer, those who wanted to take photographs and hug us more than once. We dare not attempt to pack the roof rack for our helpful guests would use it as an excuse to outstay their welcome and get under our feet on the wet concrete.

We were so late leaving. We turned to face our home and the sun was behind us and the sixteenth century house in front and we saw that there was a rainbow rising behind it and climbing towards heaven before bending magnificently over the valley to root itself in the opposite hillside. We felt akin to Noah. Our faith is as strong, our ark as welcoming, our voyage through life well chartered and there is a good Captain on the bridge.

There came upon us the complete tiredness of having coped. As long as one must keep going there is energy to do so. Only when the job is done and there is opportunity to stop does tiredness overwhelm. We journeyed, as did Hiawatha, 'into the portals of the sunset'. The great golden ball led us all the way as far as Tarporley. It blinded us with its brilliance and awed us with its time-keeping precision. Unlike us it is never late and when it left us in darkness it illuminated the moon. The silver disc was full as we pulled into the car park of the Wrexham Travelodge, suffused with a great relaxing weariness.

Perhaps, I thought, it is time to lay down my pen for a while whilst we are all well; whilst Auntie Mary eagerly awaits her ninetieth birthday and Harry proudly steps into his seventieth year; whilst Lusky still barks from the back of the Range Rover and our old cat, Tommy, continues to thrive though he cannot always get to the door and makes a pool on the lino.

I have experienced this physical tiredness many times before when all that was immediately necessary has been done. For a moment, before sleep imprisoned me, I thought of the comfort of my sleeping bag at the end of the first day in camp with fifty children on islands too beautiful to describe. Tents had been pitched and gadgets made, wood and peat stacked and children fed. The water bin had been filled, the fire had died to embers and the silence was broken only by the proximity of the Atlantic breakers playing on the shore.

I drew the Travelodge curtains, conscious of the need for darkness and deprived of it by the full moon and I sang quietly a camp-fire song:

> Bed is too small for my tiredness,
> Give me a hill topped with trees,
> Tuck a cloud up under my chin,
> Lord, blow the moon out, please.